PRAISE FOR GRANTLEE KIEZA'S BOOKS

'Engagingly written … one of the most nuanced portraits to date'
The Australian

'Vivid, detailed and well written'
Daily Telegraph

'A staggering accomplishment that can't be missed by history buffs and story lovers alike'
Betterreading.com.au

'A free-flowing biography of a great Australian figure'
John Howard

'Clear and accessible … well-crafted and extensively documented'
Weekend Australian

'Kieza has added hugely to the depth of knowledge about our greatest military general in a book that is timely'
Tim Fischer, *Courier-Mail*

'The author writes with the immediacy of a fine documentary … an easy, informative read, bringing historic personalities to life'
Ballarat Courier

ALSO BY GRANTLEE KIEZA

Knockout: Great Australian Boxing Stories

The Remarkable Mrs Reibey

The Kelly Hunters

Lawson

Banks

Macquarie

Banjo

The Hornet (with Jeff Horn)

Boxing in Australia

Mrs Kelly: The Astonishing Life of Ned Kelly's Mother

Monash: The Soldier Who Shaped Australia

Sons of the Southern Cross

Bert Hinkler: The Most Daring Man in the World

The Retriever (with Keith Schafferius)

A Year to Remember (with Mark Waugh)

Stopping the Clock: Health and Fitness the George Daldry Way (with George Daldry)

Fast and Furious: A Celebration of Cricket's Pace Bowlers

Mark My Words: The Mark Graham Story (with Alan Clarkson and Brian Mossop)

Australian Boxing: The Illustrated History

Fenech: The Official Biography (with Peter Muszkat)

Award-winning journalist Grantlee Kieza OAM held senior editorial positions at *The Daily Telegraph*, *The Sunday Telegraph* and *The Courier-Mail* for many years and was awarded the Medal of the Order of Australia for his writing. He is a Walkley Award finalist and the author of twenty-one acclaimed books, including recent bestsellers *The Remarkable Mrs Reibey*, *The Kelly Hunters*, *Macquarie*, *Banjo*, *Mrs Kelly*, *Monash*, *Sons of the Southern Cross* and *Bert Hinkler.*

HUDSON FYSH

GRANTLEE KIEZA

For Everald Compton, a nation builder who like Hudson Fysh has encouraged so many Australians to soar

The ABC 'Wave' device is a trademark of the Australian Broadcasting Corporation and is used under licence by HarperCollins*Publishers* Australia.

HarperCollins*Publishers*
Australia • Brazil • Canada • France • Germany • Holland • India
Italy • Japan • Mexico • New Zealand • Poland • Spain • Sweden
Switzerland • United Kingdom • United States of America

HarperCollins acknowledges the Traditional Custodians of the land upon which we live and work, and pays respect to Elders past and present.

First published on Gadigal Country in Australia in 2022
This edition published 2023
by HarperCollins*Publishers* Australia Pty Limited
ABN 36 009 913 517
harpercollins.com.au

A catalogue record for this book is available from the National Library of Australia

ISBN 978 0 7333 4154 0 (paperback)
ISBN 978 1 4607 1348 8 (ebook)

Cover design by Christine Armstrong, HarperCollins Design Studio
Front cover images: Hudson Fysh courtesy Australian War memorial (B02182); Biplane courtesy Mitchell Library, State Library of New South Wales (FL8581273/FL8581340); Landscape by Peter Harrison/Getty Images
Back cover image: Qantas booking office, Longreach, Queensland, ca.1920s by Cliff Postle, 1913–2004, courtesy National Library of Australia (nla.obj-151430031)
Author photograph by Milen Boubbov
Typeset in Bembo Std by Kelli Lonergan
Printed and bound in Australia by McPherson's Printing Group

Foreword

By Wendy Miles, Hudson Fysh's daughter

My father was an extraordinary man who helped to create and shape one of Australia's most celebrated businesses, the airline Qantas.

I loved him dearly, and at the age of 92 I still treasure the great influence he had on me and my family, and the principles of honesty, loyalty, love and respect that were the cornerstones of his own life.

His children and grandchildren all called my dad 'Hud' or 'Huddy', and we remember him as a good and kind man, with a wonderful sense of humour who devoted his life to bringing Australia closer to the rest of the world through air travel.

Qantas was launched with two tiny biplanes in the remote country of western Queensland, and in 1922 Dad flew the company's first scheduled passenger flight from Longreach to Cloncurry.

He was a far-sighted man and he remained at the controls of Qantas until 1966, guiding it through the Great Depression, the savage years of World War II and into the jet age.

Hud was born in the days of the horse and cart, but he was still the chairman of Qantas when the airline began negotiations to buy its first Jumbo jets, aircraft that revolutionised the way that Australians travelled.

His achievements are all that more remarkable given his tough start in life.

Hud's childhood was difficult and confusing, and for a long time he bore the scars of his parents' unhappy marriage.

His schooling suffered, but the determination he developed playing sport propelled him through his early challenges in life, whether it was on the deadly ravines of Gallipoli or in the skies over Palestine where he became a decorated war hero.

That same determination helped Hud turn a little bush business into one of the biggest and most respected airlines in the world.

Yet throughout his life, he was an extremely modest and unassuming man.

One of my father's lifelong passions was reading as it opened whole new worlds of adventure and learning.

I hope you enjoy this book and the story of my father Hudson Fysh, who was a truly great Australian.

Sydney, June 2022

Fysh is met by his wife Nell and children John and Wendy at Archerfield Aerodrome in Brisbane after a flying visit to London on behalf of Qantas in 1937. *State Library of NSW FL520053*

Preface

Hudson Fysh was a very determined man. He was very shy in those early days, but he was a man of courage and determination, and he had good vision … this vision of service to the public which he was determined to carry through.

JOHN FYSH, HUDSON'S SON[1]

IT WAS A CLEAR, CRISP day beside the serpentine Brisbane River in 1971 as Sir Wilmot Hudson Fysh,[2] strode purposefully towards the corner of Creek and Adelaide streets. Dressed in a dark suit and tie, the still spritely seventy-six year old was walking along a path he had trodden more than fifty years earlier, long before his hair had turned white and he had developed the need for thick black-framed glasses, and long before he and his aeroplanes had conquered time and distance to make Australia one with the outside world.

This tall, lean and unassuming colossus of the aviation world turned left, then walked through a wooden door with a frosted glass sign that in gold script proclaimed 'Gresham Hotel'. With its colonial architecture and iron lacework, the Gresham had been a landmark in the Queensland capital for almost a hundred years, and it was the site where an Australian icon was born.

Fysh took his place in a chair at a wooden table in the middle of the famous old pub's smoking lounge and, looking down the barrel of an Australian Broadcasting Commission camera, he began to outline in clipped, modulated tones the story of the airline Qantas and how it had come to life in that very room. He told the documentary makers about how he had survived the horrors of Gallipoli and aerial combat over Palestine, and how the little bush airline he and his friends had launched in 1920 had grown into an international powerhouse that was the envy of airlines around the globe.

By 1971 Fysh was a retired business giant who spent most of his time playing with his grandchildren, hitting the fairways and wading through trout streams. But when he had first visited the Gresham in 1920, he had been a novice pilot and the nervous sidekick to his mercurial pal Paul 'Ginty' McGinness,[3] a small, dashing fighter ace from Victoria who was 'game for anything'.[4] Back then, Fysh let McGinness do the talking as they assailed a prosperous grazier named Fergus McMaster[5] with their sales pitch. They asked him to help finance a small business running a couple of cheap aircraft in the outback that could service remote sheep and cattle stations as aerial taxis. McGinness and Fysh could also fly around the western parts of Queensland and into the Northern Territory, looking for the best paddock to put down and asking local squatters if they'd like to pay for a joy ride. Very few people in the outback had seen an aeroplane, and even fewer had flown in one. That day in the Gresham's smoking lounge, all three men knew there would likely be years of struggle ahead for their little venture.

The devil-may-care McGinness was champing at the bit for a challenge, but like most of the adventurous fighter pilots who tried aviation careers after the Great War, he did not have the temperament for the day-to-day order and stresses of business life.[6] Fysh was different: calm, quiet and methodical, he spent ten years living in western Queensland, where his two adored children were born and where, in the heat and dust, he steered the infant Qantas through turbulent times.

The first aircraft Fysh flew for Qantas cost £450 and carried one nervous passenger in an open cockpit. Under his stewardship for almost half a century, the airline grew into a company that in 1967 paid $123 million[7] for its first four 747 Jumbo jets from Boeing; each one could carry hundreds of passengers in comfort,[8] and would make global travel affordable and accessible for future generations of Australians. By 1971, when Fysh revisited the Gresham, Qantas had come to embody the spirit of Australia, with its symbol of the flying kangaroo, and it set the benchmark for service and safety around the globe.

Born in Tasmania in the age of the horse and cart, Fysh saw them overtaken by the motor car. He started his working life as a farmhand and jackaroo, then survived Australia's baptism of fire on Gallipoli. He came through aerial combat in dogfights with German aces, delivered secret messages for Lawrence of Arabia, and became an old man in the age of heart transplants and trips to the moon. Along the way he transformed from a shy, awkward teenager into a war hero and titan of Australian business. Never motivated by money or fame, but rather by what was good for his country, Fysh was at the heart of it a simple, humble man who lived in rented accommodation for most of his life, rarely owned a car, and was known as 'Hud' or 'Huddy' to his family and friends.

Because of his efforts at the controls of Qantas, Fysh lived his final years at a time when Australians routinely flew on his airline's machines to the furthest corners of the globe in a single day. Few apart from Fysh could have foreseen the airline's ultimate success from its humble beginnings.

Chapter 1

Henry Reed was a good man. He loved God and good work and good people. His sympathies went out specially to the poor and friendless; and to be the means of ministering to their temporal and spiritual welfare was for many years the joy of his life.

Salvation Army founder William Booth writing about Hudson Fysh's grandfather[1]

As HUDSON FYSH WAS dodging Turkish bullets on Gallipoli, a powerful image always gave him wings. It was the same mental picture that later helped sustain him while he was escaping German machine-gun blasts in the skies above Palestine, and which drove him on through the darkest times in the early days of his fledgling airline, when outback dust storms tore into his pair of flimsy aircraft. Fysh would visualise his grandfather Henry Reed[2] and recall stories of the sharp-eyed, fiercely determined young Yorkshireman. Whenever Fysh thought of Reed, he would remember that within him too was the courage and tenacity to survive any trial.

Reed was born late in 1806, in Doncaster in England's north. He was the youngest of four children of Samuel Reed, an eccentric postmaster,[3] and his wife, the former Mary Rockliff. Samuel died when the boy was small, and he remembered nothing of his father except the tolling of the church bell to mark his passing.[4] At thirteen, after a solid grounding in the Wesleyan faith and a smattering of education, Reed was apprenticed to a merchant in Hull; he learnt the basics of shopkeeping from the floor he had to sweep morning and night to the top of the shelves he routinely stacked with arms that grew longer and longer over the next seven years. It was in the counting house where his talents really shone, and the wide world beckoned.

When Reed had just turned twenty, he bade his widowed mother goodbye, promising to return a rich man, and left Gravesend aboard the 350-ton ship *Tiger*, seeking to make his name and fortune on the far side of the world. The cargo included fine English soap, as well as choice hops from Kent and brewing utensils that were much in demand for the free settlers in Van Diemen's Land,[5] then best known as the end of the line for transported British convicts. The voyage was trying, and Reed was in the cheapest class, steerage. But when he wasn't praying for deliverance, he was studying all he could about the way the skipper and the crew handled a large vessel, how they navigated by the stars and steered to make the most of whichever winds were blowing.

The *Tiger* finally sailed up the Derwent on 13 April 1827. After a confinement of four months in cramped quarters, the long and lanky young man took in the surrounds of the fledgling Hobart Town, with the majestic Mount Wellington in the distance. Then he stretched his legs in the best way possible, walking two hundred kilometres across the northern part of the colony to find work. For company he had another passenger named Vallance and a double-barrelled shotgun. The cannibal bandit Thomas Jeffries, who had eaten at least one of his four murder victims, had been hanged recently in Hobart Town, but bushrangers and escaped convicts still roamed the area with murderous intent. Meanwhile the so-called Black War was raging; in reality, it was a series of massacres of the local Aboriginal population and the occasional retaliation by their kin.

After a week of cold nights as winter approached, Reed and his companion trudged into Launceston unharmed – though on his return journey to England, Vallance was murdered by pirates when made to 'walk the plank'.[6] Reed presented a letter of recommendation to Launceston's first legal practitioner, young John Gleadow,[7] who was combining his legal career with a successful venture as a merchant in the growing town of two thousand people. Launceston had become an important export centre for the colony's northern pastoral industry, and Gleadow gave Reed a job in his store. With a head for figures and a readiness for hard work, Reed was soon running his own trading concern and looking towards expansion.

The local bounty hunter John Batman[8] befriended Reed, and the young Yorkshireman was a witness at Batman's marriage to the convict Eliza Thompson[9] at the recently built St John's Anglican Church in Launceston.[10] Batman convinced Reed that there was wealth to be made in land acquisitions and agricultural improvements using convict labour, and in January 1828, just nine months after arriving in the colony, Reed took hold of 260 hectares at the Nile Rivulet, near Deddington, not far from Batman's property. The land was about thirty kilometres south-east of Launceston and came under a grant from Governor George Arthur. While Reed benefited from Batman's guidance, one of their neighbours, the renowned colonial artist John Glover, described Batman as 'a rogue, thief, cheat and liar, a murderer of blacks and the vilest man I have ever known'.[11] Batman came down hard on any protests from the Indigenous people over the influx of white settlers. He brought in Aboriginal trackers from New South Wales to support what he called 'roving parties' to drive away and sometimes kill the first peoples of the area,[12] and Governor Arthur admitted that Batman 'had much slaughter to account for'.

To obtain his land grant Reed declared his assets at £605 and 7 shillings, calculating his worth based on possessing £385 cash, twenty-five thousand needles at 35 shillings per thousand, a case of forty-eight hats valued at 45 shillings each, four sixty-yard pieces of the fabric bombazine worth two shillings threepence per yard, twenty-four bags of shot at 18 shillings per bag, and his double-barrel gun worth £20.[13] With two assigned convict servants to do his heavy lifting, Reed quickly turned a small fortune into a large one, but to save money during those early days, he and his workers wore boots of rough greenhide and trousers made from kangaroo skins.[14] By the age of twenty-four, Reed had established himself as a prosperous landowner, farmer and merchant, and he named his property Rockliff Vale after his mother's people. He soon acquired more land and thirty more convicts to toil on it.[15] In what he later called a 'sinful life' he worked hard and played hard, gambling on billiards and cards, and backing the racehorses he acquired.

He began his shipping ventures by chartering the *Britannia* with Launceston merchant and banker James Henty for a trading voyage to the settlement that the Henty family had established on the Swan River in what is now Western Australia. Once when returning to Launceston from Sydney on a small vessel called the *Hetty*, Reed let his impatience get the better of him. Annoyed as the skipper waited for more favourable winds at the mouth of the Tamar, a long tidal estuary, Reed set off for the sixty kilometres to port alone on a dark and stormy night, rowing the ship's small boat. He'd made it through Whirlpool Reach and into the icy Supply Bay when a violent gust overturned his craft. Clinging desperately to the keel for five terrifying hours in choppy waves and evil squalls, Reed drifted helplessly about the wide bay. Fearing he would freeze to death before he drowned, he tore off his clothes, said a quick prayer and, though a poor swimmer, began flailing through the cold, black waves towards the distant shore. Fatigue and the conditions overcame him, though, and exhausted, he felt himself sinking as the water rushed into his mouth and nose. 'God be merciful to me,' he prayed. Then his cold feet hit the riverbed, and with joy and relief he realised he was standing in neck-deep water on the gravel spit of Swan Point. Slowly but happily, the naked survivor waded ashore at midnight and knelt to give thanks for his deliverance.

Reed continued in what he called his 'sinful' ways, and though it seemed that everything he touched turned to gold,[16] there was a deep longing in his heart for spirituality. He bought three ships – the brig *Henry*, followed by the *Socrates* and the whaler *Norval* – and kept them sailing in busy trade, often under his command. He also established a whaling station at Portland Bay in what is now Victoria, and later sold it to the Hentys, who had started grazing more of their sheep in that part of Australia.

Before long Reed's vessels, laden with Vandemonian wheat, wool, kangaroo hides, sealskins, whale oil and wattle bark, were travelling between Launceston, Hobart and Sydney, and eventually on to London. They carried home to Australia such valuable commodities as kauri timber from New Zealand and sugar from Mauritius, and the voyages of the *Henry* helped open the way for the European settlement of South Australia.

IN 1831 REED, NOW A RICH young man based at his mansion Macquarie House in Launceston,[17] decided to make his first return visit to England. His voyage on the *Bombay* was life changing – and almost life ending. In the middle of a savage winter off Cape Horn, two of the Indian crewmen froze to death. A vicious gale threatened to tear the ship apart, and Reed's whole life flashed before him. Weeping bitterly, he promised to serve God for the rest of his days. Though there would remain a constant warring within him over his 'sinful' inclinations, he became a zealous evangelist, perhaps overcompensating for having accumulated so much wealth.

When he finally arrived in London, he made important business alliances to sell Australian wool. In an all-important personal alliance, he married his seventeen-year-old cousin Maria Susanna Grubb.[18]

Back in Launceston, Reed became a superintendent of the new Sunday school opened by the Paterson Street Methodist Church. He was also a director of the Bank of Australasia, and he loaned John Batman £3000 for expansion plans on the mainland. Batman claimed to have negotiated a purchase there of more than 240,000 hectares with eight Wurundjeri elders in exchange for tools, blankets and food; he said he had chosen a site on the northern bank of the Yarra River, which would be a good place for a village.

Reed travelled to the site that would become the colonial outpost of Melbourne, saying he wanted to find some means of preserving 'the natives there' from the 'destruction' they had suffered in Van Diemen's Land.[19] He well knew Batman's ways. At the time there were only three huts: Batman's, John Fawkner's, and one belonging to a ship's hand.[20] In Batman's hut, Reed preached to a congregation composed of Batman, the bounty hunter's brother Henry, three of Batman's Aboriginal trackers from Sydney, and William Buckley, the infamous 'wild White man', a giant escaped convict with a terrifying appearance who had spent more than thirty years living among the local Indigenous people. Reed told his congregation about the mercy of Jesus and the sanctity of all life, and he prayed with them every day. The Aboriginal trackers understood a little English, and Reed preached the Gospel to them constantly – whether they liked it or not. 'No doubt,' he declared proudly, 'this was the first time the Gospel was ever proclaimed in Victoria.'[21]

He even journeyed up the river beside the huts and lived with a group he called the 'Yarra Yarra' people in a vast wilderness for some time, hoping to bring them to God; later he wrote that his work there 'was blessed'.[22] Returning to Launceston, he ministered to all who would listen, including convicts awaiting execution. In 1837, he prayed for days alongside four of them who were soon to be hanged together in Launceston Gaol. One of them, John Hudson, was being put to death for striking a cruel guard who had been tormenting him.

Although much of Reed's energy was now devoted to evangelism, he continued to reap plentifully from what he sowed in the business world too. His ships were soon carrying Vandemonian settlers along with their sheep, cattle and horses across Bass Strait to the Port Phillip region, though Batman did not see much of Melbourne's development, dying there of syphilis aged just thirty-eight.

By the time Reed was forty-one he had appointed managers to navigate the multiple income streams he had created beside the Tamar. He sailed with his wife and growing family for England, where he lived for the next twenty-six years, developing further trade connections and drawing ever closer to God.

AS REED WAS ESTABLISHING his headquarters in Britain, another enterprising Yorkshireman, Sir George Cayley,[23] was developing an industry of the future: perfecting the design of his gliders, a passion that had consumed him for almost all of his long life. A brilliant engineer, he had been fascinated as a child by the flight of birds, and in 1799 he revealed his plans to give humankind wings, engraving an image of a primitive aircraft on a silver disc.[24] He experimented continuously with his ideas, and in 1809 he published a defining treatise, *On Aerial Navigation*, in which he outlined the three elements required for powered flight: lift, propulsion and control. In 1853 at his imposing Yorkshire home, Wydale Hall, near Scarborough, he supervised the first recorded flight by a person in an aircraft. Cayley's coachman, John Appleby, was left ashen-faced by his short journey on his master's glider; after this truly groundbreaking event, he told the 79-year-old inventor, 'Please, Sir George, I wish to give notice. I was hired to drive, and not to fly!'[25]

A few months later, in 1854,[26] Henry Reed purchased a large farmhouse and about thirty-five hectares of land in Tunbridge Wells, Kent. He then hired one of Britain's leading landscape designers, Robert Marnock, to create an earthly paradise complete with a two-hectare lake. While the work was underway, Reed's wife Maria – the mother of their eleven children – died, aged just forty-six. Despite this, the magnificent Dunorlan Park, with its Italianate mansion built entirely of Normandy stone and requiring eleven servants to run it,[27] was completed in 1862. Over the entrance Reed displayed his family crest: a sheaf of wheat above the motto, 'Nothing without the cross.' Soon the now fifty-six year old, sporting a long grey beard worthy of an Old Testament prophet, married for the second time;[28] he tied the knot with a devout 35-year-old church worker, Margaret Frith.[29] Over the next seven years she bore the last five of Reed's children, and their third, Mary Reed,[30] would be Hudson Fysh's mother.

Margaret had grown up in a home known appropriately as 'The Cross' just outside Enniskillen, Ireland. She came from a long line of devout Christians. John Frith, born in 1503 as the son of a Kent innkeeper, could have risen to great riches in the material world, but he was just thirty when King Henry VIII had him burnt at the stake in Smithfield, London, for heresy after he preached that neither purgatory nor transubstantiation could be proved by Holy Scripture.[31] A young apprentice tailor named Andrew Hewitt was chained with the Reverend Frith, and it took two agonising hours for them to die because the wind kept blowing the flames away from their charred flesh. When one of the priests watching the execution admonished the weeping spectators not to pray for the prisoners because they were no better than dogs, Frith managed a smile and prayed for the Lord to forgive his tormentor.[32]

A century and a half later, three brothers from the Frith family fought for the Dutch prince William of Orange, later England's King William III, against Catholic forces at the Battle of the Boyne in Ireland in 1690. As a boy Hudson Fysh was told that all three had been huge men for their times, at more than 190 centimetres tall, and that they had lived by the motto, 'A Frith never turns back.'[33]

Henry and Margaret Reed turned their mansion and its gardens over to the Lord's work, inviting local churchmen to hold open-air services under the magnificent beech trees. Reed donated large sums of money to William Booth as the Wesleyan minister enlisted soldiers for Christ into his new Salvation Army, which was taking England by storm, and Reed once opened the grounds of Dunorlan Park to host a gathering of 1400 of Booth's troops. Booth called his benefactor a man of 'unswerving integrity, great courage, inflexible will, and tireless energy'.[34] Reed also financially supported the preaching work of the China Inland Mission and the East London Christian Mission, and he helped to establish churches in the East End as well as schools in the slums around Bow Common.

Despite his spirit of self-sacrifice, some fellow worshippers found fault with the opulence of Dunorlan Park. One servant called it an architectural monstrosity that represented 'everything one might expect from a man with too much money and too little taste'.[35] After one too many churchgoers complained that its lavishness reflected a reverence for the God of Materialism rather than the God of Christianity,[36] Reed auctioned off the building and land, then he and his family moved into a smaller but no less stately home at Harrogate; he called it Dunorlan Villa.

He continued to preach to congregations and large Christian gatherings throughout the north of England and throughout Scotland. He bought three cottages in Leeds to house the aged poor, and in his home town of Doncaster, he bought ten.

Realising his twilight was fast approaching, Reed returned to Launceston in 1873 with his large family. He brought servants on the voyage as well as a cow to provide fresh milk for his youngest children, including five-year-old Mary. In the colony that was by then known as Tasmania, he bought the magnificent home Mount Pleasant – situated on almost fifty hectares overlooking Launceston and the Tamar – then had it rebuilt. Hudson Fysh remembered it as 'one of the finest residences in northern Tasmania and always the delight of us children when we visited there'.[37] Reed also made a luxurious summer residence in Mount Villa on the Wesley Dale farm at rugged Mole Creek, in the upper Mersey Valley of Tasmania's north. But building congregations remained his chief task until his dying day, and he helped establish the New Guinea Mission, providing the group with the steamer *Henry Reed*; as a result, New Britain's Henry Reed Bay was named in his honour. Years later, part of his fortune was used to build Launceston's Henry Reed Memorial Christian Mission Church and the nearby Dunorlan Cottages to provide free housing for destitute elderly women.

Henry Reed had fathered sixteen children and established a Tasmanian dynasty built on property, shipping and trade, approaching his business ventures with the same zeal that he brought to his life as a philanthropist and evangelist. Having turned his initial stake of £605 seven shillings into what would be many millions today, he died at Mount Pleasant in 1880, just short of his seventy-fourth birthday. He left his children and grandchildren a lifelong example of pluck and grit.

MARY REED WAS TWELVE at the time of her father's death. Despite the sadness of losing him, she pressed on through life with a resolve as strong as his. She received some schooling in England, but the finishing touches came at the Presbyterian Ladies' College in Melbourne, where she distinguished herself as a pianist. As a young woman she sailed for London; inspired by her father's faith, she preached the Gospel and dispensed charity in the slums. In 1888 she went to China as a missionary, but she returned to Launceston two years later after chronic asthma forced her home for a period of convalescence.

While Mary continued her father's zealous ministry, Lawrence Hargrave,[38] a Sydney inventor, was soaring as the acknowledged world leader in aviation research. Born beside the Thames in Greenwich, Hargrave left London in 1865 for Sydney, where his father had been the NSW attorney-general. Hargrave spent years absorbed by various adventures: mining for gold, exploring the wilds of New Guinea, and – most importantly, after being inspired by George Cayley's writings – wrestling with the idea of creating aircraft propelled by engines. After experimenting with thirty-six different designs, Hargrave developed a three-cylinder rotary engine in 1889. Four years later, he moved his wife and children to Stanwell Park, between Sydney and Wollongong; he owned property and coalmines in the area but was searching for something he believed to be far more important: the answers to the mysteries of flight. He started experimenting with box kites.

That same year, Mary Reed was flying high too. Having reached the age of twenty-five, she had inherited a small fortune of £15,000 from her father's estate in 1893. She was in London at the time, working as the voluntary secretary of the China Inland Mission. A striking-looking woman with large almond eyes that seemed to search the souls of everyone she met, Mary married a handsome Tasmanian, Wilmot Fysh,[39] at the Congregational Church in Bow, London.[40] He was a dapper twenty-seven year old with a handlebar moustache and a head for figures. Like Mary's father, Fysh was a Launceston merchant, but he was a spectacularly unsuccessful one despite his powerful connections; they included his uncle Philip Fysh,[41] twice Premier of Tasmania and one of the driving forces behind the Federation of the Australian colonies.

The Fysh family had been farmers in Norfolk. In 1859, Wilmot Fysh's father Frederick Lewis Fysh[42] accompanied his older brother Philip from London to Melbourne on the *Bombay*. After working for the shipping firm L. Stevenson & Sons for ten years, Philip was offered the chance to run its new Hobart office. In 1862, as the Tasmanian economy tanked, Philip bought his employer's wholesale agency and changed its name to P. O. Fysh & Co., general merchants. He made it the leading wholesale business in Hobart, and in 1866 Fred Fysh travelled across Bass Strait from Melbourne to run the Launceston branch. When Philip began his long political career, Fred Fysh started his own business, selling clothing, fabric, bedding and perfume and eventually enlisting the help of his son Wilmot.

Mary's new husband had none of the business acumen – or luck – that her father had possessed. They were very much in love, though, and within a few weeks of their marriage they were expecting their first child and sailing home to the warm embrace of their families in Launceston.

While Wilmot and Mary Fysh were making their way across the waves back to Tasmania, a German aviator named Otto Lilienthal was riding gusts of wind near his home in Berlin after having built a fifteen-metre-high artificial hill to launch his gliders. And on 12 November 1894 at the beach near his home in Stanwell Park, Lawrence Hargrave and his assistant James Swain anchored four box kites to a pair of sandbags, then tethered a trapeze to the honeycomb-shaped kites. As the breeze blew up and the kites rose so did Hargrave, soaring on the trapeze to the astonishing height of almost five metres – although it felt like heaven with the box kites floating overhead, the blue Pacific Ocean on one side and the lofty Bald Hill behind him. Hargrave said that controlling his flying craft was a matter of utilising the wind after the mode adopted by the 'albatross, turkey buzzard, vulture and other sailing and soaring birds'. His work was soon being celebrated by the press in New York, Chicago and London.

Hargrave's research had been huge news at the 1893 International Conference on Aerial Navigation in Chicago, an event organised by Octave Chanute, a French-born American railway engineer. Chanute sang Hargrave's praises to the world, writing in his 1894 book *Progress in Flying Machines*: 'If there be one man, more than another, who deserves to succeed in flying through the air, that man is Mr Laurence [*sic*] Hargrave, of Sydney, New South Wales.'

The news of Hargrave's achievements probably did not reach Wilmot and Mary Fysh at their stylish Gothic revival home, The Gables, at 52 High Street, East Launceston, but they had other reasons to celebrate. It was at The Gables on 7 January 1895 that Mary gave birth to their first child, a baby boy the couple named Wilmot Hudson Fysh. The 'Hudson' was in honour of Hudson Taylor,[43] the Yorkshireman who had founded the China Inland Mission and spent fifty-one years preaching the Gospel there.

Hudson was born into a loving family that was quickly becoming downwardly mobile. Amid a financial depression that bit Australia in the 1890s, their money was dwindling, but inventiveness and drive was in Hudson's DNA. His first childhood memories were of being pushed on a swing at the rear of The Gables when he was about three years old. He felt himself rising further and further in the air, then coming down and rising again, higher still. It seemed as though he became weightless – he felt like he was flying.

Chapter 2

As he sped over the crowd 100 feet up, the aviator released sheafs of coloured handbills which the crowd eagerly sought as souvenirs. Encircling the area several times the aeronaut gradually ascended to 4000 feet, looped the loop again and executed a long volplane with the machine riding as steadily as a hawk in the stiff breeze.

Newspaper report on the flight of Maurice Guillaux at Geelong, with Hudson Fysh in the crowd.[1]

HUDSON FYSH'S boyhood in the sylvan setting around Launceston played out against a backdrop of lightning-fast developments in the science of aviation, and the real prospect that one day soon humankind might touch the sky in powered machines. By the time of his birth, a pair of bachelor brothers, Orville and Wilbur Wright, were exploring human flight at the home they shared in Dayton, Ohio. They experimented with kites, corresponded with the Chicago aviation expert Octave Chanute, and read every report they could on Lawrence Hargrave's tests in Australia and Otto Lilienthal's experiments in Germany. Fysh had his feet and mind firmly on the ground, though; instead of spending his formative years watching the rapid progress in aviation, he spent them despairing at the disintegration of his father's business and the downfall of his parents' marriage.

A year after Hudson's birth, Mary delivered twins: a son, Frith,[2] whose name honoured her mother's pious ancestors, and a daughter, Geraldine.[3] Hudson was only about three when his family moved from The Gables to a new spacious double-storey home, Ketteringham, situated on a hectare in St Leonards, which was then a village about six kilometres from Launceston, on the North Esk River. The stately house had been built of handmade convict bricks in 1843. Amid the willows and golden river sands of a country idyll, Hudson watched his mother supervise the development of the property, renovating the house and outbuildings, overseeing the construction of the stables and tennis court, and creating sublime gardens. Some of the plants were for the kitchen table but others were for pure pleasure, including magnificent roses with masses of the brilliant white Frau Karl Druschki variety that bloomed on a bed built from tons of rich silt.

Two more siblings were born at the new home – Peggie in 1899[4] and Graham[5] in 1903 – and their mother had a growing staff to help her: a full-time cook and a housemaid, a gardener-coachman, and a nurse for the children. The gardens always needed work; kerosene lamps had to be refilled and cleaned, while the candlesticks were frequently replaced; and the many fireplaces needed constant cleaning. When the children were sick they were allowed fires in their rooms, and sometimes there was a big fire in the hall.

Hudson inherited his mother's long, thin face and kind, searching eyes. Though Wilmot thought his wife fanatical, Mary instilled in her children the Christian tenets she had inherited from her father, teaching them the importance of honesty, fair play, charity and common sense. There were family prayers with the servants every morning after breakfast; except on Sundays, when the family and staff trooped up to the local Methodist church.

Sometimes they would travel into Launceston to listen to hymns sung in the Christian Mission Church that Grandfather Reed had built, and Mary insisted the children learn a psalm or hymn each morning to be recited over a Sunday lunch that was always cold; there was to be a minimum of cooking or any other work on the Sabbath. Only religious books were to be read on the Lord's day of rest, and games were banned, though the children were encouraged to take long walks into the countryside, up Abels Hill or down to the river.

Fysh's parents, Wilmot and Mary. Their marriage was stormy and doomed. *State Library of NSW FL8576691, FL8576692*

Fysh was raised with his pious mother's admonition to 'never go down in life, always rise and better your position'[6] but Wilmot did not heed that advice.

He did not share Mary's zeal for Bible teachings, either, and she began to look upon him with increasing scorn.[7] The family was only keeping up appearances through Henry Reed's money and Wilmot was quickly draining Mary's bank account to keep his business afloat.

Although Mary's relationship with Wilmot was increasingly strained, she would often help him in his business, then known as Fysh, Scott and Co. on Launceston's Paterson Street. Hudson would accompany his mother and watch his father as he sat perched on a high counting-house stool, adding figures in a thick dog-eared ledger. The boy was astonished by his father's flair for arithmetic; how he could glance down at long columns of numbers and tally an accurate total in his head. But while talented with basic figures and kind and gentle in an aloof, remote way,

Wilmot was, according to Fysh, lacking in common sense and 'completely overshadowed' by Mary's intellect. [8]

In his preschool years Hudson had two overriding ambitions: to be able to read and to own a gun. But when he was about five and staying at the Reed family's holiday cottage Dawlish in George Town on the Tamar, he rode in a motorcar for the first time, marvelling at the noise of the combustion engine and the euphoric rush as it accelerated away without the means of a horse. The exhilarating experience created a lifelong love for all things mechanical.

OTTO LILIENTHAL'S DREAM of becoming the first person to make a powered flight had died with him in 1896 when he suffered a broken spine in a crash, uttering as his last words: 'Small sacrifices must be made.' Others were willing to risk their lives by following his example. In Paris, the French lawyer Ernest Archdeacon and engineer Gustave Eiffel founded the Aéro-Club de France. Meanwhile, a debonair Brazilian coffee plantation heir named Alberto Santos-Dumont became the world's most celebrated action-adventure hero, piloting his cigar-shaped airships in the skies above Eiffel's great tower.

Little wonder that Hudson began dreaming of flight when he was about six, a year after his great-uncle Philip Fysh – then the Tasmanian agent-general in London and the proud owner of a white beard that looked half a metre long – had established the Fysh crest of a hand holding a flying fish above the motto *Nitor in Adversum* (Strength in Adversity). As a child Hudson would imagine himself lifting off from the top of the stairs at Ketteringham and circling the house effortlessly, his arms and legs fully extended, before flying through his bedroom, then out through the window and around the tall fir trees in the garden. He said that this fantasy gave him a 'delicious sensation of gliding free and triumphant round the countryside'.[9]

These boyish flights of fancy came down to earth with a thud, though, in the turbulence of his parents' marital disharmony. Wilmot's retail enterprise flopped; Fysh later recalled that his father 'was no businessman, and he finally had to close down after having borrowed all he could from his relatives including my mother'. Fysh watched his mother and father quarrel until the inevitable parting. Wilmot went to a small dairy farm near Launceston, but he could never make more than a bare living from milk and cream.[10] Under the terms of their legal separation – a rare event for the time – Hudson and his brother, Frith, went to live with their father on the struggling farm while their sisters, Geraldine and Peggy, and the baby, Graham, stayed in considerably more comfort with their mother.

Fysh never forgot that traumatic day in 1903 when a horse-drawn cab arrived at Ketteringham to take him and his things to his father a few miles away. Only eight years old, he howled, stamped his feet and refused to go, until finally he was persuaded to climb into the cab with the explanation that it would help his mother.

Until they were given bicycles, Hudson and Frith walked each day to the Launceston Grammar School, five kilometres away. But Wilmot did his best to help his young sons find happiness in their new surrounds. He bowled ball after ball to them on a cricket pitch scraped out of bare earth, and gave them their first kick of an Australian Rules football. He also taught them to fish. Under the willows beside the North Esk, Hudson hauled up blackfish and eels that lived in deep holes along the river, and he developed an enduring passion for fly-fishing after hooking a 'really big fellow' in the tumult around the great revolving waterwheel of the St Leonards flour mill. Hudson also learnt to shoot, though the kick of the shotgun he first fired sent him flying backwards and bruised his shoulder.

The shy boy made animal friends on his father's farm. The memory of them stirred him all his life – especially of the greyhound Speed, a real goer with a docked tail. Speed became the boy's all-time favourite dog, though he enjoyed chasing rabbits with a fox terrier named Dick. Like his mother, Hudson suffered from asthma and fevers, and he missed Mary terribly; when the little boy was sick or melancholy, Speed would camp beside his bedroom window, and when Wilmot would allow the dog a brief visit inside the house, he would lay his head on Hudson's bed and whimper his support. Once Hudson shut Speed in a lucerne box in the stable, and the dog ate his way out through the wooden bars, causing all sorts of consternation in the household.

Not even the love in his dog's big brown eyes could cheer Hudson when he felt especially lonely, and many times in the first year after his parents' break-up he ran away from the farm and back to Mary's loving arms.

As the dissolution of the marriage had been a legal separation rather than a divorce, Wilmot was prevented from remarrying. It made him increasingly morose, bitter and lonely, and Hudson could never reconcile himself to the other women who shared the small cottage at his father's farm.[11]

After Hudson's final escape back to Ketteringham, he was interviewed personally by the fabled Tasmanian jurist Sir John Dodds,[12] and he was allowed to stay permanently with his mother. Hudson's relationship with his father never really recovered, though Frith stayed on the farm.

In later years, memories of Hudson's boyhood centred around Mary: her soft cool hand on a hot forehead when he was sick, and the beautiful soothing music she produced at night from her beloved Bechstein baby grand piano, which would have the children crying out 'more, more, more' in chorus from their upstairs bedrooms. The smell of the bush stayed with Fysh always, as did the waning golden light of dusk on the Esk, and the smell of the horses and all the harness in the coach-house of his childhood home.

IN THE YEAR THAT Mary and Wilmot Fysh separated, Orville and Wilbur Wright arrived at the windswept fishing town of Kitty Hawk, North Carolina, with the ambitiously titled Wright *Flyer I.* They had been experimenting with their manned gliders in the sand dunes there since 1900. The town was 1100 kilometres east of the Wrights' home in Dayton, but the wide open spaces and strong steady Atlantic breezes were ideal for the brothers' test flights.

The *Flyer I* was built from spruce and ash, with wire struts and unbleached muslin stretched over the wings to make it smoother through the air. The machine had a wingspan of twelve metres and weighed 275 kilograms, and – most importantly – it had an engine: a four-cylinder, twelve-horsepower motor that the brothers believed would facilitate the first ever powered flight. The engine drove a pair of handcrafted 2.5-metre propellers that sat facing backwards behind the pilot. The Wrights had designed their machine so that the pilot would lie prone, facing forward, in the centre of the bottom wing. The *Flyer* was too heavy to be launched by wind gusts, so the brothers built an eighteen-metre wooden monorail they called the Grand Junction Railroad.

On 14 December 1903 they enlisted the help of men from the nearby Kill Devil Hills Life-Saving Station. The group carted the machine and its rail track to the incline of a sand dune known as Big Kill Devil Hill. Orville and Wilbur tossed a coin to see who would be the first person in history to fly under the power of an engine. Wilbur won – but after the *Flyer* raced down the Grand Junction Railroad, it stalled and finished nose deep in the sand three and a half seconds later. The brothers spent three days fixing their craft and then decided to try again in front of five witnesses, one of whom, John T. Daniels, was given Orville's camera to record the moment.

In the shadow of Big Kill Devil Hill, on the icy winter's morning of 17 December 1903, and with a gale of more than twenty knots blowing salty mist off the grey Atlantic, Orville climbed onto the *Flyer*'s lower wing and lay face down. With the engine roaring and the propellers spinning, the *Flyer* headed along its wooden runway at 10.35 a.m., Wilbur holding one wing to keep the *Flyer* stable, sprinting along beside it as fast as his thick coat and trousers would allow. Orville pulled back on the elevator lever, and the *Flyer* rose into the teeth of the arctic gale.

Shaking all over, Daniels took the snapshot some people called the photo of the century as Orville Wright briefly left the surface of the earth. The first powered flight lasted just twelve seconds before the *Flyer* came down into the soft sand on its wooden skids.

AT THIS TIME, young Hudson Fysh was still more interested in the visits to Ketteringham of his grandmother Margaret – or Granny Reed as he called her – than he was about the arrival of powered flight. A coachman in livery and silk top hat would drive Henry Reed's widow to Ketteringham at regular intervals, usually when one of the children was sick. She came with the delicious medicinal treat of calf's foot jelly, and exquisite hothouse grapes nurtured by the Mount Pleasant gardener.

Granny Reed would host great family gatherings for Christmas dinner at Mount Pleasant; twenty or so relatives would take their seats in the dining room for a sumptuous silver service feast as their ancestors gazed down from portraits. Hudson was already proving himself to be an outdoors child rather than a student, and he was right at home spending hot Christmas Days in Mount Pleasant's lush gardens, or splashing about in its artificial lake after a game of tennis or bowls on the manicured lawns. Sagging trestles held a weighty array of Christmas gifts – cricket bats, tennis racquets, rocking horses and watches for the children – and each would also receive one of Granny Reed's gold sovereigns, enough for Fysh to buy his first .22 calibre rabbit gun and still have a few shillings in change.

It wasn't all fun and games, though. Henry Reed lay buried in the family tomb on the property, and above the grand entrance to the main house were the Reed crest and motto, 'Nothing without the Cross.' Fysh and the other children were warned not to wear boots on the polished verandah, while the stateliness of the home impressed upon Fysh the importance of hard work, tradition and respect.

Fysh knew he had many talents, but his was an unsettled and often unhappy boyhood, despite the big home, the wealthy grandmother, and the occasional trip by steamer across Bass Strait where the Reeds would holiday at the Menzies Hotel in Melbourne, with its luxurious rooms and German band. His parents' break-up, his father's business flops, his own battles with asthma and many changes of school all contributed to 'an agony of shyness'. Because of his parents' separation, Fysh felt looked down upon in his community and he developed an inferiority complex that lasted four decades. He often lamented what he called 'an intolerable nervous reaction in contact with my fellows, especially in the tenfold agony of making a speech'.[13]

Once when cajoled by some girls to enter a swimming race in St Leonards, Fysh glided to victory over some rival boys but was struck dumb when he saw his young female fans preparing to drape around him a purple sash with the word 'Champion' in gilt lettering. He was so bashful that he dived back into the river and swam away, and the victory sash was sent to him later. Once he was thrown from a horse and, with his foot caught in the stirrup, dragged across the St Leonards bridge. He was lucky not to be seriously hurt but developed a morbid dread of horses. Although he was a competent cricketer, nervous tension made him stumble and fumble at football. He became so reticent that when he broke a collarbone playing the game it was three days before he told his mother that he was injured. For a long time when any visitors apart from Granny Reed came to Ketteringham, he would disappear into the bush.

He was fortunate that his mother was an inspiration, broadening her horizons at every opportunity. She travelled to Melbourne to study art photography with the renowned John Kaufmann, and her artistic nature study *The Tussock*, taken on a windswept beach in Falmouth, Tasmania, became one of her son's lifelong treasures.

Aviation continued to astonish the world. On 25 July 1909, when Fysh was fourteen, Louis Blériot – a heavily moustachioed French inventor who had amassed a fortune making acetylene headlamps for automobiles – advanced the Wrights' first flight of 1903 exponentially. He flew his sputtering, dragonfly-like Blériot Model XI across the English Channel from Les Baraques, near Calais, to Dover. In the process he claimed a £1000 prize from Britain's *Daily Mail* newspaper. His aircraft had decapitated a dog that snapped too closely at the noisy, whirring propellor just before take-off. While the *Daily Mail* proclaimed that his flight had ushered in a new era for the world, the writer H. G. Wells warned that it showed the vulnerability of Britain to air attack.[14] The military powers of Europe were already devising ways to use aviation as a weapon of mass destruction.

FYSH'S MOTHER HAD lost heavily in the Melbourne real estate crash of the 1890s but she could still find the funds to send his sisters Geraldine and Peggie to Sandecotes, a select girls boarding school in England. Two cousins also went to the Leys school in Cambridge. But Mary and Wilmot quarrelled over who would pay for Fysh's education. Mary believed that she should pay for the girls and Wilmot should pay for the boys. Caught in the middle of another of his parents' feuds, Fysh's education was piecemeal, with a heavy, though not always welcome, emphasis on religious instruction from his mother. Mary did not believe in infant baptism, so Fysh did not formally become a Christian until he had learnt enough about God and his commandments to make that commitment, aged twelve. By fifteen he had attended Launceston Grammar as both a day boy and boarder, and also Launceston's George Edmunds Academy, the Reverend Roach's school in Devonport, Georgetown State School, St Leonards State School, and St Peter's College in Adelaide.

Fysh also had private tutors, but none of his teachers had been able to impart much knowledge into a head that was focused on the great outdoors. Mary decided that he should try to make up for lost time by attending Geelong Grammar, founded by the Church of England fifty-five years earlier. The school had a record of encouraging its students to contribute to the community. Mary planned to pay for Fysh's education with dividends from shares she had in a tin mine at Mount Bischoff in north-western Tasmania. She had high hopes that Geelong Grammar's headmaster, Leonard Lindon,[15] a classical scholar of Jesus College, Cambridge, and a former deputy headmaster at Sydney Grammar,[16] could set a course for Fysh towards a successful future.

During Fysh's final years of schooling in Tasmania, Australians saw flying machines for the first time in their own skies. On sand dunes at Narrabeen Heads, north of Sydney, on 5 December 1909, George Taylor became the first man to fly a heavier-than-air machine in Australia when his engineless glider took wing before about a hundred spectators. Just four days later, Colin Defries, a London pilot, took off in a Wright Model A, with a motor, at Victoria Park racecourse in the Sydney suburb of Zetland. He reached an altitude of five metres before crashlanding as he tried to retrieve his hat, which had blown off.

Then, on 17 March 1910 in Bolivar, north of Adelaide, a nineteen-year-old motor mechanic named Fred Custance, who had never flown before but was up for anything, flew for about 180 metres, reaching an altitude of fifteen metres in a Blériot XI. He came down so hard that the undercarriage was smashed and the propeller broken.

A touch of magic was needed to control an aircraft properly, and who better to demonstrate that the very next morning than the world's greatest conjurer and escape artist, Erik Weisz – better known by his stage name, Harry Houdini. The Hungarian-born son of a New York rabbi had been in Melbourne performing as the 'Handcuff King', and he'd brought along his $5000 Voisin biplane that owed its design to the Hargrave kites. At Plumpton Paddock near Diggers Rest, about forty kilometres north-west of Melbourne, in front of assembled newspaper reporters as well as a photographer and a cinematographer, Houdini thundered across the grass for forty-five metres before lifting off. As the spectators gasped, he made another of his thrilling escapes, skimming over the treetops with inches to spare before coming in for a graceful landing. Before the day was out, he made two more flights, the longest of about three and a half minutes in which he flew in a great loop for more than three kilometres over rocky outcrops and stone fences.

Among those who saw Houdini fly at Diggers Rest was Harry Hawker, the 21-year-old son of a Moorabbin blacksmith. Hawker was a mechanic looking after five vehicles, including a Rolls-Royce owned by the wealthy grazier Ernest de Little, lord of Caramut House in western Victoria. The magical experience of seeing Houdini fly primed Hawker for bigger things.

Other Australians were busy building aeroplanes, and on 16 July 1910 at his father's farm near Kyneton, north of Melbourne, John Duigan made a seven-metre hop in his homemade biplane.

EVEN THOUGH FYSH WENT to Geelong Grammar in 1910 as the oldest boy in his class, he was 'hopelessly behind everybody else'. The school was still a relatively small institution, with just thirty day boys and ninety boarders, but it drew students from around Australia. Fysh forged lifelong friendships there. He made many crossings of Bass Strait to Geelong on the old steamer SS *Loongana* with his schoolmate Johnnie Webster, a doctor's son from Launceston with an infectious, self-deprecating laugh. He and Fysh were in the same dormitory. They played football and cricket together, and Webster was the stroke for Grammar's eight in a Head of the River rowing triumph.[17]

Geelong Grammar was an Australian version of an elite English public school, and Victoria's wealthy Western District graziers saw it as the stepping stone for their children to reach Oxford or Cambridge. Many of Fysh's schoolmates were children of Australia's landed aristocracy, families who owned large swathes of the country. Names alongside his at rollcall included Fairbairn, McCaughey, Chirnside, Manifold, Fawkner and Featherstonehaugh. Jimmy Fairbairn[18] and Charles Hawker[19] were graziers' sons who would go on to have similar experiences in life. Charles Hawker grew up on a sheep station outside Adelaide, and during his school years he spent many Sunday afternoons hiking with Fysh through the Moorabool Hills outside Geelong; Hawker would regale his shy, reticent friend with tales of life on a sprawling property and discuss all manner of other subjects with deep intellectual analysis. In later decades Fysh came to regard Hawker as 'one of our greatest Australians': a thinker, a scholar, a man of 'unswerving character and the crusader for what he thought was right'.[20]

On Saturdays, small groups of boys ventured into the countryside at the crack of dawn, walking, riding bikes, or rowing down the Barwon. The excursions primed Fysh for the future as he revelled in the fresh air and countryside. He was also invigorated by Grammar's sporting activities, including the swimming events before breakfast throughout the summer. His childhood illnesses became mere memories despite the compulsory cold shower every morning, even during Geelong's winters when howling winds coming off Port Phillip ripped through his crowded dormitory. He grew into a tall, athletic seventeen year old with a wiry physique and an enquiring gaze, ready to absorb whatever life lessons he could from those around him.

Though not an exceptional sportsman, Fysh was a good swimmer and won the open backstroke in the school baths. He also managed to win a place in Grammar's First XVIII Australian Rules football team, and he opened the batting for the First XI cricket side in his last year at the school but without much success. Jimmy Fairbairn's cousin Beau[21] was a far better batsman, going on to play county cricket in England for Middlesex after Fysh watched him crack 120 against Wesley College on Geelong's pitch.

Fysh won prizes in History and English, and one of his essays was so impressive that his teacher Arthur 'Jarpo' Morris – who had been at Geelong since 1889 – initially suspected that some passages had been copied from a book. Fysh remained 'on the slow side'[22] in other classes, though, finding it impossible to catch up after a somewhat misspent youth; he wasn't surprised when he scored just 12.5 per cent on a Latin exam and 26 per cent on a French one. The commercial class, which taught young men business principles, was much more in his line.

Geelong Grammar nurtured Fysh's love for sports after he was sent there in 1910. Back row, second from the left, he was a standout in athletics, cricket, Australian rules football, and rowing. *State Library of NSW FL8644113*

He was not at the school long before Leonard Lindon retired as headmaster after sixteen years to be replaced by the Reverend Francis Brown.[23] The Bristol-born hat maker's son had won a mathematics scholarship to Oxford, and he believed strongly in chapel services, scripture lessons and confirmation classes for his students. Fysh and his schoolmates called their new headmaster 'The Crow' because of his austere appearance and initial remoteness.[24] But Fysh flourished under Brown's leadership, and the headmaster took an interest in the shy boy despite his modest academic performances.

Australia had just introduced its Universal Service Scheme as the only English-speaking country with a system of compulsory military training. Fysh became a senior cadet, at first baulking at the tough discipline but coming to relish the camaraderie and spirit of his fellow trainees.

He could appreciate aesthetics as well. Encouraged by his mother's love of art photography, he joined the school camera club and had his winning photographs published in Grammar's magazine. His study of the old school tower was one of the last to be published before the building became a jam factory when the school shifted to Corio Bay in 1913.

That move coincided with Fysh's swansong at the school, as he rejected appeals from Charles Hawker to stay, along with their headmaster's promise to make him a prefect. After he had started studying wool classing at the Geelong scour, he'd realised he did not have the application or desire for a university education. He obtained work as a jackaroo on the 8000-acre (3237-hectare) merino sheep property owned by his uncle Henry Reed in Logan, twenty kilometres south of Ketteringham. Fysh finished Grammar with no formal qualifications, but as he sailed home across Bass Strait he realised he was doing so with a stronger constitution, a large array of friends, and a measure of self-confidence that had been absent when he arrived.

He was still afraid of horses, and it took a year and a broken wrist before he felt comfortable in the saddle, but from his first week at Logan he relished the work. Rising before dawn to immerse himself in the tough chores, he tried to keep up with the older workers as he cranked away with a hand pump to drain underwater tanks for cleaning. Then he pitched in with the harvesting and loading of sheaves into a horse-drawn dray. He toiled alongside his cousin Mac McCarthy, a lad his own age 'and a very fine and loveable character'.[25] The eager eighteen-year-olds were paid ten shillings a week when the gun shearers were making ten times that, but still they delighted in the hardy outdoor life. Fysh helped with the burning off of grasses to prevent bushfires and the catching of breakaway animals spooked by the flames. Sometimes he drove a small team of four bullocks pulling a cart that carried stone for the foundations of a new shearing shed. He learnt how to work a sheepdog, to gather and care for sheep, and to shear the cross-bred wethers before killing them, cutting the carcasses into pieces for weekly distribution to the other workers. The Tasmanian winters were so cold that one morning Fysh had to light a fire to warm the frozen crowbar before it could be used to dig post holes. But despite all the hard work there was still time for rabbit and hare hunting, with a pack of eight dogs of all sizes and breeds.

There was also time for Fysh to serve as a reserve with the 26th Light Horse Regiment. He went into Easter camp with his horse in 1913 and 1914 at Ross, on the road to Hobart. It was at military training in the languid Tasmanian countryside that he became familiar with machine guns.

The storm clouds of war grew darker every day over Europe, as rival powers England and Germany flexed their military muscles with their respective allies joining a rapidly escalating arms race. Winston Churchill, Britain's First Lord of the Admiralty, sensed the importance aviation would play in warfare and during 1912 and 1913 he made many flights, even though his wife Clementine pleaded with him to keep his feet on the ground after his instructor was killed in a crash.[26]

When Fysh left the Logan property in 1914 and returned to Gordon College in Geelong[27] to study for his wool-classing diploma, he paid 22 shillings sixpence a week for his food, lodgings and washing at Mrs Brearley's boarding house. He continued his military training on the mainland with the 70th Ballarat Infantry, unaware that he would soon need all of it.

FYSH HAD HARDLY ARRIVED back in Geelong when the fifty-year-old Archduke Franz Ferdinand, heir to the Austro-Hungarian throne, and his wife Sophie arrived in the Bosnian capital of Sarajevo to inspect army manoeuvres. Europe was now a powderkeg of militaristic tension, and the Balkans were a flashpoint. Dressed in the finery of an Austrian cavalry general, the Archduke sat with Sophie in the back of an open-top 1910 Graf & Stift Double Phaeton limousine when it stalled outside Moritz Schiller's food store near Sarajevo's Latin Bridge. A sickly nineteen-year-old assassin named Gavrilo Princip opened fire with a Browning automatic pistol.

Just a week after the assassinations in Sarajevo, Fysh saw the flight of an aeroplane in person for the first time. On 4 July 1914, he was among the thousands who defied a bleak Victorian winter's day at the Geelong racecourse to watch a visiting French aviator, Maurice Guillaux, fly his £2000 Blériot XI 'Looper' stunt machine in a strong south-westerly wind. After a band played 'La Marseillaise', the dapper, dashing Frenchman, rugged up with a woollen hood, thick muffler and leather coat, covered a circle of the course, then made Fysh and the rest of the crowd gasp by pulling off an almost perpendicular dive and some loop-the-loops.[28]

Twelve days later, at 9.12 a.m. on Thursday, 16 July, Guillaux took off from the Melbourne Showgrounds. He was heading for Sydney's Moore Park with Australia's first airmail – a sack containing 1785 souvenir postcards and a few letters – along with Australia's first air freight – some Lipton tea and O.T. lemon squash bound for Sydney's Commercial Travellers Club. The flight was originally meant to be made by the American stuntman Arthur Burr 'Wizard' Stone, who seemed to spend more time in hospitals than in the air, but with his young Bundaberg mechanic Bert Hinkler[29] watching on, Stone had crashed while testing his new aircraft beside Maribyrnong Road in Sunshine.[30] As Stone lay in hospital with all his teeth knocked out, his top lip all but severed and his back, neck and jaw seriously damaged, Guillaux chugged off in an aircraft with an open cockpit. He soared over mountains he had never seen before, with only a compass and the railway line to guide him.

After two days of flying, Guillaux landed at the Goulburn racecourse on 18 July to refuel and to nurse the aches in his wrists from wrestling the controls. He stopped in Liverpool, on Sydney's outskirts, then circled around Parramatta and Manly in torrential rain. Just before three o'clock, ten thousand people were assembled at the Sydney Sports Ground to see a rugby union Test match between Australia and New Zealand. As the storm passed, Guillaux finally saw the bonfire marking his landing spot in the adjacent Moore Park. When he landed, rugged up as though he'd flown in from Antarctica, he was met with massive cheering from the crowd pouring out of the sports ground, then more cheering when he raised the sack containing Australia's first airmail high above his head – though the package could have got there sooner on the train.[31]

Despite the hoopla around this historic flight, aviation as a potential livelihood still had no place even in Fysh's imagination. On the point of completing his wool-classing course, he had taken out his union ticket ready to class his first shed near Swan Hill on the Murray River in October 1914. But nothing could turn back the inevitable march of war in Europe. On 28 July, Austrian troops – backed by Germany – invaded Serbia, which Russia had promised to defend. As in a bar-room brawl, one slugger after another waded into the fray.

Chapter 3

There have been scenes of awful slaughter, with heaps of dead and wounded and ghastly wounds and long lines of stretcher-bearers with their gory burdens, but men march cheerily past and take up positions for attack or defence with the certain knowledge that many of them will share the same fate.

MELBOURNE'S GENERAL JOHN MONASH WRITING HOME TO HIS WIFE[1]

WORD HAD JUST REACHED Victoria of the Austro-Hungarian Empire's murderous march into Serbia,[2] when Captain Wilhelm Kuhlken sailed the newly built 6000-ton German cargo steamer SS *Pfalz* into Port Phillip Bay on 30 July 1914. The ship had a state-of-the-art refrigerated hold loaded with German lager,[3] and Kuhlken planned to return to Bremen with Tasmanian apples and South Australian pears. He kept a close eye on the reports of the troubles at home, knowing that as Great Britain limbered up for the biggest fight humankind had ever seen, Germans would be in hot water if they lingered too long in enemy ports.

On 1 August, Germany declared war on Russia, and the nation was soon also at war with Luxembourg, France and Belgium. In Melbourne, Prime Minister Joseph Cook's Federal Cabinet voted to offer Great Britain the Australian fleet and twenty thousand men if needed.[4] At 7.45 a.m. on Wednesday, 5 August, Captain Kuhlken ordered his crew to head for a safe haven in South America. The ship steamed towards Port Phillip Heads and freedom just as British Prime Minister Herbert Asquith, in London, announced 'that a state of war exists between Great Britain and Germany'.[5] As the *Pfalz* arrived off Portsea at 10 a.m., Australia was also now at war. The gun crews at Fort Queenscliff were guarding the western entrance to Port Phillip, thirty kilometres from Fysh's lodgings in Geelong. Another crew from Fort Nepean fired a 45-kilogram projectile as a warning across the German ship's bow – the first Australian shot of the Great War. The *Pfalz* was turned around, and Kuhlken and his crew, all German naval reserves, were arrested. They spent the next four years in Australian prison camps.[6]

All the young recruits on the compulsory forces, including Fysh, were called up immediately and sent to Queenscliff to dig trenches around the artillery batteries there. In an atmosphere of unprecedented patriotism, Fysh's mother sent him a telegram imploring him to do his duty, and he decided to volunteer. Fysh joined the new nation's cavalry, the Australian Light Horse, after an uncle, Willie Grubb, agreed to help the youngster's application by donating a horse. At a parade in Queenscliff, volunteers were asked to take three steps forward, and Fysh took those steps. He was among the first of more than 400,000 men who volunteered to fight in what would become the Great War.[7] More than a third of Australia's male population of the eligible age of eighteen to forty-four[8] signed up. Recruiting officers turned back boys in short pants and bare feet as young as thirteen, and men as old as seventy-one; some who were rejected stumbled away crying with shame.[9] Back in Tasmania Fysh arrived with his horse at the Pontville army camp, outside Hobart, on the morning of 25 August 1914. C squadron of the 3rd Light Horse Regiment was being formed.[10] Fysh and five other riders were told that places were limited but they could earn their spurs if they passed their riding tests. He watched one horse after another baulk at the first jump and send their riders skidding off the slippery military saddles, which took time to master. Having spent time in those saddles during his days in the reserves, Fysh powered his horse flat out with a light rein towards the first jump and went over with a smile on his face. He was given the regimental number 415 but was still so bashful, even about the name Hudson, that he usually introduced himself as Bill, a name that many of his old friends still greeted him by decades later.

His details were officially recorded on the enlistment form as:

Age: 19 years 8 months
Height: 5ft 10in [178cm]
Weight: 11 stone [70kg]
Chest measurement: 33 inches [84cm]; expanded 37in [94cm]
Complexion: Fair
Eyes: Blue
Hair: Brown
Trade: Wool classer
Next of Kin: Mary Fysh (mother)

Fysh listed his religious denomination as 'Protestant'. Many of his new comrades seemed frightfully sacrilegious to the reticent teenager, who had lived a somewhat sheltered life. In the bell tent he shared with other recruits, he was struck dumb by their bawdy talk and lurid tales of sexual conquests.

A month later, Fysh's painfully thin younger brother Frith listed his father as his next of kin rather than Mary. Frith had stayed on his father's farm, Glen Esk, after Hudson's departure and had worked in clerical positions on Launceston newspapers after leaving school. Most recently he had been farming in New Norfolk, north-west of Hobart. He showed the recruiting officer a letter that Wilmot had written, granting Frith his 'fullest consent to go to the war'.[11] Frith joined the army at Pontville, aged eighteen.

THE FEDERAL GOVERNMENT gathered together as many ocean liners as it could to transport soldiers for overseas deployment. One of them, leased from P & O, was designated HMAT A2 *Geelong*, and it made its first voyage as a troop transport on 22 September 1914, less than a month after Fysh enlisted. The *Geelong* carried 440 soldiers from Melbourne to Hobart, where it collected another 912 soldiers, including Fysh and his 3rd Light Horse Regiment. Fysh had a small flask of brandy with him, a gift from an uncle who told him it might give him comfort in times of need. He formed a close friendship with 24-year-old Joe Radnell, who had enlisted a week earlier in Hobart after a stint as an underground miner in Bendigo.

The ship left Hobart, bound for Albany, Western Australia, on 20 October 1914. Fysh found himself still listening to the bawdy tales of the more worldly-wise recruits, but this time while he was trying to sleep, not on the hard floor of a tent but in a string hammock, swaying to the rhythm of the waves. He arrived in Albany during a rainstorm on 26 October.[12] Five days later, as a bright red dawn broke, the 1st Australian Imperial Force and the New Zealand Expeditionary Force – together known as the Australian and New Zealand Army Corps or ANZAC – left King George Sound off Albany in a breathtaking convoy of thirty-eight troopships, including ten from New Zealand. Smoke billowed from their funnels, and there was fire in the bellies of the men: 21,500 Australians and 8500 New Zealanders. Together they had twelve thousand horses bound for England and the war in Europe, although many would never make it that far.[13] There were 25 female nurses, too. The poet and war correspondent Banjo Paterson, sailing on the *Euripides*, called the awesome spectacle of manpower and machinery 'the most wonderful sight that an Australian ever saw … thirty thousand fighting men, representing Australasia … under way for the great war'.[14]

From the deck of the *Geelong*, Fysh watched the Australian coast fade away until it was enveloped by darkness. Soon there was no sight of land, and nothing but the calm everlasting blue of the Indian Ocean all around him, like a sheet of shimmering glass. The convoy stretched for twelve kilometres, and the Japanese cruiser *Ibuki* was one of four warships enlisted to give the transports safe escort. Every few nights Fysh had a stint on guard duty, four hours on, four off, often watching a freshwater tank. Eight days into the voyage he saw one of the escort ships, the HMS *Sydney*, dash away from the convoy under billowing plumes of smoke; he later learnt that the *Sydney* had hunted down the German raider *Emden* and forced it to run aground in the Cocos Islands.

There was disease among the troops, their constipation eased by laxatives with explosive consequences. Deaths were inevitable even before any of the men heard gunfire, some succumbing to pneumonia and typhoid. Fysh watched the sad, sombre burials at sea: the bodies slid over the ship's rail and plopped into the ocean while the chaplain offered a prayer. The voyage was murder on the horses, too, as they stood or rocked about for days on end in their stalls.

The men and women in the first convoy thought they were going to England and then across the Channel to fight the Germans in France. But with a looming Turkish presence threatening British control of the Suez Canal, the ships were ordered to disembark their troops in Egypt. Fysh's voyage from Hobart ended in Alexandria on 3 December 1914, after forty-three days. He marched to camp along the streets of the ancient city, past yelling crowds with their donkeys, camels and mules. The troops passed through a red light district in what the Australians called a country of 'sand, sin, sorrow and syphilis'. The bashful Fysh had never seen a naked woman until an obese Egyptian sex worker, standing in a doorway as the Light Horsemen tramped by, lifted her dress around her neck and yelled out to the men, 'Come on Australia, me welly good.'[15]

The Light Horsemen travelled by train from Alexandria to Cairo on a journey that took the best part of seven hours. They then marched fifteen kilometres south to Maadi on the east bank of the Nile. It was here the men were reunited with their mounts. Of the 7843 large-boned Australian Waler horses on the convoy, 224 had died on the journey, but that was more than a thousand fewer than expected. Shaky and stiff in their legs and joints, they were led around for a week and ridden lightly for a while until they had recovered sufficiently to be readied as war horses.

Eucalyptus trees had been planted around Maadi 30 years earlier to strengthen the banks of the Nile,[16] and they reminded the young men of the homeland some would soon die for. Fysh's 3rd Light Horse Regiment joined the 1st and 2nd Regiments to form the 1st Light Horse Brigade. Together, they found Cairo a source of wonder and enticement. A fast electric train carried them from Maadi to Cairo's Bab al-Louq Station, from where these hardy young men, in their broad-brimmed hats decorated with emu feathers, would stroll to the local Australian Comfort Committee. Here they were provided with all manner of useful items, from buttons, shoelaces and chocolate to dried soup and pocketknives.[17] In Cairo there were the pyramids to climb and photograph, markets with all kinds of exotic treats, and cinemas showing the latest American and British silent films, as well as the Cairo British Recreation Club and the Cairo Zoo. For some of the men, there were the brothels.

But the city was a mere distraction from the deadly serious business of war, and the Light Horsemen moved to a new base at the Heliopolis racecourse, just north of Cairo. They spent most of their time there digging trenches and engaging in war games in the desert hills. It was springtime, with the fruit blossoms out and the trees covered in new green. Flies were everywhere, and occasionally hot desert winds whipped up the sand into clouds. One Australian soldier died from a fall off a pyramid, and others – breathing in mouthfuls of dust – died from pneumonia. The thousands of others sharpened their bayonets and greased their Lee-Enfield .303 rifles.

BRITAIN WANTED CONTROL of the Dardanelles waterway, which would allow its ships to take the Ottoman capital at Constantinople. If Turkey fell it would keep the Suez Canal safe, while opening another supply route to the Black Sea for Russia in the fight against Germany. But on 18 March 1915, the French battleship *Bouvet* and the British battleships *Irresistible* and *Ocean* were sunk in a failed campaign to shell Turkish forts guarding the Dardanelles. As a result, Sir Ian Hamilton – the British commander of the Mediterranean Expeditionary Force – decided on a land assault upon Turkey's gnarled and twisted Gallipoli Peninsula, even though some of his advisers thought it would be madness to try to launch infantrymen onto steep, rocky cliffs that were heavily fortified.[18]

This was to be Australia's baptism of fire as a new nation, but the excitement at the chance to fight for their country soon gave way to 'great frustration and heart-burning'[19] among the Light Horsemen as the terrain was far too steep for cavalry. So while Fysh and his comrades were left to unsaddle their horses and cool their heels, they watched the Australian infantrymen, chosen for the first landing, march out of Heliopolis to the chest-thumping rallying call of 'Australia Will be There'.

The Anzacs already had a reputation as larrikins defying authority when on Good Friday night, 2 April 1915, they rioted in Cairo's red light district. On a street called Haret el Wasser as many as 2500 tanked-up Australians and New Zealanders ran amok among Cairo's brothels in what became known as the Battle of the Wazir. Many of the Australians were seeking revenge for having caught venereal diseases that would be logged on their service records forever; some were just angry that their beer had been watered down with what many were sure was not water. Several brothels were torched and local firefighters assaulted. Fysh was among a huge crowd of onlookers jostling and fighting for a better view as bed frames, mattresses, tables and clothing were hurled from windows several floors up and then set alight in the street. A general's car was overturned in front of the luxurious Shepheard's Hotel. Police fired warning shots with their revolvers, and a detachment of the Lancashire Territorials formed up in front of Shepheard's with fixed bayonets to restore order.

By mid-April more than two hundred vessels were being readied opposite Gallipoli at Mudros Harbour, on the Greek island of Lemnos, as Britain and her allies prepared the greatest sea invasion in history. In the sky a hundred kilometres away, Wing Commander Charles Samson of the Royal Naval Air Service spied on the southern half of the Gallipoli Peninsula while biplanes from the world's first aircraft carrier, HMS *Ark Royal*, covered the northern sector.[20] The Anzacs prepared to land on a 300-metre pebbly shoreline under the rise of Ari Burnu. The troops called it Z Beach, but it was soon to be known as Anzac Cove. A few kilometres away, fifteen thousand Turkish soldiers stationed in the village of Bigali were readying cannon fireballs that exploded in the air and burnt everything beneath.

The first wave of the dawn landing at Gallipoli was planned

for 25 April 1915 with 1500 men, 500 from each from the 9th, 10th and 11th Battalions of the 3rd Infantry Brigade under Edinburgh-born Lieutenant Colonel Ewen Sinclair-Maclagan.[21] Another 2500 Australians were to be close behind, followed by 8000 more men from the 1st and 2nd Brigades with about 4000 New Zealanders later in the day.

At 3.30 a.m., in the ghostly quiet of that dark, still morning on the Aegean Sea, steamboats started their dangerous journey due east, towing rowboats loaded with soldiers like lines of wriggling snakes. As the boats neared the shore at 4 a.m., the first faint glow of a lemon-coloured dawn allowed the Anzacs to distinguish between hills and sky. The calm sea glistened like a sheet of oil, but as the Anzacs scrambled into the water near the shore, bullets splashed all around them in the half light. Most of the soldiers jumped into water up to their hips, some up to their chests, but others misjudged the depth and drowned under the weight of their 36-kilogram packs. More and more Anzacs followed. By noon, most of Sinclair-Maclagan's 3rd Brigade, the first to hit the beach, were dead or incapable of fighting on. The landings ceased temporarily at 12.30 p.m. as the Anzacs gathered their wounded. By nightfall, 754 Australians and 147 New Zealanders had died on Gallipoli that day.

Two weeks later, after the first Anzacs had dug into their precarious positions on Gallipoli, the 1st Light Horse Brigade received orders to join them. Fysh later recalled a sense of relief that they would not be denied a place in the fight, though they would attack as infantrymen without their mounts, well aware of the carnage that had greeted the first landing. They embarked from Alexandria on 9 May. Those who survived on Gallipoli would not see a woman for seven months; those who did not would be buried in rough graves thousands of miles from home.

For Fysh, there was a short voyage across a smooth Mediterranean and then Holy Communion at midnight in a crowded, dimly lit cabin as the young men thought about their families at home, and of the likelihood of an imminent death and burial on a foreign shore. By the dawn's eerie light on 12 May, seated in a barge powering across the Aegean, Fysh felt the knots in his stomach become agonisingly tighter as a sense of turmoil surrounded him – he was scared almost to death and wasn't ashamed to admit it. But the Turks had been pushed back from their defensive positions, and the Light Horsemen made it onto the beach without the fiery, bloody welcome that had greeted the first Anzacs there.

As Fysh crawled up a near-vertical ravine, his heart began racing faster than it had ever beaten before when he felt the terrifying, almost surreal sensation of men shooting at him. There was the deadly rattle of machine-gun and sniper fire from the heights as the Light Horsemen clambered their way up the peninsula's main artery, Shrapnel Gully, so named for the Turkish shells constantly exploding above the invaders. The bombs sent down showers of lethal metal, tearing apart the bushes around the Anzacs. Fysh dodged and ducked from one sandbag shelter to the next as bullets kicked up dust around him.

Finally he reached a spot under Courtney's Post, and the men were ordered to dig in on the steep hillside covered with stunted pine and shrubs. John 'Simpson' Kirkpatrick,[22] a 22-year-old stretcher-bearer with the 3rd Australian Field Ambulance, ferried wounded men downhill on his donkeys to the hospital boats waiting at the beach. Fysh marvelled at Kirkpatrick's stamina and courage as his faithful beasts of burden brought the wounded through that mad, twisted country day and night.

It was near Courtney's Post three days after Fysh arrived on Gallipoli that a Turkish sniper shot Australia's most senior officer, Major General William Bridges, through the femoral artery in his right leg. The wound became gangrenous, and Bridges died on board a hospital ship on 18 May. The following day Turkish machine gunners shot John Kirkpatrick through the heart.

FYSH SPENT HIS FIRST FEW days on Gallipoli scurrying from hole to hole like the rabbits back home when his dogs were after them. He was every bit as afraid and felt just as helpless, living in conditions more inhospitable and dangerous than he had ever imagined back at Ketteringham. He would rarely taste fresh vegetables for the next seven months, and he later vividly remembered the torpedoing of HMS *Triumph* off Gaba Tepe on 25 May because it resulted in fresh onions washing up on the beach.

At regular intervals the Turkish batteries in the Dardanelles would blast the Anzacs with 'Jack Johnson' bombs, so named after the African-American boxer who until recently had held the world heavyweight title. The bombs would whistle a warning seconds before their arrival and send the men diving for cover as they exploded with an almighty roar in clouds of black smoke. Lice and dysentery became intolerable, while the medical officers would give men a handful of rice to help with their stomach ailments. Toilets were hastily erected over slit trenches under canvas supported by forked boughs – but men had to avoid them in daylight, regardless of nature's strident calls, after the first of Fysh's comrades to be shot was hit in the leg by a sniper's bullet while relieving himself.

After a few weeks on Gallipoli, a warm July meant a chance for Fysh to escape the grubby clothes he wore day and night. With lookouts posted, he and some comrades were able to have an occasional day at the beach, bathing naked in the calm, refreshing waters of the Aegean. The men were able to delouse themselves under the hot beachside sun, inspecting every inch of their discarded uniforms and running lighted matches along the seams to destroy the lice and their eggs. The Turks were always doing their best to destroy the Anzacs, too, and often the beachgoers would have to race from the shallows or the stony shore as shrapnel bursts arrived from 'Beachy Bill' or 'Asiatic Annie', the names the Aussies gave Turkish gunners.

The beach was also the scene for the manufacture of improvised Anzac bombs to counter the seemingly limitless supplies that the Turks were forever hurling at Fysh and his comrades in the close confines of trench warfare. The Australians used the open spaces of the beach to put together their own weapons of mass destruction, weaving together explosives with whatever metal was handy, including bike pedals, and packing them into jam tins loaded with a few centimetres of fuse. Sometimes these bombs were tested in the shallow waters of the Aegean, which invariably produced more than a few fish dinners as a welcome change from bully beef and hard biscuits.

The men of the 3rd Light Horse Regiment were ordered to take a defensive role throughout much of the Gallipoli campaign, and they were absent when 372 from their brother regiments, the 8th and 10th, were killed or wounded in a bayonet charge on Turkish machine-gun posts along the Nek, a narrow stretch of ridge, on 7 August. Many of the young Australians were cut down before they could get more than a few yards from their trenches. On the same day, two hundred men of the 1st Regiment attacked at Dead Man's Ridge, with 148 killed or wounded. From a vantage point in reserve near Sniper's Nest, Fysh had a distant view of the battle of Lone Pine between 6 and 10 August, when the Australians suffered more than two thousand casualties in a rare victory on Ottoman soil.

Fysh had become a machine gunner, and during the Anzacs' August offensive he was assigned to Quinn's Post, the most advanced position of the Anzac line and the bulwark against the Turks breaking through to the heart of the Allied camps. The Turkish trenches were just fourteen metres from the Anzac defences, and there was constant tunnelling by both sides in an attempt to blow each other to kingdom come. Being a machine gunner induced in Fysh what he later recalled as 'a wonderful feeling of heartfelt but guilty relief'[23] that he would not be in the first line of soldiers to climb from the trenches to attack the enemy strongholds. Those men had big white patches sewn on the backs of their uniforms so that the second line of Anzacs could distinguish them from the enemy after they went over the top.

There was always heavy fire from both sides at Fysh's position, and once a Turkish bomb sliced through the Anzacs' protective wire netting. A small man from the 2nd Light Horse didn't hesitate to pick up the sputtering bomb and hurl it straight back. Fysh's heart rate had just slowed down when it took off again at yet another fizzing sound, this time right by his feet. No one in his trench had seen the bomb come over. They dived desperately for the ground as it burst with an ear-splitting roar, sending pieces of jagged metal in all directions. Fysh felt his forehead go numb, then a burning pain – one piece of shrapnel had just missed his eye. Partially blinded, he stumbled to the forward field ambulance station with blood streaming from his head. A bullet splinter had ricocheted off the parapet, and he thanked heaven it was only a splinter. He was given a thin strip of sticking plaster for the deep wound and immediately sent back into the line. Another shrapnel fragment lodged in his water bottle, like a hot metal plug, and other pieces scarred the wooden butt of his .303 rifle, which he would later proudly display to the Light Horse Commander General Harry Chauvel on an inspection tour.

Bombs, bullets and illnesses weren't the only hazards for the Anzacs. The balmy summer of July and August soon gave way to bitter cold and snow as the battle for Gallipoli became a bloody stalemate. The Turks could not be pushed back any further, and the Anzacs suffered enormous casualties trying to inch their way forward on what became a freezing, foreboding terrain.

THE LIGHT HORSEMEN WERE ordered to move up to Pope's Hill, a razorbacked ridge above Anzac Cove that offered a favourite spot for Australian snipers. On the way, Fysh shuddered as he brushed against the feet of a dead Turk sticking out from the side of a trench. Then he gagged at the sight of maggots that crawled on the unburied dead lying around the bloodstained sandbags of the Australian parapet. Bullets whizzed past him in his trench, and he kept his head down as he and the other men put the barrels of their rifles above the ground and shot in the presumed direction of the Turkish gunfire seventy metres away. Fysh was in a state of panic when the sun came down on his first day on the hill, and he was trembling by the time two men climbed past him and sprinted into no man's land to retrieve the body of a mate. They dodged gunfire to drag the corpse back into the trench, pulling it in over the living.

Before long Fysh was a dab hand at midnight burials, carrying the dead down the steep gorges on stretchers for internment in a cemetery that would have been a target during daylight. Exhausted from the grim and gruelling journey, he would help dig the grave and then roll the body into it and bow his head as the chaplain offered rushed, hushed prayers. Invariably distant machine-gun fire and explosions would give a grim accompaniment to the proceedings and bathe the chaplain's white robes in an eerie red glow.[24]

Chapter 4

Gallipoli and the Sinai Desert gave me an appreciation of the ordinary man – which I am myself – that lasted right through life.

HUDSON FYSH ON ONE OF THE LESSONS HE LEARNT IN WAR[1]

BY SEPTEMBER 1915, Fysh and the other Anzacs were living, in his words, like rats in holes, hanging on to a tiny piece of hillside with the enemy in front and the beach a few hundred metres behind. There were weeks of stalemate, long periods of boredom punctuated by moments of sheer terror when gunfire and bombing resumed. Fysh often cooked for himself over a small fire where he sheltered, using empty tins as pots and crushed army biscuits as thickeners, sometimes complementing his bully beef rations with the wild thyme growing all around. Occasionally there were treats sent in food parcels from home, such as chocolate, raisins, condensed milk or Quaker Oats, but fresh food was scarce, and the worsening cold was soul-destroying. Still, Fysh refused to crack open his uncle's flask of brandy, fearing even more desperate times ahead.

The Australian war correspondent Keith Murdoch sent a 25-page, 8000-word damnation of the Gallipoli campaign to his new prime minister, Andrew Fisher, calling it 'a costly and bloody fiasco', and 'undoubtedly one of the most terrible chapters in [Australia's] history'.[2] Murdoch blamed aristocratic English buffoons: 'the red feather men' among His Majesty's top military brass. Appointments to the general staff, Murdoch said, were made from 'motives of friendship and social influence' rather than competence.[3]

Fysh woke one glacial morning to find a large, jagged piece of ice between him and the next digger as they huddled through the night on their steep hillside. Frostbite chewed into the Allied troops, especially the Nepalese Gurkhas, and many were evacuated to Lemnos with black and mummified feet. Fysh was spared the worst of it, bolstered by a regular dose of thick, fiery military-issue rum that warmed him from head to toe. The rum was prized, hoarded and traded, and men were glad to have teetotallers in their midst.

On 22 November 1915, Britain's supreme commander Lord Kitchener reluctantly agreed to evacuate Gallipoli, and his government approved the plans on 7 December. By then Fysh and his comrades had not tasted meat for three months. A conference of the commanders decided that the Anzacs would bluff their way out of Gallipoli in stages, a plan conceived by Brigadier General Cyril Brudenell White.[4] Two-thirds of the army would leave in trickles, with the other third to sprint to boats waiting on the beach in the dead of night.

The evacuation began after sunset on 13 December. By 16 December the Anzac garrison had been reduced from forty-one thousand to twenty-six thousand, and the nights of 18 and 19 December were chosen for the final mass evacuation with the last dash of two thousand men at 2 a.m. on 20 December. The Turks bombarded the beaches as usual on the night of 19 December, but the weather was perfect for the Anzacs' final escape: the sea and air calm, the sky cloudy, foggy and dull, with a little drizzle, so that everything in the distance was dim and blurred. By 8 p.m. there were only five thousand Anzac troops in the whole of the peninsula, thinly holding the front line against 170,000 of the enemy. But the Allied gunfire was staged to confuse the Turks into thinking there were many more.

By midnight, groups of six to a dozen men crept down the gullies to the beach like so many rivulets flowing into the main river. The last man in every case was an officer, closing the gully with a previously prepared frame of barbed wire,[5] or lighting a fuse that an hour later would blow up a tunnel to prevent the Turks from giving chase. The marching lines arrived at Gallipoli's Brighton, Anzac, Howitzer and North beaches almost at the same instant, ghostly figures in the dim light. They then marched onto their respective jetties, sandbags deadening the sound of their feet. They boarded the 400-man motor barges known as 'beetles', crowding in together, the general pressed up against the private. The only sound was the hum of the engines.

At 1.55 a.m., the last man got away, leaving behind only decoy guns, water draining from tin cans tied to the triggers. The Anzacs had withdrawn forty-five thousand men, along with mules, guns, stores, provisions and transport under the noses of the Turks.[6] The evacuation was the most successful operation of the whole Gallipoli campaign, which had resulted in the deaths of more than eight thousand Australians and almost three thousand Kiwis.

On a ship's table teeming with bugs, Fysh tried to get some sleep as the vessel zigzagged across the Mediterranean to dodge the German U-boats. When he and his comrades finally marched back into camp outside Alexandria, their dishevelled appearance was greeted not with a brass band but wild laughter from crowds looking on. Some of the evacuees had not seen women for months and were eager to make up for lost time, and venereal disease raged. In Alexandria, Fysh faced his first 'short arm' parade, with the regiment's medical officer leaving nothing to chance. The men lined up in double rows waiting for the command: 'Unbutton flies!' As man after man offered up his private parts for the medical officer's inspection, the parade was conducted to a chorus of belly laughs.

After a few days in Alexandria, the Light Horsemen left in open cattle trucks for Cairo and their horses at the Heliopolis camp. Almost to a man, they gave Alexandria a farewell salvo, opening fire on the immense flocks of wild ducks that made their home on the city's vast lakes.

On his first night back in Cairo, Fysh slept between white sheets at a Salvation Army hostel. It was the first time he had slept in a proper bed since enlisting near Hobart sixteen months before. But after a bright and merry Christmas, his Light Horse

Regiment was on the move again, riding out of Heliopolis on 31 December 1915 bound for Wadi El Natrun and the clouds of pink flamingos that lived in the marshland below sea level there, sixty kilometres north-west of Cairo.

There were sightings of wild ibex goats in the desert on the way, but the Australians were more interested in the threat of a raid by Senussi tribesmen on the Kataba Canal, a prime source of fresh water for Alexandria and the Allied troops there. The threat passed, and the Light Horsemen moved on to Girga, two hundred kilometres south along the Nile, where wandering desert tribes were sparking alarm. Fysh took time to rest and recuperate, lazing in the sun, swimming daily across the world's longest river and back, taking his camera as he went fossicking through the hills for ancient tombs, and watching the crowds line up for their turn outside a brothel offering 'black velvet'.

At the end of May 1916, the 1st Light Horse Brigade was ferried over the Suez and rode on forty kilometres east to the British Garrison at Romani, built among the shifting sand dunes of the Sinai Peninsula. Romani was an important defence against the Turks taking the Suez Canal, and two thousand Light Horsemen joined the thirteen thousand infantrymen of the British 52nd Division. The Australians were being paid six shillings a day and could barter with locals for unfamiliar food and small luxuries, while the British paupers based there were on just a shilling a day and had to make do with army rations.

Most of the water in Romani was foul, but there was shade and succulent fruit under the date palms, and the red sunsets were breathtaking. Reconnaissance missions in the area would sometimes require twenty-four hours in the saddle with only a brief rest for the horses to drink brackish water as they carried men and equipment totalling 120 kilograms. The water was so rotten that Fysh finally opened his uncle's flask of brandy. In the dry desert heat, it tasted like heavenly nectar.

Occasionally, to give his horse a little respite, Fysh would jump off his mount and walk beside him, plodding through the sand. At night men were sometimes so exhausted they would get off their animals only to fall asleep holding the reins.

AIRCRAFT TECHNOLOGY had advanced at the speed of light since Fysh saw Maurice Guillaux's primitive aircraft in Geelong just two years earlier. Fysh had only encountered a few enemy machines making bombing raids over Gallipoli, and he had seen them cause little damage. So, he was unprepared when he saw an enemy bomber approaching the Romani camp one morning. He watched in stunned disbelief as vague shapes descended from the machine and then caused a straight line of explosions, one after the other, that came closer and closer to him until the final bomb burst in the middle of the camp. It killed an officer, seven soldiers and thirty-six horses.

Not long after that bombing raid, Fysh arrived at the small, isolated garrisons of Qatia and Oghratina where British soldiers were stationed to protect a party of Royal Engineers on a well-digging expedition. Fysh rode up to find scores of rotting horses, killed during an attack by more than three thousand Turkish troops, including a battery of twelve machine guns.

Tonsilitis caused Fysh to spend two weeks in hospital at Port Said before he was assigned to the 1st Light Horse Brigade, 1st Machine Gun Squadron, under the command of Lieutenant Ross Smith,[7] a 23-year-old South Australian who had landed on Gallipoli the day after Fysh. Smith was about 175 centimetres tall (5 feet 9 inches) and described as wiry with a 'typical cornstalk build, complexion rusty, freckled, blue-grey eyes – aquiline real hard case features you could chop wood with' and 'a laconic manner, truly casual Australian'.[8] Before the war Smith had worked in the Adelaide warehouse of the merchants Harris Scarfe.

The Anzac machine gun crews would need all their rapid-fire skills late on the night of 3 August 1916, when fourteen thousand Turks, aiming to establish a position within artillery range of the Suez Canal, came charging across the desert with blood-curdling cries of 'Finish Australia'. The Turks ran into the machine guns of Fysh's 3rd Light Horse, which had established isolated outposts in the path of the main attack. Throughout the following day, the German general Kress von Kressenstein[9] ordered his Turkish troops to hurl everything they had at the Australians to get past this thin line of defence. The Turkish numbers were overwhelming, and two of Fysh's Tasmanian pals, Owen Bingham and Alf Tolman, both sergeants, were killed. Another Tasmanian, Jack Riley, was shot in the foot.

Fysh feared death at any moment as the Turks kept surging in waves, and he was glad to hear the order to retreat with the other machine-gun squadrons to the safety of the main defensive line. New Zealand reinforcements arrived, and the British infantry held the hill trenches, knowing their lives depended on it. By sunset on 4 August the Turkish attack had failed all along the line, and Fysh and the machine gunners were switched to Romani's left flank where the Turks were attempting an enveloping manoeuvre.

Fysh's commander Colonel 'Galloping' Jack Royston[10] was shot in the calf. Ordered by Major General Chauvel to have the wound treated, Royston instead earned his nickname by galloping off with strips of bandage trailing behind him before the dressing was completed. The next day, Chauvel ordered Royston to hospital, but Galloping Jack said there was too much to do and left within a few hours, the bullet still in his leg.[11] He charged back into battle, galloping up to Fysh and the other machine gun teams and bellowing, 'Fire! Fire! Why the hell don't you fire?'[12] The Turks were more than a kilometre away, but the men let out a few bursts in their general direction to keep the commander satisfied. It was said that Royston wore out fourteen horses through hard riding that day.

The Battle of Romani lasted five days and ended in failure for the Turkish attack. The British lost 202 of their men and buried 1250 of the enemy; they also took 4000 prisoners. The Light Horsemen chased the Turks forty kilometres west to Bir al-Abed where they again fought a ferocious battle, the Turks fighting back with the superior firepower of the heavy cannons that they had dragged across the desert sands on thick wooden planks.

In a brief respite from the fighting, Fysh's closest friend Joe Radnell went to collect a tin of peaches. He had eaten half of them and was bringing the rest back for Fysh when one of the Turkish 5.9-inch shells exploded near him, sending a shard of metal through his chest and killing him instantly. Fysh was struck dumb. Chauvel ordered the Anzacs to withdraw, as the horses could not go on without water, but Fysh was eager to avenge his mate. Feeling brave for the first time in the war, he protested loudly with all the curse words he had learnt from two years in mortal combat, but no amount of protesting could change the orders.

Before making their retreat, Fysh and some comrades had only minutes to bury Joe in a shallow grave on the side of a sandhill. They placed a rough wooden cross over the grave, climbed aboard their mounts and rode out. Lonely and forlorn, Fysh tried to stop his tears falling as he rode back to Romani leading Joe's horse, which bore an empty saddle.

DURING THE FOLLOWING three months the Turks retreated further east to El Arish, a hundred kilometres from Romani, while the Allies consolidated captured territory, establishing garrisons throughout. Fysh was promoted to corporal, which he thought was a rather cushy job of doling out rations and ammunition, and tending to the lead horses. The prospect of imminent death aside, it was a much healthier situation than on the Gallipoli cliffs. Fysh slept on the sand under the cloudless desert skies, scooping out a comfortable depression for his hip and with a groundsheet over him to keep off the dew.

British forces carried out patrols and reconnaissance missions to protect the construction of a railway and water pipeline, and to thwart the chance of Turkish forces mounting raids by destroying water cisterns and wells. Fysh and the 3rd Light Horse rejoined the Allied advance across the Sinai in November 1916. The British were marching across the desert on wire netting spread ahead of them in the sand.

On the morning of 21 December, Fysh was part of the Anzac Mounted Division, which reached El Arish to find it recently deserted by the retreating Turks. Fysh camped under the date palms on a Mediterranean beach. The following night he rode with a force of six thousand men on a wild dusk-till-dawn gallop across the desert for forty kilometres south. The horses' nostrils flared, and the eyes of the young riders were focused on battle as shrapnel from the Turkish mountain guns rained down all around.

Then, on 23 December 1916, in the Battle of Magdhaba, the

Anzac Mounted Division launched its attack against the well-entrenched Ottoman forces defending a series of six temporary fortresses. With aircraft providing reconnaissance, the Anzacs galloped as close to the Turkish front line as possible, then dismounted to attack with bayonets under the covering fire provided by artillery, and by Fysh and his comrades working the machine guns.

Fysh charged from one position to another, bringing ammunition to the machine-gun teams. He saw the man working the number one gun shot in the chest by a bullet that first tore through the gun's breech block. Fysh thought for a moment about taking the man's place on the trigger but candidly admitted later that he was suddenly gripped by 'fear, terrible fear'.[13] He convinced himself that his duty was to go back to the horses and the ammunition depot, and for a long time after that he felt he had 'squibbed' it as twenty-two men around him died.

Finally, at around 4.15 p.m., the Turks surrendered with ninety-seven of their ranks dead and three hundred wounded. The Allies took 1282 prisoners.

The next day was spent burying the dead, and burning everything that would burn. Fysh and his comrades rode back to El Arish for the second consecutive night in the saddle with a day's bloody fighting between. Many of the wounded Australians were dragged on makeshift stretchers behind camels as dust clouds rose all about. Some of the horsemen were so spent that many fell asleep in the saddle, their mounts veering off from the main column as though riderless only to be herded back by comrades who were still awake.

Two weeks later, Fysh was back in the saddle for the Battle of Rafa as the Allies aimed to complete the recapture of the Sinai Peninsula. The Egyptian Expeditionary Force's newly formed Desert Column attacked the 2000-strong Turkish garrison at El Magruntein on 9 January 1917 with the aim of capturing it in daylight or retreating in the night for want of water. The garrison was housed on rising ground surrounded by flat grassland. It was a day of intense fighting, and Fysh was kept constantly on the move, burning up the desert sand supplying belts of .303 calibre bullets for the machine guns. At one stage he took cover from enemy fire behind a rise hardly large enough to hide a mouse when he noticed a wounded man lying exposed with bullet strikes erupting in the earth all around. The knots of fear in Fysh's stomach that had plagued him at Magdhaba struck again – but as he trembled, something deep inside forced him to bolt out into the open and drag back Lieutenant Clem Harris, who had been shot through the ankle.

Finally, encircled by Australian Light Horsemen, New Zealand mounted riflemen, mounted British yeomanry, cameleers and armoured cars, the Turks waved the white flag right at dusk. They had suffered 200 dead and 168 wounded, and 1600 were taken prisoner, while 71 on Fysh's side would never go home.

THE NEW YEAR, 1917, brought a commission as Fysh was promoted to second lieutenant with the machine gunners. He was replacing Ross Smith, who had volunteered for the Australian Flying Corps as an observer helping to defend the area around the Suez Canal. Fysh was again plagued by self-doubt, worried about whether he was up to the job and self-conscious about how his mates would react to him becoming an officer, now that another young Tasmanian Ernie Reeman was appointed his groom and batman. But with the Allies mounting an attack on Gaza in an effort to take Jerusalem, there was little time for reflection.

Ross Smith and the rest of Australia's young airmen had become crucial to the Allied assaults in Palestine. In the early days of the war in Europe, the military aircraft were little more than large kites with putt-putt motors; they were used for observation like the balloons of earlier times. In the first few days of the war, pilots of rival scouting aircraft threw nothing more savage at their foes than an upturned finger or brandished fist. But before long pilots and observers were throwing rocks at enemy aircraft, then using shotgun blasts, grenades, and sometimes even rope and grappling hooks to jam propellers.[14] The combat between rival aircraft became so fierce that their aerial battles were called 'dogfights'. Aviators who shot down at least five of the enemy machines were called 'aces'.

Australia's only military aviation base, the Central Flying School, had been established at Point Cook, south-west of Melbourne, just before the outbreak of war, with two flying instructors and five flimsy training aircraft. The Australian Flying Corp's first complete flying unit, No. 1 Squadron,[15] left Australia for the Middle East in March 1916. When they arrived in Egypt, they were still novices at flying, let alone aerial combat, but they were eager to learn quickly from instructors provided by Britain's Royal Flying Corps.[16]

Initially, the Australians were equipped with cumbersome two-seat B.E.2 biplanes designed by Geoffrey de Havilland before the war and built by Britain's Royal Aircraft Company. Powered by Renault engines, the B.E.2s had good stability, which was essential for aerial photographic reconnaissance missions using the primitive plate cameras of the time. But the stability was coupled with poor manoeuvrability, and with a top speed of just 116 kilometres per hour the machines were no match for the faster and more agile German Fokkers, Aviatiks and Albatros Scouts supporting the Turkish armies in Palestine. By 1917, Anthony Fokker – a young Dutchman born to a coffee plantation owner in what is now Indonesia – was turning out thousands of aircraft for the Kaiser's war effort. Fokker had perfected his 'interrupter gear', which allowed pilots to fire bullets through a spinning propeller, and his aircraft had effectively become deadly accurate machine guns on wings. They were also capable of carrying large quantities of heavy bombs.

After the Allied victories at Romani, Magdhaba and Rafa, the Ottoman Army was pushed still further east, and Fysh and the Light Horsemen prepared to fight in a combined assault on the enemy stronghold of Gaza, the gateway to taking Jerusalem. In the air, the Australians had a key role in destroying Turkish supply lines, and they quickly gained a reputation for extraordinary daring and courage as they flew in the face of fear every day.

On 19 March, a week before the Allies' mass attack on Gaza, a pilot from Britain's No. 14 Squadron was shot down during a bombing raid on the rail line between Junction Station and Tel el Sheria. Turks were known to have hacked Allied airmen to death.

Despite the enemy forces closing in on the damaged machine, Ross Smith, now an Australian Flying Corps observer, and his pilot Reg Baillieu[17] swooped onto difficult ground to rescue their comrade, who had set his machine on fire lest it fall into enemy hands.[18] As the Turks charged towards the stranded airmen and bullets began whistling around the Australians, Smith drew his revolver and held them off while Baillieu loaded the Englishman into the Australian machine. Then Smith, still firing at the Turks, leapt onto the overloaded aircraft as all three made their escape. The two Australians were awarded the Military Cross.

On the following day, as if to go one better, 22-year-old Frank McNamara[19] – who had trained to be a schoolteacher in Victoria – dropped three bombs from his single-seat Martinsyde aircraft onto the Turkish rail line only to have a fourth explode prematurely. The blast hit McNamara like a sledgehammer, and the shrapnel left one of his thighs shredded.[20] He was heading back to his base at El Arish[21] to get urgent treatment for his wounds when he saw Captain Douglas Rutherford, a former salesman from Rockhampton, in deadly distress. Rutherford had been on the same bombing mission, but his two-seater B.E.2 had been hit by anti-aircraft fire, and he had crash-landed near a Turkish base.

Despite his wounds, the rough terrain all about, and the Turkish cavalrymen on the move towards the scene with blood in their eyes, McNamara dived down and landed close to the stricken machine. Because there was no spare cockpit in the single-seat Martinsyde, Rutherford scrambled onto McNamara's wing and held on to the struts for dear life. But his weight overbalanced the Martinsyde, and with the leg wound McNamara couldn't control the aircraft, which turned over as he tried to take off.[22] So the pair set fire to the Martinsyde and rushed back to Rutherford's B.E.2.

As McNamara jumped into the cockpit and held off the Turkish cavalry with his revolver, Rutherford made some quick repairs. The B.E.2's engine roared, and despite some damage to its struts and fuselage the two airmen finally lifted off, McNamara dodging the Turkish bullets as a dust cloud enveloped the aircraft. McNamara lost so much blood and was in such agony that he almost blacked out several times, but somehow he managed to fly the aircraft for 110 kilometres to safety. He 'could only emit exhausted expletives' when he landed,[23] and he lost consciousness soon after. Rushed to hospital, he almost died after an allergic reaction to a tetanus injection, and doctors had to administer artificial respiration to keep him alive. Soon, though, the plucky Victorian youngster was 'sitting up, eating chicken and drinking champagne'.[24] For his 'brilliant escape in the very nick of time and under hot fire',[25] McNamara was presented with the Victoria Cross.

THE TURKS REPELLED THE Allied Forces at Gaza on 26 March 1917. A second attack was launched on 17 April, and Fysh later candidly admitted it was a time of 'intense fear' for him. At one point he raced across an open cornfield looking for better positions for his machine gunners only to find a large detachment of Turkish cavalry armed with lances and preparing to charge – mercifully, heavy gunfire kept them at bay. This was the last action Fysh saw as a Light Horseman. He admitted he was deathly afraid in every battle, but soon enough he was looking at the even more dangerous combat zone of the heavens in order to continue the fight.

After the Turks drove back the second attack on Gaza, Fysh's machine-gun squadron was camped near the beach-side base of the No. 1 Flying Squadron at Belah, as Fysh called it; its proper name is Deir al-Balah, Arabic for 'Monastery of the Date Palm'. Fysh rode over to see a boyhood chum from Tasmania, Lieutenant Eustace Headlam.[26] A star cricketer and golfer before the war, Headlam had served with the 3rd Light Horse in Gallipoli. He was now an observer with the No. 1 Flying Squadron, and he told Fysh that the Australian Flying Corps needed more men for that role.

Since its beginning the squadron had enlisted volunteers from the Light Horse, believing that skilled horsemen were the best possible material for the 'cavalry of the sky'. Good horsemen had many of the qualities of good airmen, with their sense of balance and their cool heads in a crisis. The dextrous hands and instinctive touch required to ride a fast horse were seen as precisely the skills a pilot needed when handling an unpredictable machine, bucking and swaying as they did in the wind and under enemy fire.[27] But once again Fysh's self-doubt raised its head, and he wondered whether he was brave enough to be an airman.

Headlam had endured his share of narrow escapes as a flyer. Only a few weeks earlier[28] he was acting as an observer in a reconnaissance mission being conducted over Tel el Sheria by the Flying Squadron's commander Major 'Dicky' Williams.[29] Anti-aircraft fire hit the motor of a Martinsyde being flown by Lieutenant Adrian Cole,[30] and he came down behind Turkish lines. The day before, Cole's actions in attacking six enemy aircraft threatening the Light Horsemen around Gaza had earned him a Military Cross. Now he was contemplating shooting it out with a nearby patrol of Turkish horsemen, when Williams and Headlam arrived like guardian angels in their two-seater B.E.2e. Not wanting the Turks to capture his damaged Martinsyde, Cole fired a flare from his Very pistol – a flare gun – into the petrol tank,[31] just as Williams dragged him on top of Headlam in a machine designed for one passenger. Cole dropped his Very, and Williams was furious, bellowing that the Flying Corps was short of them; he ordered Cole to retrieve it. The nearby Turkish horsemen were apparently so stunned by what they were seeing that they held off firing at the Australians until the overloaded B.E.2 lumbered along the ground and took to the air. Williams brought his airmen home, continuing a proud tradition in the squadron of never leaving anyone behind.

Others were not so lucky. A day before that escapade, Lieutenant Norm Steele had been shot down over Gaza, and he died of his wounds soon afterwards. Only a week or so before Fysh arrived at the airfield, Lieutenants Gerald Stones and Joe Morgan had also been killed in action.

Despite the dangers, Fysh accepted Headlam's invitation for a ride on a B.E.2e piloted by Flying Officer Stan Winter-Irving,[32] a young grazier from one of the wealthiest families in Victoria.[33] There was the ever-present threat of German flying machines appearing to spit fire over the desert, but Fysh revelled in the ride of his life. As he sat in the forward of the two seats, his nervousness gave way to euphoria while Winter-Irving powered his aircraft down the runway and lifted off. The surrounding countryside and bell tents of the airmen quickly grew smaller and smaller as the pilot pushed the machine into the wild blue yonder. Fysh felt the same sense of dreamy astonishment he had fantasised about as a boy when he imagined himself taking off through his bedroom window and floating around his mother's garden. He now gazed about him in all directions, looking at the world from on high for the first time in his life. Fysh put in an application to become an aerial observer.

There was no such excitement for Fysh's brother Frith. He had crashed in a heap on the Western Front after first serving in Egypt. Frith had become a driver in the field artillery before falling ill on the Somme.[34] He began to strip weight from his already emaciated frame, and he suffered badly from night sweats and had difficulty breathing. After doctors diagnosed his condition as tuberculous pleurisy, he spent months in English hospitals before being invalided home.

While Fysh's application to be an aerial observer was being considered, seven Australian machines bombed Ramleh before dawn on 23 June, destroying two German aeroplanes on the ground. Three days later, eight more Australian aircraft bombed the Turkish Fourth Army headquarters at the Mount of Olives in Jerusalem.[35]

One of Fysh's last acts as a Light Horseman was to rebury his dear friend Joe Radnell in a proper grave. *State Library of NSW FL7583095*

Fysh wanted desperately to be part of the action. Major Williams had told Fysh not to get his hopes up, as there were many applicants for the No. 1 Squadron. But Williams liked the look of the lad and found a spot for him at a gunnery course in Aboukir (now known as Abu Qir), along the beach north-east of Alexandria, that would teach Fysh the finer points of aerial combat.

There was just one duty Fysh needed to perform before he left the Australian Light Horse: a favour for an old mate. He had asked permission from his commanding officer to ride out and rebury his friend Joe Radnell in a more fitting grave. Permission was refused, the situation around there being so dangerous, so a few days later Fysh gathered some of Radnell's other friends and set off under the cover of darkness across the desert using a compass to find the way to a rough wooden cross. The remains were dug up, placed on a ground sheet, and carried to firmer ground for reburial. Fysh said a last goodbye to his friend and bade farewell to the 3rd Light Horse Regiment. The war was changing direction, and Fysh was reaching for the sky.

Chapter 5

I was a much quieter, fearful type, who counted the odds. Perhaps the two of us together added up to much more than either of us individually. I was … an introvert. McGinness was an extrovert if ever there was one.

HUDSON FYSH ON HIS WARTIME PILOT AND BUSINESS PARTNER PAUL MCGINNESS[1]

HUDSON FYSH WAS perched precariously in the front cockpit of Lieutenant Ron Austin's[2] B.E.2e, waiting to power down the narrow desert airstrip for his first mission as a member of the Australian Flying Corps. It was 3 August 1917.

The fragile construction of the flying machines astonished Fysh: they were cloth, wire and wood wrapped around treacherous, noisy engines that he regarded as 'cranky, finicky things never to be trusted'.[3] Fysh and Austin were dressed in fleece-lined top boots, heavy leather coats and gloves to cope with the high altitude that made them feel frozen despite the desert heat below. Their machine had the British red, white and blue roundel painted on its sides and under the wings to quickly identify it. The machines they would oppose were marked by black Iron Crosses. Just three years after Fysh saw an aircraft for the first time in Geelong, he was about to fly on one into an artful trap.[4]

The German air wolves, led by Oberleutnant Gerhardt Felmy[5] in his black-tailed Rumpler aircraft, had been feasting upon Australian pilots and their ponderous, ill-equipped machines, and Fysh knew they would do their best to make his first aerial mission his last. He was terrified – the Germans had recently killed several of Australia's best young pilots.

Fysh was officially seconded to the Australian Flying Corps on 6 July 1917.[6] Two days later, near Gaza, the Germans attacked three Australian aircraft on a scouting mission. Captain Charles Brooks, an Englishman attached to the Australian squadron,[7] was killed when the wings of his Martinsyde folded and its tail fell off, sending his machine crashing to the ground like a stone. Felmy, who was involved in that dogfight, then went after the B.E.2e of Tom Taylor[8] and his observer Frank Lukis.[9] With two well-directed bursts of machine-gun fire, Lukis forced Felmy to shear off before he headed safely back to base, but the German then went after the third machine, a B.E.2e piloted by Claude Vautin. He was driven down by the superior manoeuvring power of Felmy and another German pilot. Forced to land, Vautin was taken prisoner.

Two days later, Fysh learnt that Felmy had sent a message by an aeroplane letter drop on the Australian airfield. Felmy told his rival pilots that Brooks had been buried with military honours. The German added that Vautin was quite well, and that Felmy had shown the youngster from Perth the sights of Jerusalem – but could the Australians send Vautin some of his clothes and toiletries? Felmy enclosed two photographs of Vautin and himself side by side; on the back of one of them, Felmy wrote, 'For my adversary – friends – specially for Murray Jones the upper-sport.'[10] There was an extraordinary chivalry among enemy airmen, and Felmy had great respect for the Australians, especially Captain Allan Murray Jones,[11] who had been a pharmacy student before the war. He had become a daring and aggressive airman,[12] putting on one of the greatest displays of aerobatics Fysh ever saw when they were in Aboukir. Murray Jones flew over the Turkish lines to drop Vautin's clothes, some kit and letters from home. The airman descended as low as sixteen metres and released the parcel, then circled the ground, returned the enemies' hand-waving, and flew home. No shots were fired.[13]

But relations between the forces were more often deadly. On 13 July 1917, just a few days after Murray Jones dropped off Vautin's clothing and a week after Fysh had joined the corps,

Working the rear machine-gun for pilot Sydney Addison, Fysh earned a reputation for deadly accurate firepower. *State Library of NSW FL8644356*

Bendigo-born pilot Archie Searle[14] took a German machine-gun bullet through the head. His British observer G. L. Paget also died in the subsequent crash.

In 1917, every Australian pilot knew they were outgunned by the Germans in Palestine. Most of Fysh's work was involved in reconnaissance, watching enemy troop movements and trying to stop the German planes from spying on Allied activities. When travelling in the B.E.2 it was usually 'nose down for home in a hell of a hurry with bracing wires screaming' if they spied hostile aircraft.[15] The B.E.2s did not have machine guns with interrupter gears like the German aircraft in which the pilot could fire through the spinning propeller. Instead, the Australian observer used his aircraft's .303 machine gun mounted on a swivel. Operating it was fraught with danger as the observer had a limited field of fire, and in the heat of combat there was always a chance of accidentally blasting off his own delicate wings.

Fysh's first mission with Ron Austin on 3 August came about after a reconnaissance patrol reported a damaged enemy aircraft abandoned on the ground near Beersheba. Four Australian teams of airmen, including Austin and Fysh, were dispatched from Belah to bomb it, so that it could not be repaired and two others flew as escorts to guard against a surprise attack. Just as they arrived over the enemy machine and began to drop their bombs, two German Albatros Scouts appeared, guns blazing – the grounded aircraft was actually a dummy used as bait to draw the Australians into a deadly trap. Tracer bullets tore through the sky as though the enemy pilots were firing blindly into a flock of ducks. Fysh fired back in his first dogfight, and despite the slow, cumbersome nature of their aircraft, Ron Austin and the other Australian pilots managed to weave away from their faster enemies and make it back to Belah.

CONDITIONS WERE much better for air crews than for the Light Horsemen, and at first Fysh thought that becoming part of No. 1 Squadron was like 'going to the war in an armchair'. Though there was always the chance of being killed by enemy fire or crashing to their deaths in the unreliable aircraft, Australian airmen lived in more comfort than anything Fysh had seen so far while fighting for his country.

Belah was in a picture-postcard setting beside the Mediterranean, and the squadron's base was set among great date palms, huge fig trees loaded with tantalising fruit, and grapevines with fat, luscious clusters. Rather than cooking for himself in a foxhole, Fysh now had access to an officers' mess with a menu of well-cooked meals, a dining table with cutlery and napkins, and – if the airmen were willing to pay – wine with their dinner. On the sand dunes above the beach where the airmen relaxed off duty, the local Arabs set up nets to trap exhausted birds that had just made their long flight over the Mediterranean. The Australians in Belah often dined on quail. Before their dawn patrols, an orderly woke them for a hot drink and a snack. There was even a piano for night-time singalongs to entertain those who had survived the day's dogfighting. Fysh had many friends in the squadron too, including another pal from Tasmania, Jack Butler,[16] whose cousin Charles Bean was Australia's official war correspondent.

Rather than sleeping in the sand as he had done as a Light

Horseman, Fysh now slept on a camp bed in a bell tent he shared with another Tasmanian, Lieutenant Oliver Lee.[17] Or at least Fysh did, until Lee and Ron Austin went to the aid of the British major Alfred John Evans after a bombing raid on the town of El Kutrani.[18] Evans had to land his Martinsyde in the desert near Kerak after experiencing engine trouble, and Austin broke a wheel pulling up next to him. The three officers had no alternative but to burn their machines and give themselves up to Arab tribesmen, who in turn handed them over to the Turks as prisoners of war. This was an especially cruel blow for Fysh, who had just loaned Lee £5.

Not long after Fysh survived his first mission on the B.E.2e, he was ordered to take off from Belah in another one to dump a load of propaganda leaflets urging the Turks to surrender – and depicting them luxuriating in British camps if they did so, with soft beds and scrumptious meals. Fysh was sitting on a large pile of the leaflets as the lumbering aircraft lurched off the runway under the extra weight. The machine rose barely fifteen metres when the Renault engine stalled and the aircraft crashed, though Fysh and the pilot picked themselves out of the wreckage unhurt.

Every week there were usually two or three scouting missions over enemy lines. Before each mission, all of the airmen wondered whether their dinner would be a last supper, and many of them drank heavily. After a bender that involved duels with flare guns and the destruction of an officers' mess, one of the Australians strayed into the desert on his way home and died of exposure.

Fysh had been with the squadron only a few weeks when he saw firsthand the aerial dash of his old Light Horse commander Ross Smith. On 1 September 1917, Smith and Les Ellis[19] were flying single-seat Martinsydes over Beersheba when they gave chase to two German aircraft. Smith and one of the Germans flew head on at each other, with Smith firing his machine gun mounted on the top wing. He brought one of the enemy down but was hit in the process. Back at their base, Fysh watched the wounded airman smash through the telephone wires bordering the camp. He still managed to land his plane safely. One bullet had creased the top of his head, and another had cut a trough along his cheek.

Later in the war Smith became the aerial chauffeur for the British officer Colonel T. E. Lawrence, or Lawrence of Arabia as he became known: a slightly built, red-haired Oxford don who had worked in the area as an archaeologist before the war and was tasked with uniting Bedouin forces against the Turks and Germans. Smith once famously interrupted his breakfast with the desert commander to jump in his machine with his observer 'Pard' Mustard,[20] 'climb like a cat up the sky' and shoot down an enemy aircraft before returning within a few minutes. Lawrence recalled that Smith jumped 'gaily out of his machine', swearing that 'the Arab front was the place [to be]'.[21] The plucky South Australian hardly had time to finish his still-warm sausages and tea before he was airborne again, helping to destroy another enemy attack. Later in the campaign, Fysh would hand-deliver a number of messages to Lawrence as the Englishman marshalled his Arab forces.

IN DECEMBER 1917, the No. 1 Squadron received the first of their Bristol Fighters, making their previous machines war-worn relics. Although the pilots were still exposed to the elements in open cockpits, the Bristol biplanes, with their 190-horsepower Rolls-Royce engines, could climb to ten thousand feet (three thousand metres) in eleven minutes. Fysh came to call these new F.2 machines 'the cavalry of the clouds',[22] as the Bristol effected a complete change over the capabilities of the squadron. Handled properly, they were revolutionary weapons with double Lewis machine guns on Scarf mountings in the rear, and a Vickers machine gun firing forward through the propeller.

The first duty was reconnaissance behind enemy lines. In that role the observers were in charge, studying the maps and directing the pilots where to go. When it came to flying the machine and to combat, though, the pilot was the boss, and it was the duty of each crew to prevent the enemy getting back to their base with photographs and scouting information. The airmen were to engage and destroy enemy aircraft every time the opportunity occurred.

The airspeed indicator fitted to the squadron's F.2s registered only up to 160 miles per hour (about 260 kilometres per hour).

Once when Fysh was riding in one with the squadron's new commanding officer Sydney Addison,[23] they attacked an enemy aircraft over the Jordan Valley. The airspeed needle moved round until it could go no further, and Fysh estimated they were doing something like 320 kilometres per hour. The aircraft was shaking so violently that he expected the oscillating tailplane to be torn away at any moment. But they remained airborne and shot down their target.

German aircraft would climb to great altitudes, and Fysh often found himself flying at sixteen thousand feet (almost five thousand metres) without oxygen. Despite thick coats and boots he would sometimes feel like 'a dead icicle', his thumbs so frozen that he couldn't pull the triggers on his two machine guns. Still, on 25 January 1918, Fysh was part of a raid made by six aircraft on Kerak, about twenty-five kilometres east of the Dead Sea. A direct hit was made with a fifty-kilogram bomb on a military building, and 'a considerable number of troops' rushed out from multiple exits into a big open square, only to run into a shower of bombs dropped from machines hovering overhead.[24]

Two months later, Fysh joined missions bombing the Hedjaz Railway and a large viaduct south-west of Amman in heavy rain and even heavier ground fire.[25] At noon on 27 March he was operating the rear machine guns on Addison's Bristol Fighter when they found three groups of Turkish cavalry, of about 250 men each, in the hills south-west of Kutrani. They flew low over the ancient Crusader castle of Kerak; seeing Turkish cavalry in the courtyard, they dropped an eleven-kilogram bomb right in among them. The airmen then pursued the horsemen over the plains with machine-gun fire.[26] Later, when Lawrence of Arabia visited Fysh's squadron at their new base in Ramla[27], Lawrence told Fysh he had been at the castle in disguise when the bomb dropped, and it had created carnage.[28]

On 2 May 1918, Fysh was part of an attack that drove down three Albatros Scouts at Amman,[29] and two weeks later he flew in the bombing of the Kutrani Railway Station.[30] Flying at ten thousand feet on 23 June, he and Addison attacked four Albatros Scouts over Bireh and drove them near to the ground before the fight was abandoned in low clouds. One enemy machine was shot down, another was shot up and had to break off the combat, and the remaining two landed but were damaged by machine-gun fire. One pilot was wounded.[31] Fysh was such a crack shot that he could hit an enemy machine at four hundred metres when the accepted range was just one hundred.[32]

Pilots and observers usually patrolled in pairs. The second aeroplane would escort the main reconnaissance machine and guard his tail from above, remembering the mantra, 'Beware of the Hun in the sun.'

There were no parachutes in the desert war, partly because of a belief that they might encourage crews to bail out too quickly. But fire in the air was terrifying, and some men jumped to their deaths rather than stay with a burning machine. Fysh once watched from the ground[33] as the pilot John Walker[34] and the observer Harry Letch[35] took on a German LVG aircraft and made the fatal mistake of attacking from above and behind in full sight of the German's rear gun. They took a fusillade, which set their Bristol on fire. As plumes of black smoke billowed from their aircraft, the Australians jumped at 8000 feet (more than 2500 metres), and Fysh and others watched stunned as the two helpless airmen rolled over and over before crashing into the ground.

The Australians had just experienced a tragedy three days earlier. Lieutenants Ernie Stooke[36] and Lou Kreig[37] had taken off on a routine test flight for a newly fitted engine. Soon after lift-off, the engine failed and the aircraft crashed into a moving locomotive beside the airfield. Both petrol tanks exploded on impact. The aircraft and the two Australians were blown to pieces with the crackle of exploding ammunition.

One of the flying maxims Fysh learnt early on was that there was safety in speed and safety in height. Slow speed resulted in stalls that killed many pilots, even experienced ones; height compensated for mistakes, enabling a pilot to recover from the stall or to find a safe landing place in case of engine failure, something that happened frequently. Whenever possible, Fysh and other observers avoided flying with some of the less capable pilots. One British airman stalled his Bristol Fighter after take-off with fatal results; another crashed into Addison's tent while he was sleeping and seriously injured Addison's knee.

Another time, after an attack on Turkish grain boats sailing up the Dead Sea, Fysh and his pilot had a forced landing amid the dust, flies and extreme heat of the Jordan Valley, four hundred metres below sea level and close to what was left of the ancient city of Jericho. One of the less-skilled pilots flew a relief B.E.2e out with an engineer dispatched by stocky little Arthur Baird,[38] the squadron's flight sergeant, whose friendship with Fysh would last for decades. Fysh was instructed to return as a passenger in the B.E.2e, but seeing dark clouds gathering over the Jerusalem hills, and knowing the pilot's limitations, he said he would wait. The pilot spent months in hospital after crashing on the return flight; the front seat left vacant by Fysh was obliterated.

THE SMALL AND WIRY Paul 'Ginty' McGinness became Fysh's favourite pilot. Fysh regarded McGinness as an airman 'full of dash and adventure', and said that in his seven confirmed victories – scored when the enemy was shot down – McGinness was hit only once, when a bullet went through the tail of his machine.[39]

McGinness was the grandson of an Irishman who came to Van Diemen's Land in 1839 before crossing Bass Strait for Victoria and settling around Port Fairy. A year younger than Fysh, McGinness grew up on the prosperous Riverview property at Framlingham, near Warrnambool, and could ride and shoot with the best of them from a young age. The property had a governess, a Chinese cook and a housekeeper. As a boy McGinness was taken to an exhibition of the Hargrave box kites in Geelong and declared that if they could fly, so could he. Back at Riverview, he began jumping off buildings with homemade wings made from bedsheets and umbrellas that turned inside out.[40] Once he even hitched a sail to his back and tried to become airborne, pedalling furiously down a hill only to crash in a heap.

McGinness was about to celebrate his ninth birthday when his father James died from complications with gallstones and jaundice in February 1905, during flooding when there was no access to medical treatment. The boy grew up quickly as a natural leader and born fighter. He was the class clown at Framlingham State School and had regular meetings with the headmaster's cane. Once he was thrashed for placing a coiled snake he had killed onto the doorstep of an elderly neighbour, knocking loudly and then running off to watch chaos ensue.[41]

When he was fourteen, McGinness became a boarder at St Patrick's College in Ballarat. He revelled in the school cricket and football games, while not so much in spelling or other schoolwork. Just four years later, his mother Catherine gave him permission to enlist in the Light Horse in Melbourne at the outbreak of war; he took his own horse, Silk, to the Broadmeadows training camp and then on to Egypt.[42] He was then an eighteen-year-old larrikin with spunk.

At Gallipoli he was with the 8th Light Horse, who were massacred in their charge on the Nek – he was one of the few from a group of 150 to survive the assault in the first line of attack. Bullets tore through the edges of his uniform as he ran towards the enemy guns. One skidded off the double webbing that held his pack, and another skidded across his hip, knocking him off his feet. His head hit the ground hard, and he fell unconscious. Because he was lying in a shell crater, it appeared to the Turks that he was dead. When he came to, bloodied and badly sunburnt, he stayed perfectly still until it was dark, then he crept back to the Australian lines.[43] His bullet wound caused him kidney problems for the rest of his life.[44]

Rapid promotion followed this incident. When McGinness was awarded the Distinguished Conduct Medal in April 1916 for 'gallantry and devotion to duty in the field', he was already a sergeant leading his troops in action during the operations at Jifjaffa, a Turkish outpost sixty kilometres from the Suez Canal.

McGinness's big brother Jim died on 1 October 1916 of wounds suffered in the fighting at Mouquet Farm in France. Unaware of that tragedy, McGinness wrote home to his mother two weeks later to say that despite being deaf in one ear from the shelling on Gallipoli, he hoped to follow his childhood dream of becoming a pilot. He'd applied for a transfer to 'the flying core', he said, because aviation, though 'only in its childhood', offered a possible profession for him after the war.[45] He also wanted a crack at Felmy after the damage the little German had caused to rival pilots and in raids against the Light Horsemen on the ground.

On 7 August 1917 McGinness was selected for flight training, and he learnt to handle aircraft at Heliopolis and then Aboukir. Deafness made it hard for him to master Morse code or to listen to instructions through headphones while flying, but he was a natural at working the rudder and ailerons. For a time he was even seconded to Lawrence of Arabia's top-secret X Flight wing at Aqaba.

On 16 March 1918 he joined Fysh at the No. 1 Squadron. At that time, McGinness stood 172 centimetres tall and weighed no more than 60 kilograms.[46] He had a forceful, engaging personality, a thin face, wavy fair hair, and grey eyes that saw far into the future when the life expectancy of airmen was measured in days and weeks, not years. He was as outgoing and garrulous as Fysh was reticent, but his mother had assured the military that he was 'without a vestige of swank'.[47]

In their first dogfight together, they and their escort encountered four Albatros two-seaters over El Afule. Fysh fired a long burst at one aircraft as it flashed past but could see no effect, when McGinness made a sharp, wheeling manoeuvre and was on its tail, diving fast, his front gun flat out. The Albatros returned fire, and hot lead rocketed past Fysh's head. Then the Albatros, seemingly out of control, surged up in a towering loop with McGinness sticking to it and blasting away with his machine gun through the propeller. Instantly Fysh and McGinness were upside down. Fysh's head spun. The inverted machine seemed to hit three hundred kilometres per hour. They were still upside down when McGinness shot the two-seater at the top of its loop, the Germans plummeting straight into the ground, their machine exploding.

Fysh had never met anyone quite like this cocky country boy who was ready to conquer the world. When the pair were given leave in Cairo, McGinness rode a magnificent white stallion in races for the Arab ruler Prince Feisal, then raced through the streets on a gharry, a horse-drawn carriage used for public transport. Even the dour Fysh had to laugh at the way McGinness would bribe the gharry-wallah to let him climb on the horse's back and then, with the dumbfounded man in the passenger seat, charge down Cairo's narrow streets using a cane to knock off the flat-topped, brimless tarboosh hats of the pompous local gentlemen.[48] McGinness romanced some of the Australian nurses in the city, and he and Fysh went swimming together, though Fysh was always self-conscious about his long, skinny legs.[49]

On 13 July 1918, McGinness and Fysh attacked a convoy of two thousand camels and five hundred cavalry south of Amman, machine-gunning it and sending the poor panicked animals in all directions.[50] The young airmen were becoming masters of mayhem.

On 31 July they attacked retreating Turks after the Battle of Nablus. At Beisan they and another Bristol Fighter machine-gunned a train and wagons at the station; after creating chaos there, they put a force of two hundred cavalry into a mad stampede. They flew on north to Semakh, where they chased several hundred troops about the railway station yard and aerodrome. Some of the enemy shot back, but most sprawled on the ground or threw themselves into ditches as a dump of flares exploded and started a fire. The Australians had no ammunition left to attack the sailing vessels on the Sea of Galilee, or the two German aeroplanes they flew over at Jenin aerodrome. But that mission helped to produce a thousand new maps of the area to aid the planned offensive of British General 'Bull' Allenby, commander of the British Empire's Egyptian Expeditionary Force.[51]

Just two days later, McGinness flew straight at an Albatros involved in a duel with an Australian flyer. He and Fysh fired heavily into the German aircraft until it crashed in a fiery heap. The Australians then fired seven hundred rounds into the hangars of El Afule aerodrome, and at rollingstock and troops in the railway station.[52]

Fysh had what he called the tensest thrill of his life on 14 August, flying with McGinness over Jenin aerodrome. He looked up to see a formation of six German Pfalz scouts two thousand feet above the Bristols that McGinness and another Australian were piloting. A climbing race began for the height advantage. When the enemy realised they were being out-climbed, they dived in formation at the two Australian machines. McGinness deftly dodged them, and Fysh counterattacked with his machine guns, splitting the formation. All six Pfalzes were shot up and forced to land.[53]

MANY OF THE AUSTRALIAN fighter planes had been donated by wealthy benefactors. The pilots were not especially superstitious and flew many dangerous missions on Friday the thirteenth, but there was always a degree of reticence among them when they had to fly a Bristol presented to the squadron by a firm of Sydney undertakers.

From mid-August, the Bristol that McGinness and Fysh flew most often was C-4623 – also known as Australia No. 20 – which had been paid for by the wealthy grazier Sir Samuel McCaughey[54] (pronounced McKackie) and his brother John. Irish-born Sir Samuel, then eighty-three, had acquired more than a million hectares of Australia, and Fysh and McGinness shot down three enemy planes in their first mission on his aircraft.[55]

Once in a long dogfight with a German two-seater over the Jordan Valley, Fysh fired nearly the whole of his 1200 rounds until McGinness was almost out of fuel and had to hurriedly turn for home. McGinness landed with an empty tank in a grain field just short of Ramla aerodrome, but in customary style he pulled it off without causing damage.

Illness forced Fysh to miss McGinness's most lauded performance of the war, and not being part of a tremendous chase made the young airman feel even worse. Bowden Fletcher[56] had taken Fysh's place behind the guns as the observer for McGinness when two Bristol Fighters defeated eight German machines and destroyed four of them on 24 August. McGinness was awarded the Distinguished Flying Cross the next day.

Fysh was back behind the machine guns a week later, and they were loaded with a new supply of armour-piercing and Pomeroy explosive bullets. Preparations for the Battle of Megiddo – also known as the Battle of Armageddon, the final Allied offensive of the Sinai and Palestine campaign – were underway. General Allenby's air commander 'Biffy' Borton[57] had flown a massive Handley Page 0/400 biplane bomber, with a thirty-metre wingspan, from England to Alexandria to play a key role in the battle. At the time, the flight was the longest in history. Borton handed the machine over to Ross Smith.

On 30 August 1918, Fysh was awarded the Distinguished Flying Cross and a citation that read: 'His Majesty The King has been graciously pleased to confer the above reward … in recognition of gallantry in flying operations against the enemy.' It described Lieutenant Fysh as 'a skilful observer, conspicuous for courage and determination, whether engaging the enemy in the air or attacking ground targets. He has taken part in numerous combats resulting in loss to the enemy and has inflicted serious damage on hostile camps and aerodromes'.[58]

Fellow observer Clive Conrick wrote in his diary that Fysh 'came in for a lot of congratulations' from all at the squadron as he was 'a very popular and unassuming bloke', later remarking that while he was usually mild-mannered, Fysh was one of the 'Tigers' of No. 1 Squadron, known for always looking for a scrap in the air while some preferred to fly 'Strategicals'.[59]

On the afternoon of 31 August, Fysh and McGinness and another two-man crew were flying at fourteen thousand feet near the Ramla aerodrome. They were alerted by their own anti-aircraft bursts that enemies were close by. The Australians powered in among the white puffs made by the Australian guns, and they challenged a pair of two-seat LVG reconnaissance aircraft. The enemy had to be prevented from returning to their aerodrome with photographs of Allenby's preparations for the campaign. McGinness put Fysh in position under one of the LVG's tails to blast away and send it into a vertical dive. The Australians watching the duel from the aerodrome applauded as the German pilot and his observer crashed head first inside the Australian lines a few miles from the Ramla airfield. When the engine failed on the other Bristol, McGinness attacked the second German machine from below. Fysh let them have it with both barrels, shooting them down too. The fuselage and wings on the port side of the enemy machine fell to earth in one direction while the two starboard wings floated off in another.

A cheer went up as Fysh and McGinness taxied home and pulled up in front of their hangars. The pair went out to examine the wreckage of the first LVG and found papers on the German officers showing they had been paid in gold the day before. But the gold was missing, and Fysh presumed the Light Horsemen had arrived before them. One of the Light Horsemen came to Fysh and handed him a cigarette case belonging to one of the Germans. It remains with Fysh's family today.[60] It was a sobering experience to attend the funerals for the two German officers – Unter Offizier Hans Vesper and Leutnant Händly of Fliegerabt 301 – in Ramla's military cemetery.

In spite of their contrasting personalities, or perhaps because of them, Fysh and McGinness had shown they formed a perfect working relationship. On 14 September they took part in the destruction of a German Rumpler aircraft near Jenin, and the enemy rarely took to the skies again over that part of the front.[61]

For a couple of days Fysh and McGinness were able to relax in Ramla, drinking wine from the vineyards at Rishon LeZion and enjoying bathing parties to Jaffa and the luscious oranges grown there. Fysh was delighted to meet up again with his old schoolmate Johnnie Webster, who had joined England's Royal Flying Corps. His squadron was based near the Australians, and Fysh would often go there to play cricket. The English pilots told thrilling stories about Webster coming in from missions with his Royal Aircraft Factory R.E.8 shot to pieces, and how he carried aerial combat 'to an extreme of bravery'.[62]

On 16 September 1918, Australian, British, Arab and Gurkha troops under the command of T. E. Lawrence began destroying railway lines that supplied the Turkish forces. Three days later, at 1.15 a.m.,[63] Ross Smith, piloting the giant Handley Page 0/400, launched the Battle of Armageddon when he dropped the massive machine's full load of bombs on the telephone exchange and railway station in Al-Afuleh. Four hours later, Fysh and McGinness were in the air, scouting on troop movements.

On 21 September the Australians made six raids on Turkey's demoralised 7th Army as it retreated. The squadron dropped three tonnes of bombs, and it fired twenty-four thousand machine-gun rounds into the bedraggled infantry and cavalry. Then the British squadrons attacked with another three tonnes of bombs and twenty thousand machine-gun rounds. The fleeing Turks soon found the steep mountain passes completely blocked ahead and behind by overturned motor-lorries and horse-wagons. Men deserted their vehicles in a wild scramble to seek cover, and many were dragged down precipices by the maddened animals. Some men cut horses from the wagons and rode in a frenzy down the road to Jisr ed Damieh. According to Australia's official history of the war, 'the panic and the slaughter beggared all description'; the long, winding, hopeless column of traffic was so broken and wrecked, 'so utterly unable to escape from the barriers of hill and precipice, that the bombing machines gave up all attempt to estimate the losses under the attack'. The airmen were sickened by the massacre, one of the worst episodes of wholesale destruction in the entire war.[64]

Three days later Fysh and McGinness were continuing to harass the retreating enemy as they crowded in on the road to Samaria at Deir Sheraf. The Australians were flying low and took their share of hits from the desperate Turkish ground fire. Fysh then saw a British R.E.8 zoom in underneath his aircraft and open fire on a group of Turkish infantry on a hilltop. Suddenly the R.E.8 dived vertically – hit by the Turks below. It crashed and erupted in a sheet of flame. The dead pilot was Johnnie Webster.[65]

What was left of Turkey's 7th Army were taken prisoner on the edge of the plains of Armageddon. Fysh and McGinness were involved in raids on the Damascus aerodrome, and by 30 September the remaining Turkish forces in the area were so exhausted that they sat on the roads nearby, heads down, too tired to move as the Australian planes swooped on them. The Australians finally showed mercy and held their fire until the beaten soldiers were taken prisoner.

With the war all but finished, McGinness and Fysh were flying over the German aerodrome in Aleppo, at an extremely low altitude, when McGinness saw a German shell on the ground opposite the hangars. Much to Fysh's horror, McGinness landed and jumped out of the Bristol, then grabbed the shell and put it in the aircraft as a souvenir. He took off before the stunned Germans realised what had happened.[66]

THE ATTACK ON ALEPPO was the last raid by the Australian Flying Corps. Fysh's squadron saw out the final days of the war at Haifa, where he and McGinness shot grouse on Mount Carmel instead of the enemy.[67] The armistice with the Turkish forces was signed on 31 October 1918, eleven days before the Armistice in Europe. By then 168 pilots and observers had seen active service with No. 1 Squadron: nineteen were killed, two died, twenty-three were wounded and twelve were taken prisoner. In 1918, No. 1 Squadron had destroyed twenty-nine enemy aeroplanes, while fifty-three were driven down, many of which crashed. By contrast, the squadron lost twelve aircraft that year through enemy action, and eight through accidents and engine failures.

Fysh had spent a year and a half as an observer and rear gunner, and for a lot of that time had been desperately anxious to get his pilot's wings. The now Colonel Williams had promised all the observers that they could eventually earn their wings, but such was Fysh's importance as an eagle-eyed scout and machine gunner that his applications to enter flying school were rejected until the war was all but over. Finally, Fysh was among a small group of observers – including 'Pard' Mustard and Les Sutherland[68] – who took a train back to Aboukir, the beachside delights of the Mediterranean, and flying lessons on the two-seat eighty-horsepower British-built Avro 504K.

Fysh had an instructor sitting in the rear cockpit behind him as he was taught take-offs, turns and landings before he moved on to spins and simple aerobatics. Then, after nine hours and fifty minutes of dual flying, the instructor hopped out and told Fysh, 'Off you go.'

Fysh's first solo take-off was exhilarating: 'up in the air alone, with a feeling of being as free as the air'. He felt he was the master of the sky, turning, diving and soaring whichever way he chose. Then his heart began racing at the thought of landing by himself.

He had seen many crashes and feared he might not be able to control a speeding machine as it neared the ground. But he made it.

On his second solo flight, Fysh had risen to a hundred feet and was heading over Aboukir Bay when he misjudged the Avro's tricky fine-mixture adjustment lever, and choked the rotary engine. It cut out. There was no beach for a landing, only a steep cliff on one side, the Mediterranean on another, and ahead a mass of mounds and excavations from diggings around what was said to be Cleopatra's Palace. In a split second, Fysh chose to land on the land, no matter how rough. The Avro hit the top of a hillock, and his oil-soaked safety belt snapped. He was thrown out onto the hillside. Still conscious, he watched the aircraft roll down into a pile of wreckage.

One of the other students, flying over and seeing the crashed Avro, decided to quit the flying course there and then, but Fysh walked back to his base and immediately climbed into another aircraft to try once more. He performed two more take-offs and landings with an instructor before going solo again. He also spent thirty minutes on a B.E.2e, making one landing.[69]

Fysh next flew a tiny scouting aircraft, the Sopwith Pup, learning to counter the severe torque of its rotary engine that caused many experienced pilots to take off at right angles. At the training school in Helipolois Fysh lied to his instructor that he was experienced in handling the difficult French-built Nieuport Scout, on which the throttle was controlled by a button on top of the joystick. After a ragged take-off, Fysh had an Avro sit on his tail and stick to him as though they were dogfighting. He put the Nieuport into a spin, and try as he might, he couldn't get out of it. The out-of-control machine plummeted thousands of feet as Fysh fought for control. Down and down, round and round he fell. But despite the dizzying sensations, his training kicked in: 'Centralise all controls and stick right forward!'[70] He finally righted the machine.

BY THE END OF THE WAR Ross Smith had been awarded the Military Cross twice and the Distinguished Flying Cross three times. He was then chosen to be copilot with his old commander 'Biffy' Borton in the Handley Page 0/400 on a pioneering long-range survey mission from Cairo to Calcutta (now Kolkata). The flight began on 29 November 1918 and finished on 10 December. Smith took two air mechanics with him: Sergeants Jim Bennett[71] and Wally Shiers.[72] The British Air Ministry wanted to extend the Empire's air routes, and Smith and his team travelled by ship from India to the Dutch East Indies, scouting out potential airfields.

On 15 January 1919, while Fysh was still learning to fly, the monstrous Handley Page V/1500 biplane bomber, with its four huge Rolls-Royce engines and wingspan of thirty-eight metres, reached Karachi thirty-three days after leaving England. The aircraft was an even bigger version of the 0/400 and the Handley Page company, which had specialised in building heavy bombers during the war, announced that soon it would begin regular flights to Calcutta and then on to Australia, with passengers enjoying the comfort they were used to finding in Pullman railway carriages, complete with armchairs and electric lights.[73]

But Australia's prime minister, Billy Hughes, a small, pugnacious scrapper with prominent ears and a sandpaper voice, wanted Australian aviators, rather than British pilots, to be the first to make any flight between England and Australia. Known as 'The Little Digger', Hughes had spent Christmas 1918 at Cobham Hall in Kent, England, which was being used as a convalescent hospital for army officers. The commandant was Australian-born Florence Bligh, then Lady Darnley, the wife of former England cricket captain Ivo Bligh. She was credited with creating the Ashes cricket trophy, and she also had an interest in aviation. On Boxing Day, she persuaded Hughes to have a conversation with Australian patients who argued that their government should sponsor a race for Australian airmen who wanted to fly home.[74] They suggested Hughes could match a £10,000 prize that Britain's *Daily Mail* newspaper was offering for the first flight across the Atlantic Ocean. He told reporters that many Australian airmen were keen to beat Handley Page in making the first flight to Australia.

As talk of an Australia-England air route intensified, an Australian company called Aerial Services Limited – whose shareholders comprised businessmen of 'Melbourne, Sydney, and Adelaide' – floated plans to establish a fleet of flying ships to operate between Australia and 'The Far East'.[75] A ground survey party under the company's managing director Reginald Lloyd, using four Indian motorcycles with sidecars,[76] left Sydney on 31 January 1919 to survey an air route across Australia with hopes to link it with Calcutta, Port Said or Baghdad. The aeronautical expert for the expedition was a Frenchman, Jean Claude Marduel,[77] who had migrated to Australia in 1908. Before volunteering for the Australian Flying Corps, Marduel had taken over a flying school in Ham Common, at Richmond, north-west of Sydney.[78] The previous instructor was Maurice Guillaux, who had shown off his monoplane to Fysh five years earlier in Geelong. In 1917 Guillaux had been killed testing a French fighter plane at Villacoublay.

AFTER THIRTY-FOUR hours and fifty minutes of solo flying, Fysh graduated as a pilot on 28 February 1919. He and McGinness finally left Egypt on the *Port Sydney* on 5 March for a comfortable month-long trip in bunks instead of the hammocks they had slept in on the way to war. After four of the worst years humankind had ever endured, Fysh thought about the comrades he had left buried in foreign soil, and then thought about the life he could now live after it had been spared so many times in combat.

On 19 March 1919, Acting Prime Minister William Watt announced on behalf of Billy Hughes that the Commonwealth Government would offer a prize of £10,000 for the first successful flight from London to Darwin.[79] Competitors were required to supply their own aircraft, and the flight was to be completed within 720 consecutive hours by midnight on 31 December 1920. The aircraft and all its components had to be constructed within the British Empire. These and other rules and regulations for the contest were drawn up, at Billy Hughes's request, by the Royal Aero Club of the United Kingdom.

Fysh and McGinness learnt about the prize on their way home. The young pilots knew they were among the lucky ones who had survived the years of carnage, so it was with a soaring sense of lightness and freedom that they arrived back in Australia ready to spread their wings as widely as they could.

Chapter 6

These two daring young aviators, who, having distinguished themselves in the fighting realm are out to do their utmost for the future of peace and commercial flying.

Newspaper report on the great overland trek by Fysh and McGinness in 1919[1]

THE £10,000 prize for an England-to-Australia flight had Fysh and McGinness enthralled when they parted ways at Tasmania's Barnes Bay after their long voyage. McGinness continued on the *Port Sydney* until the ship reached Melbourne, while Fysh and some of the other Tasmanian airmen spent ten days at the quarantine station on Bruny Island. They were released on 17 April,[2] and after the SS *Bass* brought them to Hobart's Ocean Pier shortly before 11 a.m., they were met by hearty cheers from a large gathering of friends and relatives on the wharves; among the welcoming party was Fysh's nineteen-year-old sister Peggie. A public address heard that Fysh and the others had all done their duty, not only by fighting against the enemy at the front but also by submitting to the quarantine regulations to protect those in Tasmania fighting the new enemy of Spanish influenza, a global pandemic that would kill as many as one hundred million people.[3]

Four and a half years before, Fysh had paraded through the streets on horseback as a Light Horseman. Now he was driven round the city with the other returned servicemen in motorcars, preceded by the Claremont Military Band. Fysh and his sister then took the train north to Launceston, where there was a tearful reunion on the little platform at St Leonards with their mother and Fysh's siblings Graham and Geraldine.

Ketteringham was as Fysh remembered it, and there was much pumping of his hand and back-slapping at a civic reception in Launceston for the returned troops. The celebrations were muted, though: Frith Fysh, having been invalided out of the Western Front, had been in hospitals for more than a year, and his lungs were all but destroyed.

Despite having been in the thick of the horrors of war and having fought bloody, fiery battles thousands of metres above the ground, Fysh remained a bashful young man in front of crowds, and he refused to attend a war medal presentation at the Mechanics Institute. It took all the courage he could muster to meet with Johnnie Webster's sister Rowena in the sitting room at the back of the Launceston Hotel and tell her about the final moments of her brother's life as he'd fought for king and country.

The Great War had killed more than sixty-two thousand of Australia's youth and left almost three times that number physically and psychologically damaged,[4] including those left behind. Not long after his return, Fysh came out of the Launceston Hotel and was walking down Brisbane Street when he spied a sweetheart he had as a seventeen-year-old schoolboy. She did not notice Fysh, and this made him glad because he felt it saved her from embarrassment. She looked forlorn and haggard; she had married while Fysh was at war, and he surmised that her husband was inflicting domestic violence after coming home with psychological battle scars – or, in that time of unprecedented patriotism, that her husband had low self-esteem after failing to enlist or see action.[5]

As a way of forgetting the horrors he had witnessed, Fysh immersed himself in resurrecting fond memories by visiting old friends and old haunts. He bought a tiny second-hand Trumbull car for £125 and played golf at Kings Meadows. But times were tough all around, with thousands of returned servicemen out of work.

Despite the wealth of his late grandfather Henry Reed, Fysh needed money and his mother, having lost so much to her husband, was not ready to part with any of what was left. Fysh had lost interest in wool-classing, and there were no funds to buy land. He thought about rejoining Australia's air force operating out of Point Cook, and considered becoming a commercial pilot flying around the remote areas of Tasmania or between the vast sheep stations owned by the Dalgety pastoral company on the mainland. There might have been a few bob to be made in rabbit skins or in mining for Tasmanian osmiridium, a precious metal used in fountain-pen tips. But Fysh knew that none of those plans were likely to give him anything like the £10,000 Prime Minister Billy Hughes was offering for the great flight from England to Australia. McGinness thought the same thing – but buying and outfitting an aircraft to make the longest flight in the short history of aviation would require a small fortune.

The Great War had enhanced technological growth out of necessity. Guns were now more accurate, bombs more destructive, and aeroplanes faster, more powerful and more reliable than anyone had imagined just five years earlier. By 1919, many aircraft had 200-plus horsepower engines and superchargers, aluminium was replacing wood for the frames, and parachutes and oxygen supplies were becoming common for pilots. The world was looking towards safe, reliable and regular international air transport becoming a reality as pilots and the aviation industry pushed the envelope further and further.

McGinness had turned twenty-one while at war and had come into a large inheritance from his father, but he didn't have the funds to mount a serious challenge in the air race. Then he had an epiphany: he thought the best aircraft for the race would be a Bristol Fighter like the one he and Fysh had flown in Palestine, though with a more powerful engine and a greater fuel capacity for the long haul. He planned to be the chief pilot in the front cockpit with Fysh and the former flight sergeant Arthur Baird, as their engineer, in the rear compartment; McGinness imagined there would be enough room for both of them without the bulk of the double machine guns and their mounting. He wrote to Fysh reminding him that in Palestine their favourite machine had been a Bristol Fighter donated by the millionaire Sir Samuel McCaughey. Maybe McCaughey and his brother John would help two decorated airmen, who had risked their lives for Australia, to create aviation history.

With what Fysh later remembered as McGinness's 'typically energetic and enterprising style', the young man drove up to Yanco, in the Riverina district of New South Wales, in his two-seater Ford to plead his case personally with Sir Samuel McCaughey. It was the first time McGinness had been out of Victoria and not had to fight a war. The genial McCaughey, born in County Antrim eighty-four years earlier, was battling the kidney ailment nephritis but gave McGinness a cordial reception. He had become a celebrated philanthropist, giving away sizeable chunks of his fortune while he still had the strength to write the cheques. After McGinness showed McCaughey the bona fides of the two young pilots, their war service records, and their success upon the Bristol Fighter that McCaughey had sponsored, the old pastoralist agreed to back them in buying an aircraft suitable for the month-long flight from England. He gave McGinness a handwritten letter confirming it.

McGinness immediately wired Fysh in Launceston with the good news, telling him to hightail it to Melbourne to begin preparations for what would be the most momentous flight in the sixteen years since the Wright brothers left the surface of the earth at Kitty Hawk.[6] Fysh was busy selling rabbit skins, but he put away his gun, sold his Trumbull car for a £50 profit and said goodbye to his family in St Leonards as he headed for the boat across Bass Strait. He and McGinness convinced Arthur Baird to sell the motor garage he had started in Carlton and join them. Fysh stayed with McGinness at Riverview as they planned out a route for the flight, then the three booked their passage to London where the great race was to begin.

THE FIRST OFFICIAL ENTRANT for the flight was the Queenslander Bert Hinkler after medical authorities ruled out the four officers who had first suggested the event to Billy Hughes. A product of the sugarcane fields around the Queensland town of Bundaberg, Hinkler had started his aviation career with a homemade glider before becoming the mechanic for the American stuntman Wizard Stone. He had then flown combat missions for the Royal Flying Corps, shooting at Zeppelins over England's north before undertaking missions over the Italian Alps.

The race rules gave competitors thirty days to make the trip, and Hinkler had amended his original optimistic timetable of ten days to a more manageable seventeen. He told Britain's Royal Aero Club, which was overseeing the air challenge, that he wanted to make an immediate start because the monsoons in India would prevent flying for three months. He assured the club that there was no chance he'd end up like another Australian airman, Harry Hawker, who had just been plucked out of the North Atlantic Ocean with copilot Mackenzie Grieve after their bold attempt to be the first pilots to cross that vast waterway had ended in a near-death experience.

It was almost an anticlimax when, two weeks later, John Alcock and his navigator Arthur Whitten Brown descended from the gloomy mist onto a bog on Derrygimlagh Moor near Clifden in Ireland, at 8.40 on the morning of 15 June. The pair were flying a twin-engine Vickers Vimy bomber, with a wingspan of twenty metres, which had left Newfoundland sixteen hours and twenty-seven minutes earlier. On their 3000-kilometre journey they had encountered the same poor visibility that had plagued Hawker's machine and, as snow fell into their cockpits, Brown had to keep climbing out onto the wing, sometimes at 3500 metres, to scrape ice from the engine's air intakes. Not only did Alcock and Brown collect a £10,000 prize from Winston Churchill, but the King knighted them as well.

Hinkler told the Aero Club that his tiny Sopwith Dove would carry 260 litres of petrol, enough for ten hours of flying each day at a cruising speed of 145 kilometres per hour. All he now needed was an official starter to check him off from Hounslow Heath and for a monitor to check his arrival in Singapore. If he didn't make a start soon, his little machine would be overtaken by bigger, more powerful aircraft. His fears were soon realised when another Gallipoli veteran, Charles Kingsford Smith, and his mate Cyril Maddocks – who had spent the months after the Armistice giving joy rides around England in war-surplus de Havilland DH.6 planes – announced they would start for Australia in mid-June, flying a Blackburn Kangaroo anti-submarine biplane.[7]

ONLY WEEKS AFTER THEIR prospects had apparently taken off into clear air, Fysh and McGinness hit a wall of turbulence. While Fysh was at Riverview, McGinness rushed into his room to wake him after reading the morning paper. Sir Samuel McCaughey had suffered a fatal heart attack at Yanco on 25 July.[8] The lifelong bachelor was buried the next day in Narrandera, leaving an estate of £1.6 million. There were large bequests to his brother John and legacies to his station managers and employees, and to orphans, churches, schools and universities, hospitals, needy returned servicemen, war widows and their children.[9]

There was nothing for Fysh and McGinness. The two stranded airmen rushed to Sydney to plead with McCaughey's solicitor. Anxiously, they showed him the old man's handwritten promise, but while the lawyer gave the young pilots his sincere regrets, he said the letter was worthless. McCaughey's beneficiaries were not willing to part with any of their windfall.

McGinness and Fysh returned to Riverview, then McGinness went to Melbourne to formally resign from the military. At the Melbourne Barracks, he ran into Major General James Legge,[10] Chief of Australia's General Staff and the man blamed for the 1916 bloodbath at Pozières that cost Australia almost seven thousand casualties, and where young Mac McCarthy, Fysh's cousin and farm workmate in Tasmania, had perished. For many months Legge had been working on ways to defend Australia against the possibility of a Japanese attack, and he was a strong advocate for building a powerful Australian air force.[11] He was now in charge of establishing an air route across the country, as the air race competitors were required to travel from Darwin, their point of arrival in Australia, to collect the £10,000 cheque in Melbourne, where the Federal Parliament was still sitting.

As decorated airmen, Fysh and McGinness were well known at the Department of Defence in Melbourne. Legge asked McGinness if he was interested in surveying the route north from the Queensland outpost of Longreach to Darwin, in the Northern Territory, a journey of 2200 kilometres through forest, scrub and desert where there were virtually no roads. The few rough tracks were designed for bullock wagons rather than fragile automobiles with thin and delicate tyres. What could possibly go wrong?

But McGinness jumped at the chance for the ultimate off-road escapade – a sense of adventure was alive in him and bursting to be free. He nominated Fysh to go with him and wired his mate at Riverview to meet him in Melbourne. Fysh couldn't wait for the chance to see the wilds of Australia and would later say that 'nothing on Earth could be worse than Gallipoli. Anything ... after that was not a problem.'[12] If they could not be part of the great race in the air, they could still be part of it on land. And they would get paid for this, too: £15 a week (about four times the average wage) at a time of high unemployment.

The pair were reassigned to the Defence Department on special duties and given uniforms. Legge sent them instructions, including a cable from Alfred Milner, the Secretary of State for the Colonies in London, a man totally unfamiliar with northern Australia and its wild, untamed country. Based on maps from half a century earlier, Milner suggested the most convenient route for the airmen after they reached Darwin would be Pine Creek, Bitter Springs, Borroloola, Westmoreland, Gregory, Hughenden, Clermont, Dingo, Rockhampton, Maryborough, Brisbane and Sydney. The route did not take into account the thick forests and big rivers that made it totally unsuitable for unreliable aircraft that needed open spaces for emergency landings.

Legge had been contacted by the renowned explorer and surveyor David Lindsay[13] of the Northern Territories Administration, who advised that a better route would be from Darwin straight down the middle of Australia, following the plan outlined by Reginald Lloyd's survey by motorcycle. After five months of touring from Sydney, Lloyd's party had arrived in Darwin on 10 June 1919, having passed through Charleville, Longreach, Winton, Camooweal, Brunette Downs, Newcastle Waters, and Katherine, though Lloyd's proposed air service to the 'Far East' failed to get off the ground.[14] Legge chose two other airmen who had served in Palestine, Major Rolf Brown and surveyor Lieutenant Arthur McComb, to map out a route that the airmen could fly south from Charleville in Queensland to Melbourne, and to pick out areas where the aircraft could land on level ground. Legge told Fysh and McGinness that for their northern expedition they could hire a driver-mechanic and travel by car from Longreach along the coast of the Gulf to the Katherine River railhead and then by train to Darwin. Along the way they were to find places suitable for aerodromes and forced landings, and to mark them on maps and field sketches for easy identification from the air. Legge told them that the landing grounds were to be no more than 550 to 700 kilometres apart and close to settlements where there were telegraph or telephone facilities, but with the immediate vicinity clear of trees, telegraph wires and other obstructions. The landing grounds would require level and hard surfaces, at least 1200 metres by 800 metres wide, with the greatest length in the direction of the prevailing winds. The pair were also to make notes on where emergency petrol and other supplies should be sent in case the airmen needed to come down between the stops.

McGINNESS LEFT HIS LITTLE Ford in storage at army headquarters, and on 30 July he and Fysh travelled to Brisbane to prepare for an unprecedented journey. They also had to organise a specially modified Model T Ford with a 2.9-litre (177 cubic-inch) four-cylinder engine producing twenty horsepower; it would have a specially modified wooden utility tray to carry the large cases of petrol needed to cover the hundreds of barren kilometres where supplies were non-existent. A pearl lugger carried a supply of petrol to the Northern Territory cattle droving settlement of Borroloola on the coastal plain between the Barkly Tableland and the Gulf of Carpentaria. At the time, the town had one pub, one police station and a small general store.

In Brisbane, Fysh and McGinness reported to the army, then chased around town for any experts on the outback of North Queensland and the Northern Territory. The 58-year-old

Paul McGinness (left) and Fysh in Melbourne before they set off for Longreach and their great overland trek.
State Library of NSW FL8580955

Englishman Alfred Cotton, who ran the massive Brunette Downs cattle station on the Barkly Tableland, told them that Legge's proposed route along the Gulf of Carpentaria coast was sparsely populated with no telegraph communication, and with forest so dense it would be all but impossible to find large areas of level flat ground. If the aircraft broke down, there would be no help anywhere close by. Cotton warned that the First Nations people were especially hostile towards Europeans around Turn Off Lagoon, west of Burketown near the Territory border, and that there was one man in particular who lived up to his nickname of 'Murdering Tommy'.

Fysh passed on Cotton's information to General Legge but was ordered not to deviate from the official route. He and McGinness spent the rest of their time in Brisbane securing supplies, including a .303 service rifle and ammunition, a winch in case they got bogged, blankets, waterproof coverings, camping equipment and a liberal supply of Ford spare parts. They also carried tools, a good supply of white cotton shirts, six spare tyres, cans of petrol, water bottles, compasses, and maps of both the earthly terrain and the night sky for navigation. Waterbags were attached to the front of the car.[15] Fysh bought a second-hand movie camera.

McGinness was issued with an apparatus with headphones to break into telephone wires along the route – if they could find them – to send emergency telegrams or speak to army headquarters. He was the natural leader for the expedition, while Fysh, who had read the maps when flying over Palestine, was tasked with compiling reports, taking photographs and making sketches of the proposed aerodromes. He planned to write a book after the adventure and illustrate it liberally. For Fysh and McGinness, the first transcontinental car trip across the roof of Australia represented their chance to make a lasting mark on the young nation and open the way for air transport to link remote areas of the country with the world at large.

Their mission was big news. They were interviewed by a reporter from Brisbane's *Daily Mail*, who described them as 'young men, enthusiastic in their work, and possessing unbounded faith in the future possibilities of flying in Australia'.[16] The article continued,

> They expect the trip from Longreach to Darwin and back to occupy a couple of months, and the car is to be loaded fairly heavily with food. Supplied with guns and a liberal supply of ammunition, these enthusiastic aviators anticipate little trouble in adding variety to the menu. They fear no shortage of the wherewithal to sustain life. Thus onward goes the march of civilisation; the residents of these outback portions of the continent from Longreach to

> Darwin, instead of being for months and years away from the comforts and excitements of city life, will be situated on the main aerial route to the old world. It requires little stretch of imagination to realise that the time is not far distant, given a sympathetic and farseeing Government, when bi-weekly mail services will be maintained along the route to be chosen by these two daring young aviators, who, having distinguished themselves in the fighting realm are out to do their utmost for the future of peace and commercial flying.[17]

Fysh and McGinness were unprepared for the overwhelming reception that awaited them in Longreach. Bolstered by the positive press, they arrived by train on 13 August 1919 as men who would put that little town on the world map and help it establish a much closer connection with the rest of Australia. For someone who had always shunned the spotlight and public speaking, Fysh had trouble processing the euphoria and adulation their arrival created, remarking that 'it was embarrassing and not good for our characters'.[18] The atmosphere of excitement was unlike anything he had ever experienced, even in the glow of triumph after returning to Tasmania. Still in uniform, the young pilots found themselves being worshipped like gods, drawing cheers by just walking down the town's dusty streets, and there was plenty of back-slapping and hand-pumping when they attended a function for returned servicemen. They created as much fuss as astronauts from a later generation would have done if they had marched down the streets of the outback town in their spacesuits. Over cold beers the pair met many investors and graziers who they believed might sponsor their future business activities, including Ainslie Templeton,[19] the general manager of the Queensland Primary Producers Cooperative.

On 14 August, McGinness sent a wire to Legge telling him that on the recommendation of the manager of the Longreach Motor Company, they had hired 43-year-old George Gorham[20] as their mechanic for the trip on a salary of £1 a day. A wiry little Englishman with a neat moustache, Gorham always wore a straw hat. His boss told Fysh and McGinness that Gorham was 'a clever and reliable mechanic',[21] and that he had been among the first men from Longreach to enlist for the Great War.[22] Gorham was an expert on the Ford's complex transmission with its three floor-mounted pedals and a lever beside the driver's seat to control the gearing. The throttle was opened by a lever on the steering wheel.

Gorham had joined the British Royal Horse Artillery in India in 1895, the year of Fysh's birth, and he had served with them for twelve years before migrating to Australia in 1910. He had arrived in Longreach two years later to become a lorry driver with the New Zealand Loan and Mercantile Agency Company. After enlisting in Rockhampton, he was wounded in Gallipoli and France, married in England,[23] and left suffering for the rest of his life from the effects of Germany's poisonous mustard gas on the Western Front. Gorham returned to Longreach and a job at the local motor works in November 1918; he had been a battler all his life, and the tyranny of distance held no terrors for him.

The journey was delayed because a coil from the Model T and some spare parts had gone missing on the train from Brisbane,

Fysh with McGinness (centre) and mechanic George Gorham and their heavily laden Model T Ford as they embarked on a 2200-kilometre trip to Darwin through forest, scrub and desert where there were virtually no roads. *State Library of NSW FL8580927*

but once the problem was rectified Gorham declared the vehicle was ready for anything the outback could throw at it. On 16 August 1919, McGinness arranged with the Defence Department in Brisbane to have 40 gallons (182 litres) of petrol provided when they reached Burketown. The following day the Longreach Motor Company had the Model T ready. McGinness bought a large pair of pliers, extra .303 ammunition, and a copious supply of plug tobacco to use as gifts for any Aboriginal people they might meet along the way. He also bought a supply of cigars and a big bottle of brandy that he carefully wrapped and stored with the medical supplies.

The men selected a large flat area at Longreach for an aerodrome, and Fysh sketched the surrounds, reporting on the nature of the landscape, the rivers and the weather. Then, on the morning of 18 August, Gorham started the engine with a twist of a crank handle, and the three intrepid explorers launched themselves into the great unknown. Gorham was wearing his straw boater, Fysh a wide-brimmed felt hat, and McGinness a pith helmet worthy of a safari. Their heavily laden jalopy headed north towards Winton and the first stop on this unprecedented odyssey. The only proper roads ahead of them were those that ran for a few metres through the tiny settlements between central Queensland and Darwin.

Just outside Longreach they crossed a wooden bridge over the Thompson River. For the next 2200 kilometres there would be countless rivers, creeks, streams, gullies, backwaters, lagoons, crocodiles and snakes. The three trekkers would see just one more bridge for the entire journey.[24]

Chapter 7

It was decided to salt down 20 black ducks shot at Snake Lagoon, and with wallaby, these formed a food reserve.

FYSH EXPLAINING HIS DIET ON THE DRIVE ACROSS COUNTRY FROM LONGREACH TO DARWIN[1]

THE VAST EXPANSE OF Australia stretched out before Fysh as the Model T chugged on over tufts of Mitchell grass, cruising across the black soil plains of western Queensland, heading north. The men had a gruelling task ahead of them, but they were confident in the car's abilities given the reputation of the Model T as a battlefield ambulance in the war.

The track out of Longreach was firm and even, and the Ford could reach sixty-eight kilometres per hour going flat out, though that was on bitumen, not dirt. Still, with two men sitting up front and the third perched on the utility tray next to the supplies, they made good time initially, covering 160 kilometres a day and chugging through the towns of Winton, Kynuna and McKinlay. Two days into their journey they reached Cloncurry, and while McGinness revelled in the sheer thrill of the drive and the prospect of tackling the dangerous unknown ahead, Fysh sat on a rocky outcrop and in his meticulous way began sketching the site for a landing field as a mob of kangaroos bounded across it. The men obtained a quote for £500 from local workers to build an aerodrome there, which was duly approved.

There were now five entries for the England-Australia race and more in the wings, but on 21 August the Royal Aero Club announced additional regulations. Each crew would have to include a proven navigator, and it was hinted that each plane would need to have a range of three thousand kilometres because of the scarcity of airfields in India. Bert Hinkler said that if he could make a start he'd leave the rest behind, rules or no rules, but he was a solo flyer and wanted to make the journey alone. The aircraft manufacturer Tommy Sopwith, realising that Hinkler was not going to meet the race criteria no matter how much he protested, decided to rescind his loan of the little Dove aircraft. Although business had boomed during the Great War, the British government had cancelled all orders with the company in 1919, and Sopwith faced financial ruin; he needed Hinkler's aeroplane to make a sale.

On 23 August, after some minor repairs to the Model T, Fysh and the others left Cloncurry on a 450-kilometre drive north-west towards Burketown, on the Albert River beside the Gulf of Carpentaria. They negotiated one long stretch of a wide sandy river crossing by cutting down tree branches to place under the car's wheels. The next day they arrived at their destination.

Burketown's birth was bloody. Not long after the port was established in 1865, Queensland's Native Police shot more than sixty local Indigenous men after a series of attacks on horses and the killing of a grazier. When one of the Aboriginal men did not die despite being shot twenty times, a trooper finished him off by smashing in his skull. Every settler in the district was said to be 'delighted with the wholesale slaughter'.[2] In 1883 a young British woman, Emily Caroline Creaghe, passed through the region and saw forty sets of ears of Waanyi people nailed around the walls of the nearby Lawn Hill Station homestead.[3]

The adventurer Francis Birtles first visited Burketown while riding a bicycle around Australia in 1907. He painted a bleak picture. 'The usual program was races in the morning, funerals in the afternoon and debaucheries at night,' Birtles wrote. 'Bets were freely made as to whose turn it would be next.'[4]

When Fysh, McGinness and Gorham puttered into the town twelve years later, they found a sleepy village that was still cut off from the outside world – apart from a weekly mail run from

Cloncurry and a visit every three months from a steamer bringing supplies for the cattle stations.

Fysh and McGinness selected a landing field, and the locals were instructed on how to clear and drain the site. But everyone in Burketown laughed at the travellers' plan to drive to Borroloola, a further 550 kilometres north-west. They pointed out that even the intrepid Birtles hadn't made it through seven years before, when driving a Flanders touring car, and had to turn back at the Calvert River. The trio heard that a drover named Higgins had just set off to Borroloola in a spring cart with his wife, infant daughter[5] and twenty-eight horses, and they resolved to catch up with him in case they needed a tow for the toughest river crossings. They bought a roll of birdcage-wire netting that they planned to lay down on sandy patches to give the Model T a firmer base, and they jettisoned everything they could from their vehicle, electing to sleep in the open with leaves and long grass as mattresses. Then they loaded the utility tray with 47 gallons (214 litres) of petrol, packed in cases, hoping it would be enough to get them through an area as big as Tasmania that had narrow, uneven tracks over hellish terrain. There would be no telephone or telegraph.

With the Ford sagging under the great load of fuel, the trio left Burketown on 28 August bound for the police station at Turn Off Lagoon near the meandering Nicholson River, 140 kilometres away. First, they had to wage a five-hour battle crossing Gin Arm Creek, thirty-five kilometres into the journey. There was no way the Ford could get across the three hundred metres of sand on the eastern bank, so they had to offload all the cases of petrol and their other gear, and carry them across the sand, then across the water, watching out for crocodiles the whole time, before hauling their supplies up the steep western bank. Then they laid down leafy branches and the wire netting to help the Ford across the long stretch of sand into the water and, with the engine revving as hard as it ever had, willing it onto solid ground across the creek.

At dusk they made it to the galvanised-iron police station at Turn Off Lagoon. They camped beside it, listening to the crickets and night birds, and later the bats and dingoes. Spectacular lilies and myriad ducks abounded in the water. After washing off the sweat and dust from the journey, the men shot some of the birds, which the local policeman cooked in a camp oven and served with damper bread. Fysh fell asleep listening to stories of 'Murdering Tommy's' atrocities.

The Nicholson River afforded them the first and only bridge crossing since they had gone over the Thompson outside Longreach. They chugged off across the thick, unforgiving scrub towards Westmoreland cattle station only for the car to end up stuck midstream in Scrutton Creek, short of the tin house that was Westmoreland's homestead. A Waanyi boy came to investigate, and Fysh gave him a note to hand to the station manager, who promptly sent over two horses to help haul the Ford back to solid ground. The manager had never seen a motor car in that part of the country before, and he shared a meal with the adventurers, including some greens, a refreshing change from the dried vegetables they carried with them.

The following day, 1 September, they motored on in agonising slowness for Wollogorang Station, which straddled the Northern Territory border. They spent a night camping at The Springs and the next day became stuck in a creek bed just short of Wollogorang. A big Aboriginal station worker and several women with him were seconded to push the Ford back onto the track, but the man sat down to watch as the women did the heavy lifting. The next morning at the station, the man told Fysh that the women – 'Judy, Nora and Wanda' – were 'no good'. They were sleeping like they were dead, he said, because 'too much pushem motor-car'.[6]

The Wollogorang homestead was more like a fortress than a house, with heavy wooden doors. It had been made spear proof after the many skirmishes between its occupants and the local Garawa people, almost always with them coming off second best. The homestead was beside an idyllic lagoon along Settlement Creek, a waterway teeming with barramundi and mangrove jack, and its grounds had an abundance of banana, pineapple, pawpaw, tomato, pumpkin, sweet potato and cauliflower plants, all tended by a Chinese gardener and a group of Aboriginal women in uniform.

Fysh and the others left Wollogorang with some of the garden's cabbage, sweet potatoes and pawpaw, and eleven kilograms of flour from the station store. The next 325 kilometres to Borroloola would be the toughest part of their journey. Higgins the horse drover arrived with his animals the next day, and McGinness and Fysh asked him to travel behind them for a while and help them out of any more sticky situations.

Inside the Northern Territory the next morning they crossed Hobble Chain Creek before reaching Big Running Creek. Gorham was driving the Ford at speed, trying to climb the second of two undulating sandhills approaching the creek, when on the rapid descent the car swerved in the sand, narrowly missing a tree but hitting a hidden rock and getting stuck in the loose surface miles from anywhere. The situation looked hopeless, but the only damage was bending in the front axle and the radius rods that stabilised it. They took out the axle and belted it back into shape with an axe over a fallen log.

Despite the mishap, Fysh found the scene all around breathtaking, with palm trees and paperbarks reflected in the cool, clear water. McGinness shot a bustard, a large Turkey-like bird renowned as good eating; it was hung up for the night, but not high enough, and dingoes crept in after dark and made their own meal of it.

The next day there was another thirty-five kilometres of lurching along in low gear through soft sand and across difficult creeks. At one stage the trio had to carry their petrol and luggage two hundred metres to dry ground before making a hundred-metre track for the car out of branches and netting.

They finally reached the Calvert River on 7 September. Fysh was aghast at the large spread of fast-flowing water with a strong ebbtide running. Higgins arrived again with his horses as they were contemplating crossing the wide and dangerous expanse. After an anxious wait, low tide revealed a stony bar, but it was unsuitable for the heavily laden Ford, so two of Higgins's horses were tethered to the vehicle. McGinness stood in the driver's seat to steer the horses across the stone bar; instead, they turned for the sandy bottom. While Fysh photographed the scene for posterity, McGinness handed the wheel to Gorham, jumped on the bigger horse, and guided it back onto the stones and across to the opposite bank.[7]

They then powered the Ford as fast as it could go up the far bank, but as it came across a small ravine, screened by long grass, one wheel went over and the car was left dangling on the bank of a seven-metre drop. Yet again Fysh feared that the expedition would end with a wrecked machine, but they hitched the winch to the nearest tree and inched the Ford onto safe ground.

A lone dingo watched them from a high point. The Calvert was teeming with fish, and beyond it animals and birds became more abundant. The trio never encountered 'Murdering Tommy', but they met many Garawa people who offered them kindness, advice on the countryside, and assistance pushing the vehicle and finding water. Without the help of those people, Fysh, McGinness and Gorham would likely never have made it through the harsh country. One of their benefactors was smoking a pipe by the Calvert when they came upon him; he told them he was 'Kingy Paringi ... plenty big fellow' and would send word out to his people to help them further along on their journey. They gave him some flour and tobacco. He had with him two naked Aboriginal women, who before approaching Fysh's camp covered themselves 'fore and aft' with branches tied with a possum-fur string. One carried a long-dead goanna, which reeked.

A WORLD AWAY IN THE salubrious offices of the Royal Aero Club at 119 Piccadilly, London, it was announced that the England-to-Australia race would not start before mid-October, although teams representing the Alliance and Martinsyde companies were ready for an immediate take-off. Bangkok had been added to the itinerary, but landing grounds in Rangoon, Singapore and Darwin were still being prepared. Singapore golfers were requested to forgo their game while sand bunkers there were smoothed out for the aircraft.

Four teams were already approved, including Ross Smith and his crew in a Vickers Vimy. The entrants had all agreed to complete the whole flight even if took longer than thirty days.

Biffy Borton, now the officer in charge of administration at the RAF's command headquarters, strongly advised the crews against attempting overly long flights for their stages, especially on the last leg to Darwin when he expected the engines would be worn out. However, the Alliance team said they were depending upon the great range of their machine, which they claimed could cover 4800 kilometres without refuelling.[8]

It was a far different story for Fysh, McGinness and Gorham as their machine's appetite for petrol increased. As the Ford lumbered along through thickly timbered country in low gear, it was returning barely eight miles to the gallon (thirty litres per one hundred kilometres).

Once they had crossed the Calvert, the only way forward was for one man to drive and the others to walk ahead, watching out for rocks and gullies while cutting down saplings with an axe and clearing away fallen trees in their path. Then the track was lost altogether, with the trio crisscrossing cypress pine forest until they could find it again. They had to stop so many times for repairs that Higgins and his horses took the lead on the track.

The abundance of game and birds did not last, and the men became so hungry they shot two flying fox, as hundreds were flitting about the blossoms of nearby trees by moonlight. But the bony bats stunk to high heaven, so the men went hungry that night. Once, just as they had improvised with bully beef and hard biscuits during the Great War, the trio made a stew in an empty petrol tin from three parrots, a crane, a bush pheasant, and three small leatherheads. The meat was tough and stringy but not nearly as tough as the galah they ate at another time.

One night, as they settled around the campfire for another meal of stringy bush tucker under the glory of the stars, McGinness recalled writing home from the war about the potential for aviation. He'd told his mother, Catherine, that it could be a profession in peacetime. This was the spark of the idea for Qantas.

McGinness said to his companions, 'What this part of the world needs is an air service.' There were many trains running from the coast, east into the interior, but few were running north

Fysh and his companions lived off the land for weeks. Fysh photographed McGinness (left) and Gorham with dinner they had just shot. *State Library of NSW FL8641960*

to south. An aerial service could link the unconnected railheads of remote towns such as Charleville, Longreach, Winton and Cloncurry, McGinness said, and help overcome the tyranny of distance and isolation for all those living far from big cities. Air travel could also overcome the dilemma of unmade roads, unbuilt bridges, and the constant threat of flooded creeks and rivers that periodically halted transport in Australia's remote north.

There were plenty of wealthy pastoralists overseeing enormous properties out west who could subsidise transport that would be much faster than anything existing in Queensland or the Northern Territory. The prevalence of flat country all around made it easy to find sites not just for small aerodromes, but also for emergency landing areas at a time when aircraft were routinely beset by engine failures. Fysh knew the idea was a winner, but he wondered where they would find the money for aircraft.

From Warby Lagoon to the Robinson River, the trio chugged along more than sixty kilometres of soft sand with just a faint track to guide them. Higgins again loaned them two horses, which allowed them to get through the roughest parts. But then they ran short of water because the animals were consuming most of it. The Ford's radiator began to spit steam.

McGinness remembered how in his youth, Aboriginal people from the Framlingham Mission near his home at Riverview had dug in creek beds to locate fresh water. He ferreted down in the sandy bed of a dry creek for a metre and a half, finding enough to fill the radiator and their canvas waterbags. A crisis was averted, but tensions between McGinness and Fysh were growing. The leader of the expedition was starting to believe that Fysh was rather too fond of photography and not fond enough of getting his hands dirty when manual labour was required.[9] Gorham was caught in the middle as they began to squabble.

To make matters worse, the sand became like a concrete barrier in front of the car. When Gorham let air out of the tyres to give them more spread, they were wrenched off and their tubes blew out. The men tried flat tyres roped to the wheel rims, but this did not get them far. Finally, with patches all over the car's tattered rubber tyres, they reached the wide Robinson River on 15 September.

Like the Calvert it seemed impossible to breech, then the tide receded and left a stony bar with thirty centimetres of fresh water flowing over it. McGinness jumped on the back of one of the horses as it and another were tied to the Ford. With them straining in the lead, and the men shouting encouragement to each other and the trusty animals, the Model T struggled through the water and onto the start of a steep bank on the far side. The car's fuel was fed to the engine by gravity, which meant that on steep rises Gorham used a bicycle pump to force the petrol through the motor, while the others hacked away at the ground to reduce its incline. Once again they called on Higgins for assistance. When they were back on flat ground, Gorham took out the front axle to straighten it.

For two days they camped beside the Robinson. A gold rush had taken place there many years before, and all that remained of the settlement were the posts of what had been a pub. At night Fysh could hear the crocodiles in the river, so he always kept the .303 handy.[10] The trio managed to drive another twenty-seven kilometres to Snake Lagoon, which was covered in water lilies and surrounded by wildlife. The men shot a wallaby and several ducks, and after a swim to retrieve the dead birds, they spied a large snake enjoying a dip among the lilies. There was solid ground to the Fulch River, where they ran into a shifting sandbank. The Ford slipped and slid before coming to a dead stop halfway across this expanse. Facing sideways, it was just clear of a huge dead tree jutting out of sand. Two horses were again needed to pull the vehicle downhill.

Over the next two days the men continued through the soft sand, crossing along the stony bars of the narrow Wearyan and Fletcher rivers. The Model T looked like it had been in a bruising fight. A big part of each day was spent repairing it: the clutch needed constant attention, and after the car hit a tree the fan was wobbling against the radiator. The radius rods had been bent so many times that it was hard to straighten them properly, which made steering difficult on a track covered with fallen logs, that required weaving in and out of lightly wooded scrub.

In this vast emptiness, Fysh and McGinness – hot, bothered, dirty and dishevelled – were now often at each other's throats. Once a perfect team in the air, the comrades became as compatible as oil and water. The always impatient McGinness frequently displayed a hair-trigger temper that would explode and subside just as quickly; perhaps he had been traumatised by the war at a time when what we now know as post-traumatic stress disorder often went undiagnosed. Gorham was never quite sure whether McGinness was a genius or a madman,[11] and Fysh, with his methodical, cautious approach to life, was usually the target of McGinness's outbursts. Travelling less than twenty-five kilometres a day, they were well behind schedule. Fysh continued to make detailed notes, and he often ordered a halt so that he could photograph incidents and scenes. McGinness accused him of wasting time.[12]

Fifty kilometres from Borroloola and its semblance of civilisation, food became scarce, water was again in short supply and petrol was running out. The men knew that if they encountered any more long stretches of sand requiring low gear, they might not be found for days, and they were already famished, thirsty and exhausted. On the morning they left the Fletcher River, the Ford skidded into a tree, bending the front axle. The radiator was also damaged, and the spindle that held the fan was broken off. Improvising as best they could, the trio repaired the damage with wood. They continued on, even slower than before, making frequent stops to prevent overheating and to fill the radiator.

On the night of 19 September they breathed a heavy sigh of relief when they reached Feathertop Creek, only a few kilometres from Borroloola, where Higgins and his family delivered their horses. At noon the next day, a group of the local Binbinga people helped to push the Ford across a long sandy patch to the McArthur River, where the water was deep and flowing fast. Fysh took some remarkable photographs of the locals hard at work, lending many hands. McGinness would later say that despite the cruelty meted out to them by early settlers, if Indigenous people were treated fairly they would 'work their soul cases to the bone for you'.[13] Some other local men were paddling canoes and the adventurers watched their impressive efforts at spearfishing.

That night the trio camped beside the McArthur, waiting for low tide in the morning; when it came, they managed to just make the crossing under the power of the Ford's little engine, running through shallow water. On the other side of the river, Fysh paid a group of Binbinga a few sticks of tobacco to push the car through yet another long sandbank and to cut a path as the Model T made a steep climb. By evening, with a group of local women laughing at their efforts, and babies watching on, the trio were on the track again.

After a 24-day odyssey from Burketown, the men had just thirteen litres of petrol left on 21 September when they finally arrived at the four houses and one hotel that constituted the town of Borroloola. Fysh, McGinness and Gorham had crossed six tidal rivers and more than forty creeks, and they had covered 140 kilometres of the 550-kilometre journey on soft sand, needing horses to tow them for thirty of those. The Model T was almost a write-off, and the expedition was two weeks behind schedule. In England, the finest aviators in the world were making their preparations to take off from London for Australian airfields that were still unmade in an event that was headline news around the world. The three adventurers needed a saviour.

Chapter 8

The friends of Lieuts. McGinness and Fysh are justly proud of this their latest adventure, as theirs is the first motor car to traverse the Gulf route, after many failures by experienced motorists.

FYSH'S LOCAL NEWSPAPER ON THE MONUMENTAL OVERLAND FEAT BY THE TWO YOUNG AVIATORS[1]

WITH THEIR MACHINE falling apart around them in Borroloola, Fysh, McGinness and Gorham looked for someone to help them make urgent repairs. As though miraculously, the young and powerfully built Reverend Hubert Warren[2] of the Roper River Anglican Mission was in Borroloola when they arrived. For twenty years Reverend Warren devoted himself to the service of Indigenous people in Australia's north[3] and to weary travellers passing through the area. He knew something about hard driving on rough roads, having just spent four months covering six thousand kilometres from Melbourne on a preaching tour that had left his own little Ford decidedly worse for wear.

The good reverend had a blacksmith's forge behind his house, and the shoulders and biceps to suggest he used his hammer and anvil frequently. Over the next four days all four men got to work fashioning a new blade for the Model T's fan, drilling a hole through what was left of the spindle, and – with hand bellows to intensify the heat of the forge – straightening the front axle and strengthening the radius rods with a coating of tin. Warren made a new part for the speedometer out of a horseshoe, and he fixed the leaking radiator and replaced a broken rear wheel with a spare he had from his own car that he used on pastoral visits to the far-flung corners of his parish. With proper timber instead of the branches used in the bush, the men strengthened the car's bodywork and its running boards, which had sagged with the weight of spare tyres and petrol tins.

In London, Bert Hinkler was busy running around in his air force uniform, all dressed up but with nowhere to go, still hoping for a crack at the £10,000. While he hunted around for a sponsor to help him buy the Sopwith Dove, backers didn't match his enthusiasm; instead, Sopwith entered a team with a bigger, more powerful machine, the Sopwith Wallaby, an improvement of their aircraft in which Harry Hawker had made his failed bid to cross the Atlantic. Meanwhile, Charles Kingsford Smith and Cyril Maddocks had been dumped from the Blackburn Kangaroo team because of a perceived lack of experience. Fysh and McGinness had to get back on the track again to ensure the Australian airfields were ready for the aviators who were about to take off.

Big trees, uneven, holey ground, and a shortage of labour made Borroloola out of the question as a landing spot, so on 25 September – with Warren's detailed directions for the best route – the adventurers left the remote town, aiming for the train station at Katherine eight hundred kilometres away. They were now on a much better track, and after experiencing some further minor damage to the radiator, the Model T reached Bauhinia Downs the next day.

Warren had soothed some of the tensions between Fysh and McGinness, but he hadn't wanted to become embroiled in their disputes. Not wanting to travel with them, he had not told them that he would soon be driving along some of the same terrain. The Model T was parked beside a creek late on 27 September when Warren sputtered along with a torn sump and bent axle casing thanks to an unplanned meeting with a tree stump. Fysh and McGinness were in a hurry to reach Katherine, but together with Gorham they helped Warren repair the vehicle as best they could. They then drove along with him on his journey, making more repairs to his vehicle and theirs. The reverend's engine was all but done while theirs was 'going beautifully',[4] and they passed through Tanumbirini Station and Nutwood Downs. At Hodgson Downs they fashioned a new set of radiator rods from the Model T's old ones and Warren's broken ones, then they all journeyed on to Urapunga, near the Roper River Mission.

McGinness deemed the area unsuitable for a landing field, but they took local horses and rode with Warren to the mission station, with Fysh, McGinness and Gorham returning by launch to their Ford. The adventurers then took a comparatively smooth drive to Katherine to catch the weekly train service to Darwin leaving on 8 October. Fysh and McGinness chose a landing site outside Katherine but wired General Legge to say that the route they had taken by car would be disastrous for aircraft. McGinness recommended an alternate way across the Barkly Tableland.

After fifty-one days of hard driving, the trio had covered 1354 miles (2179 kilometres). The first half of the journey to Burketown had taken just six days, but they had covered less than twenty-five kilometres a day for the remaining forty-five. Apart from the streets of the small towns they had visited, they hadn't seen a metre of made road the whole way. Although there had been a good deal of rancour between Fysh and McGinness during the worst of the travails, Fysh later admitted that 'McGinness, with his enterprise and optimism, was the one who got us through'.[5]

The Model T looked like roadkill, so it rode on a freight carriage while the three adventurers took the slow train to Darwin, via the mining town of Pine Creek, and arrived in the Territory's capital on 12 October 1919. They checked in that morning at the two-storey Victoria Hotel, the first stone building in the town. The exhausted McGinness lay on his bed and saw a gecko on the ceiling, then the next thing he knew the cleaning lady was waking him up the following day – he had slept for almost twenty-four hours. Fysh slept just as long; when he woke, he started to have a good look around his new home. Darwin was still a remote outpost with no more than two thousand people: 'a strange and tough town … bare, barren and uninviting', Fysh recalled, 'full of pearlers, buffalo hunters, prospectors, sandalwood gatherers, cattlemen and adventurers'.[6]

Fysh and McGinness went to meet the surveyor David Lindsay,[7] of the Northern Territory's Lands Department, who at sixty-three was still a bronzed, powerfully built action man with a bombastic nature.[8] He had already told Legge that the route the adventurers had taken was unsuitable for aircraft, and he was amazed that Fysh and McGinness had even managed to drive through the Gulf Country. Lindsay was a controversial character who advocated for the segregation of European people from Indigenous people in the Territory, but he knew his job as a surveyor and was eager to help Fysh and McGinness with the clearing of an airfield in his town.

Although Reginald Lloyd's survey had identified the Darwin racecourse as a potential landing ground, Fysh and McGinness insisted on seeing some other areas, including a police paddock six hundred by four hundred metres, near the Fannie Bay Gaol, that had been chosen as an emergency strip. After a day's scouting McGinness wired Legge to tell him that this emergency strip could be readied for the race at very little cost, though there was a bigger site in Darwin that would cost £1700 to prepare. Fysh and McGinness then prepared a much more detailed report about their long journey from Longreach and the impossibility of using it for the air race.[9] Meanwhile, they told everyone they could in Darwin that aviation was the way of the future for all Territorians.

THE GREAT ENGLAND-TO-AUSTRALIA race began with a staggered start while the landing grounds were still being chosen. The Frenchman Etienne Poulet was ineligible to claim the £10,000 prize, but he thought he could steal the thunder from the Australians by making an epic flight to Darwin for the glory of France. On 14 October he and his mechanic Jean Benoist left Paris in a twin-engine Caudron G4.

Now realising that Lindsay's proposed route through the centre of Australia was the best way to go, Legge sent instructions to McGinness and Gorham on 17 October that they should head back to Cloncurry posthaste, across flatter but still brutal country to the south, and establish landing sites there as they travelled. Fysh was told to make the Victoria Hotel his home while he oversaw the work on the landing strips in Darwin and Katherine, ensuring both areas had the necessary petrol supplies so the contestants could refuel on the way to Melbourne.

The Sopwith Wallaby was the first of the official entrants to take off, leaving Hounslow, in London's west, on 21 October with Sopwith employee and Gallipoli veteran Captain George Matthews,[10] from Morphettville, South Australia, at the controls, and Sergeant Tom Kay,[11] from Springmount, Victoria, navigating. They carried a generous supply of Bovril and Wrigley's chewing gum, and they were wearing military uniforms under their leather fur-lined kit as though setting off on an Arctic expedition. Matthews told reporters he wasn't taking a collapsible boat as 'it would only prolong the agony' of being lost at sea. Many Australians were present for the farewell, including Harry Hawker, who arrived in a Sopwith triplane at the last moment to watch the take-off.

Forty-five minutes after the fog over Hounslow lifted, the two airmen were up in their single-engine biplane. They circled Hounslow for twenty-five minutes, then headed for France. Ill fortune followed them the whole way. Within hours fog forced them to land at Marquise, north of the French port of Boulogne. Eventually they flew on to Cologne in Germany – however, a rough landing damaged the aircraft and injured Kay, and bad weather kept them grounded for twelve days.

The day after Matthews and Kay left London, McGinness and Gorham set off south for Katherine in the trusty Model T, which had undergone extensive repairs. McGinness had brought hundreds of metres of white calico to mark out airfields, along with tobacco and a large amount of cloth, known as 'Turkey red', to pay Indigenous workers to clear ground and to give the car a push if needed.

About a hundred kilometres into the journey, McGinness found a suitable forced landing ground at Batchelor Farm. He informed Legge that the telegraph line headed south from Darwin for 142 kilometres and not only was an ideal reference point for the aviators, who flew by visual navigation, but also had a good ten metres of cleared ground either side.

The next day the Model T chugged into Katherine with a broken spring. After making repairs, and after McGinness had preached a sermon on aviation being the best method of transportation for those in the remote outback, he and Gorham left at 6 a.m. to mark out another forced landing area on the experimental government farm at Matarnka Station.

Up in Darwin, the excitement over the air race grew as hot as the summer sun there, and for the first time the town was the focus of the world's news. The local paper warned residents that if they encroached on the open space where the aircraft from England were likely to land, there would be a great danger to both the spectators and the airmen. 'An aeroplane,' the paper explained, 'may skim along close to the ground at possibly over a hundred miles per hour, and its propeller, revolving at a fearful pace, would cut a human body in two without the aviator feeling the jolt, but an aeroplane, diving into a crowd of people would be an awful catastrophe. Children should be well under control …'[12]

On 30 October, Fysh wrote to tell Legge that he had rejected Reginal Lloyd's idea of the Darwin racecourse as a landing ground. He said that it was in a depression, could not be extended, and would require extensive clearing of stumps and saplings. Fysh chose the site at Fannie Bay instead, three kilometres to the north of Darwin and only four hundred metres from the shore. As he explained to Legge,

> The greater length of the ground runs north-west, which is the direction of the prevailing wind at the present time of year. A run of some eight hundred yards can be got in this direction … The area has been cultivated with rice and other crops while a portion of it is natural grass. The soil carries a good quantity of gravel and is hard and even … The present obstructions which are being cleared are seven hundred yards of wire fencing, two banyan trees, two large earth mounds and several small trees on the edges of the site.[13]

Fysh was under tight constraints of time and money. Locals told him that 'the wet season starts in the district about the first week in December and continues till sometime in March. Very heavy rains fall ... heavy weather sets in from the north-west and storms are expected from the south-east ... A machine out in the open at the landing place would stand a good chance of being wrecked.'[14] The superintendent of public works in the Territory, W. C. Kellaway, advised Fysh that the work would have to be done by casual day labour at the rate of four shillings an hour. Fysh paid £700 to have the Fannie Bay site cleared further, including the removal of the banyan trees and two great mounds that had been brush turkey nests. He also supervised the clearing of the landing site at Katherine that required the logging of a large area of forest; he travelled down there every fortnight with the £600 payroll, staying at the Pine Creek Hotel and sleeping with the money in a suitcase under his bed and a revolver under his pillow.

At Matarnka Station, McGinness left some white calico with the manager, as well as instructions on how to use this cloth to mark out the landing area with a circle fifty metres in diameter and a large T. McGinness also left instructions on how to prepare an oil-burning choofer so that an approaching pilot could see the direction of ground winds in the thick plumes of smoke the oil produced.

The Model T then travelled on a narrow camel track for another eighty-five kilometres before McGinness and Gorham camped for the night. They reached Daly Waters on 27 October to plot out the next landing field. The site was covered with small stones that could have torn the fabric or damaged the wheels of aircraft, and McGinness gave a group of Jingili women at the station Turkey red material to make dresses after they spent a day sweeping the stones away. He gave a group of local men, who had watched from the edges of the field, plug tobacco.

The next day, McGinness and Gorham reached the eight small houses of Newcastle Waters Station. For the next six days, McGinness developed a more substantial airstrip there. Again he paid the Jingili people in goods, noting his expenditure to them as two bags of flour, twenty-four sticks of tobacco and twenty metres of Turkey red.[15]

Having wasted time hacking their way through the unsuitable Gulf Country, Fysh and McGinness were now under enormous pressure to get the airfields completed. They knew that one of the official entrants was already a week into the journey to Australia and that more would be starting soon. McGinness sent a telegram to Fysh telling him to hurry up in Darwin and Katherine, and to provide maps of the Queensland airfields for when the pilots arrived from London.

After more than a week at Newcastle Waters, McGinness and Gorham headed south, then made a sharp turn east for Eva Downs. The station was two hundred kilometres away across what station workers called 'devil-devil' country, so rutted and holey that cars were rarely seen there. The radiator boiled so much that the Model T resembled a small steam locomotive. The heat was soul destroying, but they made it to Eva Downs – and found there was barely enough water to cool their parched radiator and dry mouths. The next morning they puffed along for another seventy-five kilometres east to Anthony Lagoon, where they gave the station manager instructions on how to make another rough airfield.

On 10 November, McGinness and Gorham reached Brunette Downs and met up again with Alfred Cotton, who had warned them about the perils of the Gulf Country. Cotton told McGinness he was astonished they had made it through to Darwin across some of Australia's most inhospitable country.

JUST A YEAR AFTER Ross Smith had been blasting Germans out of the sky over Palestine, he took a shot at aviation history in Hounslow on 12 November 1919. He revved up the two Rolls-Royce 360-horsepower Eagle VIII engines powering his big Vickers Vimy, and propelled the aircraft down the grass runway as his team became the second to officially start the race to Darwin. Alongside Ross was his brother, Lieutenant Keith Smith,[16] two years older, and the mechanics Jim Bennett and Wally Shiers. Keith Smith had spent many of his war days as a gunnery instructor in England, but had been a pilot with the Royal Flying Corps and Royal Air Force.

Given the heroics of John Alcock and Arthur Whitten Brown

crossing the Atlantic in a Vimy, it was only natural that Vickers would have a crack at flying one of their aircraft to Darwin. Although the big green machine they entered had been built too late for war service, it was just the aircraft for an arduous campaign over hostile territory. It was registered as G-EAOU, which the crew said stood for 'God 'Elp All Of Us'. Ross Smith, Bennett and Shiers had previously flown from Cairo to Calcutta with Biffy Borton on the Handley Page and knew at least some of the hazards ahead. After leaving London in the Vimy, they immediately ran into snow that blasted them in their open cockpits, and they were driven up to almost 2800 metres by a blizzard. Ice shrouded their goggles, so they took them off and then endured intense pain, frozen in their seats as winds of 140 kilometres an hour flayed their eyeballs. Somehow they survived to land in Lyon.

McGinness and Gorham were on the move too, thumping along after breaking a back spring on their way to prepare a landing strip at Avon Downs. The government had no time to purchase sufficient suitable land in the area, so the Defence Department instructed McGinness to use the only Crown land available, even though the one local road ran right across the airstrip.

While McGinness was at Avon Downs sorting out this dilemma, Captain Roger Douglas,[17] a Queensland boxing champion and linotype operator at the *Townsville Daily Bulletin* before the war, took off from Hounslow on 13 November in his Alliance P2 Seabird, the *Endeavour.* Beside him was his navigator, Lieutenant Leslie Ross.[18] Their machine had a huge 450-horsepower Napier Lion piston engine and an enclosed cabin; the Alliance company had claimed it could fly 4800 kilometres without refuelling. But after the *Endeavour* climbed into low cloud, it was aloft for only a few minutes before entering a spin. It crashed into an orchard at Surbiton, just ten kilometres from Hounslow. Both aviators were killed, and the Alliance company was ruined, but the deaths were no deterrent to the other airmen.

McGinness and Gorham left Avon Downs having organised a supply of 240 gallons of petrol and five gallons of Castrol oil to be delivered there. They made a short trip east to Camooweal in order to oversee the work on that town's landing field, then on 15 November they set off for Cloncurry to end a round trip of 2229 miles (3587 kilometres). While making their way across the outback, they had often navigated by the stars in a car that many times seemed ready to fall apart. It was pouring with rain when McGinness and Gorham reached Cloncurry. They found shelter with the patriarch of the town, Alexander Kennedy,[19] a Scottish-born pastoralist who had just celebrated his eighty-second birthday and was taking a break in town from his property Bushy Park, about a hundred kilometres to the south-west. Kennedy's wife of almost fifty years, Marion,[20] made the weary travellers hot drinks and food, while Kennedy sat at the table amazed at the story of the mammoth drive around the Top End.

Kennedy had arrived in Australia from Scotland almost sixty years earlier after seeing a sign on a building: 'Wanted: Young Men for Queensland'.[21] In 1861 he landed in Gladstone, then he worked on sheep and cattle runs around Rockhampton before taking up more than 160,000 hectares at Buckingham Downs, south of Cloncurry, in 1878. With their cattle in tow, he and his family trekked more than 1100 kilometres from Rockhampton and lived in a tent until a small hut was built. The journey took eight months.

One of Kennedy's business partners, James Powell, was killed by Indigenous warriors near Cloncurry in 1884, but eventually Kennedy was running fifty thousand head of cattle on holdings of more than 3100 square kilometres. For decades he rode across what is now called Mount Isa without realising that underneath him were some of the world's richest deposits of lead, silver, copper and zinc, but in 1895, while out looking for stray cattle, Kennedy's son Jack found the copper deposit that became the Duchess Copper Mine.

When McGinness told Alexander Kennedy about his time flying in Palestine and his dream of starting an air service for outback Queensland and the Northern Territory, Kennedy thought he'd struck gold – what a difference it would make for people in the outback if medical help could arrive by air and if they could cover the vast distances quickly. He still mourned the death of his newborn daughter Euphemia back in 1882, when floods had prevented the family from getting to a doctor at Cloncurry. The canny Scot stroked his long white beard and said he'd be interested in stumping up some capital to give an aerial venture wings in western Queensland, so long as he could be its first paying customer.

McGINNESS SENT FYSH more telegrams urging him to sort out the airfields on his watch. Fysh was slow to respond, and McGinness concluded that after their ill feeling on the journey north, Fysh was now reluctant to take orders from his old comrade.[22] Eventually Fysh informed McGinness that the work in Darwin had been carried out for £700.

Gorham returned to Longreach, and McGinness got to work at Cloncurry organising an aerodrome and ordering petrol drops at various landing sites for Captain Henry Wrigley[23] and Sergeant Arthur 'Spud' Murphy.[24] The two airmen were flying their slow but reliable B.E.2e from Melbourne to Darwin in the first trans-Australia flight, checking the route the race contestants would take after arriving from London. Wrigley and Murphy had left Point Cook on 16 November and had made stops in Cootamundra, Narromine, Bourke, Charleville, Cunnamulla and Longreach.

Five days after they had set off, back in London the Blackburn company's twin-engine flying Kangaroo took off from Hounslow, piloted by Lieutenants Valdemar Rendle[25] and David Williams. The Kangaroo was a reconnaissance torpedo bomber and had seen service on anti-submarine patrol towards the end of the war. The mechanic for their team was Lieutenant Garnsey Potts and the navigator Captain Hubert Wilkins,[26] already world famous for taking the first war film in the Balkans in 1912 and for exploring the North Pole.

On 30 November, McGinness was in Cloncurry to welcome Wrigley and Murphy's B.E.2e after their two-week flight from Melbourne. They had been delayed by engine trouble, and McGinness spent a whole day helping mechanics work on their aircraft at Cloncurry's Western Motors Garage and ordering an urgent delivery of parts by train from Melbourne.

Meanwhile, Adelaide's Captain Cedric 'Spike' Howell,[27] a former Sopwith Camel fighter pilot with nineteen aerial victories to his credit, and his navigator Sergeant George Fraser,[28] a Victorian-born mechanic with Rolls-Royce, took off from Hounslow in their Martinsyde A Mark 1 on 4 December. The Martinsyde was the fastest machine in the race, capable of 240 kilometres an hour. But throughout their trip, Howell and Fraser encountered mechanical problems, and the machine suffered damage. Six days after leaving England, both men were killed in a crash off St Georges Bay, Corfu.[29] Howell's new wife wasn't far from the crash site, on a ship bound for Australia, and no one had the heart to tell her he'd died until her ship had docked in Adelaide.[30]

The first team to take off in the race – the duo aboard the Sopwith Wallaby – had reached Vienna, but George Matthews and Tom Kay had a tough time obtaining petrol there. Then fog, rain and snow kept them grounded again. Finally they made it to within 160 kilometres of Belgrade, but they were almost out of fuel when heavy fog again caused them to make an emergency landing in a field. They were immediately surrounded by armed men and arrested on suspicion of being Bolsheviks. Matthews and Kay were kept in a room about three metres square, and fed on bread and figs.

At Suda Bay in Crete, Bulgarian prisoners of war were pressed into service to get Hubert Wilkins's Blackburn Kangaroo out of a boggy landing field. On 8 December the Kangaroo took off from Suda Bay bound for Egypt, then a broken crankcase forced the crew to turn back. The Kangaroo skidded to a stop, wrecked beyond repair, against a mound of earth next to the Canae lunatic asylum.

That same day, Ross Smith and his team were in the final stages of the race aboard their Vickers Vimy, and their flight was making headlines around the world. Wrigley and Murphy finally left Cloncurry, heading for Darwin to meet them, and McGinness sent telegrams to all the airfields in the area telling them to put out the white calico circles and big Ts.

Smith had learnt to fly only three years earlier, but he had steered his crew safely through one deadly obstacle after another on their way to Darwin. From Lyon, their first port of call, they had flown to Pisa, where heavy rain made their landing field a bog, and thirty locals were needed to dig their machine out of the slush. As they cruised down the coast of Greece, the peak of an island suddenly appeared out of a cloudbank, requiring some heart-stopping evasion. At Ramadi in Persia, the weight of fifty soldiers from the 10th Indian Lancers was needed to stop the Vimy flying away unmanned during a desert storm; freak gusts pummelled the ailerons so viciously that the control wires all stretched or broke, forcing Bennett and Shiers to make emergency repairs.

After that, favourable winds sped the Vimy on to Baghdad and beyond, to the tiny port of Basra. On Sunday, 23 November, they left for Bandar Abbas on the Persian coast, bright sunlight sparkling over the varnished wings, and the polished propellers becoming, in Smith's words, 'halos of shimmering light as the two engines sang away merrily'.[31]

The next day took the Aussies more than 1100 kilometres over a mountainous landscape to Karachi, a flight made more perilous by their growing fatigue. In Karachi they learnt that the Frenchmen Etienne Poulet and Jean Benoist, with almost a month's head start in their Caudron G4, had already reached Delhi. By the time the Vimy passed over the desert sands and the jungles to land in Delhi, the Frenchmen were in Allahabad.

On 27 November the chase continued, but a defective oil gauge forced Smith to make an emergency landing in Muttra. The Frenchmen continued to Calcutta, while as Smith revved up the Vimy engines the next day in Allahabad, a belligerent bull stood defiantly on the runway in front of them until a brave little boy distracted it long enough for take-off. When the Vimy landed in Calcutta, the crew learned that Poulet and Benoist had made it to Akyab in Burma (now Sittwe in Myanmar). Just after the Vimy left Calcutta, two large hawks assailed it in a suicide mission; one bird hit a wing, while the second was torn apart by the port propeller, which – despite suffering great damage – held together.

The Caudron was still at the Akyab aerodrome when the Vimy arrived, and the chase was over. Even though Smith gave the Frenchmen another head start the next day, the Vimy reached Rangoon first, by more than an hour. The Caudron's engine had a cracked piston, and while flying over Thailand the aircraft was attacked by a vulture that shattered its right propeller. Poulet made an emergency landing on a mountain plateau, and his hopes of being part of the first team from Europe to reach Darwin were dashed.

Smith and his team powered on, cheating death time and time again. While flying through a huge cloudbank surrounding treacherous mountains between Rangoon (now Yangon) and Bangkok, Smith put the Vimy into a dangerous sideslip when he accidentally bumped the rudder bar. Between Bangkok and the Thai city of Singora, the four exhausted aviators were battered by rain with the force of hailstones. Ross had to take a break and hand the controls to his brother Keith. When they finally made it to Singora, after three hours of tag-team flying, they found that half the aerodrome was covered with water and the other half with tree stumps. Somehow Ross landed the Vimy with the only damage sustained being a broken tail skid.

In Singapore the local racetrack was too small for them to make a normal landing. The mechanic Jim Bennett climbed out of his cockpit and shimmied along the tail section of the machine, hanging on with all the strength he could summon as his weight forced the tail down sharply. Ross brought the Vimy in hard, and it came to a dead stop within a hundred metres, Bennett riding it like a runaway bronco. Later, in the Javanese city of Surabaya, the machine was bogged on a field reclaimed from the sea. It took twenty-four hours of constant work by locals to make a matting tarmac 275 metres long and 12 metres wide to provide traction for take-off.

In spite of all the privations and perils, the four Australians and their big, slow machine pressed on. By 9 December, they were at Timor. The next day, with Fysh up early in Darwin to prepare for their arrival, they crossed the Arafura Sea.

Quarantine officers had sent messages that with the Spanish flu ravaging the world, the airmen would have to be medically examined before being allowed to go among the people. But the wild stampede of the local populace to the landing field 'resembled an old-time gold rush in the west'.[32] At the front of this crowd of several thousand stood Fysh. He was officially representing the Department of Defence, even though he had grown sick of his uniform and was dressed in a light-coloured suit for the summer heat.

In the midafternoon, a speck in the sky appeared like a distant bird. The speck grew larger and larger, until the cheering locals could plainly see the green Vimy. From about thirty kilometres away, the crew saw the white guiding mark in the centre of the field that Fysh had arranged.

Watching spellbound, Fysh recalled how he and Ross Smith had risked death against the Germans every other day over Palestine only a year earlier. Now Smith was completing a flight that would change the world, as he circled the big bomber over the Fannie Bay field and his crew waved to the crowd from the open cockpits. Finally, the Vimy floated down onto Fysh's landing strip at 3.50 p.m. on 10 December 1919. The crew had covered 18,250 kilometres in twenty-seven days and twenty hours.[33]

Fysh rushed to be the first person to congratulate his old comrade, as Smith and the others climbed wearily out of the first aircraft to cross the globe. Smith's weather-beaten, ruddy face broke into an enormous smile. 'Well, we're here,' he told Fysh, and as his old friend pumped his hand, he added with a laugh, 'I didn't recognise you at first – where's your uniform?'[34]

Fysh handed Smith a bundle of telegrams that included congratulations from the King, the Australian Prime Minister Billy Hughes, and Winston Churchill, now Britain's Secretary of State for Air. Despite the strict social distancing regulations, the crowd rushed the aviators and carried them shoulder high from the ground.[35]

For Fysh, the arrival of the big biplane was one of the most moving sights of his life, a feat of courage and endurance that he regarded as perhaps the greatest achievement in aviation history.[36] It was also an unprecedented boost to the infant aviation industry in Australia, and Fysh wondered if the little outback airline he and McGinness had talked about over their lonely campfire might just get off the ground yet.

Chapter 9

The flight from England to Australia was something more than a feat of aviation; it was a milestone in the history of civilisation.
AUSTRALIAN PRIME MINISTER BILLY HUGHES[1]

THE DAY AFTER FYSH welcomed Ross Smith and the Vimy crew to Fannie Bay, George Matthews and Tom Kay made a desperate, death-defying jail breakout in their own bid to reach Darwin. Suspected of being Bolsheviks and arrested at gunpoint by a local militia near Belgrade, the hapless flyers were locked in a dungeon for four days – until they took advantage of the temporary absence of their guard and made a run for it. Hearts racing, they grabbed their passports and papers and then bolted for the Sopwith Wallaby, taking off through a dense fog to continue their journey.

Back in Darwin, Fysh verified the lead seals secreted about the Vimy to confirm it was indeed the same aircraft that had left London. On 12 December, he was back at the landing field to welcome Henry Wrigley and Arthur Murphy in their B.E.2e. They had become the first fliers to cross Australia, completing the exhausting and hazardous trip from Melbourne to Darwin twenty-six days after leaving Point Cook. They were two days late to greet Smith and his crew on arrival from London, but Wrigley and Murphy had staged another important nation-building feat, travelling 4500 kilometres in forty-seven flying hours.

General Legge also released a statement praising McGinness and Fysh for their great trek and the way that had established 'a very fine record' breaking down the distances between new aerodromes to just 400 miles.[2]

Under instructions from McGinness, and in conjunction with the Lands Department, Fysh had prepared a detailed map for use by the Great Air Race teams of the route he and 'Ginty' had devised through Queensland and the Northern Territory. Ross and Keith Smith studied the map closely, and on 13 December the Vimy crew left Darwin and headed for Cloncurry as their first stop towards the £10,000 cheque in Melbourne. While they were on their way, Buckingham Palace announced that the Smith brothers were to be knighted, and that Shiers and Bennett would have the honour of bars added to their Air Force Medals.[3]

Some of the outback settlers were literally dying to see them en route. A grazier, Hugh McMaster, spent ten weeks recovering from typhoid in Cloncurry hospital[4] and was lying in his bed there on 15 December[5] when a local bank manager, English-born Henry Gribben, was admitted with sunstroke, having waited out on the claypan in broiling heat for a chance to see the wonder of the ages. Gribben mumbled to McMaster, 'If Hell is any hotter than this, I hope I don't go there.' All the available ice in Cloncurry was requisitioned to bring down Gribben's temperature, but he died the next morning without ever casting eyes on those magnificent men in their flying machine.[6]

It was even more dangerous for the pioneers of the air. On the same day that Gribben was admitted to hospital, the now Sir John Alcock received a standing ovation when his Vimy, which had crossed the Atlantic, was presented to the Science Museum in London as one of the wonders of the scientific age. Three days after that, as Alcock was taking a new Vickers single-engine seaplane, the Viking, to the first postwar air show in Paris, he crashed and died trying to land in fog near Rouen. He had just turned twenty-seven.

Ross Smith's Vimy experienced mechanical problems with the starboard engine not long after take-off from Darwin. Four hours later, Smith made a forced landing on a dried-up swamp called Warlock Ponds. It was so hot the four airmen sheltered under the shade of the Vimy's wings and fought swarms of mosquitoes before taking off in the cool of the following morning.

Between Newcastle Waters and Cloncurry, the crew were startled by a loud crack; they discovered that the propeller damaged by a hawk near Calcutta had split apart. They brought the big machine down in a forced landing at Cobb Creek near Anthony Lagoon. The thermometer registered fifty degrees Celsius in the shade, a heat that was so intense it melted the crew's aviator goggles and the Vimy's little windshield.[7] A party of bore sinkers in two cars arrived to help the crew by providing a sheet of galvanised iron for repairs. Jim Bennett used splinters from a piece of packing case to fill the crack in the propeller, bound the blade with thin strips of the galvanised iron and secured them in place using screws from the aircraft floorboards. He covered the whole blade with fabric and painted it, then repeated the procedure with the opposite blade to ensure they rotated in harmony without vibrations.[8]

McGinness was busy trying to find the lost crew. After wiring all the stations and airfields he could, he breathed a great sigh of relief when the Vimy team were finally located. He ordered petrol and oil to be delivered to Avon Downs so they could refuel, and the big machine finally touched down at Cloncurry just before noon on 20 December, a week after leaving Darwin.[9] Everywhere they landed, the aviators were treated with all the fanfare of conquering heroes.

Fysh was under instructions to remain in Darwin, where he was supposed to meet the other crews coming from London. He spent months getting to know what was still a wild frontier town. Christmas brought the wet season with its monsoon rains, stifling humidity and mosquitoes like a black mist. Upstairs at the Victoria Hotel, Fysh had to put his suitcases on empty cigarette tins to keep the white ants away, but not before a pair of his good boots had been eaten through.[10]

To pass the time, he read voraciously, trying to fill in the gaps from his stop-start education. He also regularly visited the Kakadu area, less than a day's drive away. The young man from chilly Launceston was mesmerised by this other world of tropical delights. He explored the vast swamps and thick jungles bursting with flowers. He gazed at the herds of buffalo running through the forest or splashing about in muddy lagoons. There were big mobs of kangaroos and wallabies, and he estimated there were duck and geese in their millions. The estuary was full of barramundi and other fish as well as man-eating crocodiles.

Fysh found plenty of other ways to occupy himself. At night he relaxed with poker games at the telegraph station. In the dry winter he played cricket on a ground where a six saw the ball plonk over a cliff into Darwin's harbour. In the wet season he played at half-forward for the Wanderers Australian Rules football team, eagerly awaiting the bloody battles with the team from Lord Vestey's meatworks.

Fysh took several hard falls on the gravel surface and lost plenty of skin, but not as much as the buffalo he wrote about in a 1350-word feature article for Australia's most popular magazine, *The Bulletin*, as he nurtured his writing talents and thought about an illustrated book of the great overland trek that he and McGinness had just made. *The Bulletin*'s editor, Samuel Prior, paid Fysh £2 two shillings for his coverage of the local buffalo hunters, who started shooting immediately after the wet season in May and kept blasting until November, when the rain started again and all work in the Top End seemed to cease. Only the buffalo hides were taken, Fysh noted, while the huge carcasses were left to rot and be devoured by wild dogs and the birds that circled above a kill, ready to pounce as soon as the skinners finished their work. Fysh wrote,

> Yet buffalo beef when taken from a young animal is surprisingly palatable, and the tongue is quite a delicacy … The waste of shooting for hides only is appalling; but the buffalo is a wild, obstinate animal which refuses to be droved. Its heavy body crashes through fences as if they were matchwood … Several of the beasts are used in Darwin by the Chinese for drawing vehicles; but the old buffaloes are slothful animals, nor can they be made to hurry by any persuasion. There seems to be in them a perpetual longing for their native bogs and wallow-holes. Introduced, so to speak, but yesterday, the massive beasts, in spite of wholesale slaying, now inhabit the whole of the true tropical regions of the Territory.[11]

IN CLONCURRY, McGINNESS tried to organise repairs for the Vimy, but there was no engineering workshop capable of fixing the propeller. The crew carried on, passing over Winton, flying to Longreach, dropping a packet of letters over Barcaldine, and circling over Blackall before touching down at Charleville two days before Christmas.[12]

The route south had been amended to include Sydney, but when the Vimy took off from Charleville on Christmas morning, heading for Bourke, there was a bang and a flash of fire out of the port-side engine. Ross Smith wheeled the big aircraft around and nursed it back to the airfield. The engine needed a new crankshaft among other major repairs, but there was no way the work could be done in Charleville. After considerable deliberations, the engine was shipped to the Ipswich Railway Workshops seven hundred kilometres east. A new propeller was also made there from 'nine layers of Queensland maple, stuck together with hot animal glue'.[13] But the work was detailed and intricate; the Vimy stayed grounded in Charleville for fifty-two long days.

Up in Darwin, Fysh was still waiting to see who else could cross the globe. By January 1920, the Sopwith Wallaby had made it to Sofia, Romania, but Matthews and Kay were stuck there during snowstorms.

On 8 January 1920, Lieutenants Ray Parer[14] and John McIntosh[15] left London in a war surplus Airco DH.9 sponsored by the Glasgow whisky distiller Peter Dawson. They carried with them a bottle of Dawson's finest for Prime Minister Hughes. The tiny Parer had a heart problem and wasn't supposed to fly above three thousand metres. McIntosh had not yet gained a pilot's licence but was up for anything; at just fifteen he had stowed away to Western Australia from his native Scotland, and he'd become a well-known axeman before serving as a stretcher-bearer at Gallipoli. Together, Parer and McIntosh left a trail of broken propellers, bent wings, choked carburettors and burnt-out engines on their great flight, as Ray Parer became known as 'Re-Pairer'.

Matthews and Kay, meanwhile, pressed on in their Sopwith. They spent three days in Constantinople while they fixed an engine leak with the aid of chewing gum, powdered asbestos and copper wire.

IT HAD TAKEN JUST 27 days for the Vimy crew to fly from London to Darwin, but another two-and-a-half months of delays for Ross Smith and his team to finally collect their money.

Two hundred guests, including federal and state ministers, Melbourne's Lord Mayor and heads of the naval and military forces, assembled in Queen's Hall, at Melbourne's Parliament House, at a lunch for the Vimy crew on 27 February 1920.

After the toasts of The King and The Governor-General had been made, Prime Minister Billy Hughes proposed the toast of Sir Ross Smith and his crew. Cheers rang out as Hughes told the crowd that their aviation feat was a milestone in the history of civilisation; that its significance was not confined to the flight in itself – 'daring, amazing and almost miraculous as it was.'

'As a result of the flight the world will not be quite the same again for all of us in this far off spot of Australia,' Hughes said. 'The Commonwealth's isolation is no longer with us. Wars are fomented because of misunderstanding; it is impossible to make bad blood between people who understand one another thoroughly, and this will remain one of the most glorious feats in the world's history.'

Hughes led the audience in a rendition of 'For They are Jolly Good Fellows.'

Then amid deafening applause, the prime minister handed Smith the Commonwealth cheque for £10,000. The beaming pilot told the crowd it would be split four ways.[16]

Expecting the imminent arrival of Matthews and Kay in Darwin, Fysh had commandeered the inmates from the adjacent gaol and had them clear the landing ground using sickles and scythes. He was astounded at the rate the grass grew in the wet season: within two months of the Vimy's arrival, the landing ground had been transformed from a bare surface to a paddock with eight-foot-high vegetation.

But as Fysh made final preparations for their arrival, Matthews and Kay were grounded yet again. The aviators came down with a dose of dengue fever in Bangkok before the Wallaby's run ended on 17 April 1920 with a crash into a Balinese banana plantation near Grogak. Matthews and Kay returned to Australia on a steamer with their damaged aircraft in the hold, and their pride seriously bruised.

In the DH.9, Parer and McIntosh had been almost swallowed by the volcanic crater of Mount Vesuvius in February. They had to fight off locals in a Syrian desert in March, and then scare off indigenous people they assumed to be cannibals on an island in the Irrawaddy River in Burma in April. At Moulmein (now Mawlamyine), they broke a radiator and propeller, and they were forced to crash the aircraft rather than plough into a sea of spectators on the landing field.[17]

They finally made it to Darwin on 2 August 1920, 206 days after leaving London. With them was their mascot, a small brown mouse they had found in the aircraft at Calcutta. The keen-eyed rodent flew with them through northern Australia as they followed Fysh's map, heading south towards Melbourne. But at Longreach, apparently fair sick of flying, the mouse scurried off and disappeared into the vast Queensland outback.[18]

The aviators dropped some souvenir charts over Bundaberg and other towns along the way, before they crashed at Culcairn in the Riverina district of New South Wales. The DH.9 arrived in Melbourne by rail, and when Billy Hughes tasted his Peter Dawson Scotch it had been seven months in transit. Hughes presented the two airmen with a £1000 cheque to share and gave each of them an Air Force Cross.[19]

By then Fysh was long gone from Darwin, having left in May 1920 after General Legge had told him his job was done. Before packing his things, Fysh had given his first recorded public speech, overcoming his great shyness in front of crowds to declare, 'I am quite safe in saying that within five years we will have a regular service between England and Australia. This will automatically make Darwin the first port of call in Australia on the great aerial highway.'[20]

AS McGINNESS WAITED FOR Fysh to join him, he made his home at Cloncurry's Post Office Hotel. He won friends and influenced people all around Queensland's outback, his status as a wartime fighter ace guaranteeing him celebrity status. Fysh described McGinness as 'generous, gregarious, enterprising and always on to anything new',[21] and he became a popular figure in the town. There was big money around him, with seventeen million sheep in Queensland at the time, producing wool each year worth more than £6 million.[22] One of the wealthy graziers McGinness impressed was Ainslie Templeton, who invited him to stay at his sprawling property Acacia Downs, north-east of Longreach. Another was the tall, straight-talking Fergus McMaster,[23] the younger brother of Hugh.

When Fergus was just twelve, he had helped his older brothers drive four thousand sheep from Logan Downs at Clermont over the Drummond Range to Kelso Station four hundred kilometres away near Longreach. For most of his adult life, he had been involved in large grazing leases throughout western Queensland. When he enlisted for the Great War, he was a 37-year-old widower with a five-year-old daughter, and he served in France as a gunner and dispatch rider. He returned to Australia at the end of 1919,[24] and on the day of his arrival in Sydney, still on the boat, he received an urgent telegram about his brother's serious illness.

Fergus McMaster was now looking after Devoncourt, a station sixty-three kilometres south-west of Cloncurry, while Hugh recuperated. Devoncourt carried a herd of ten thousand cattle worth £15 each, a small herd compared with those on neighbouring stations.[25] The recent theft of unbranded poddy calves was rife, and stock was poisoned or shot by thieves when crackdowns were implemented. At Devoncourt, poddy-dodgers had once stolen all the new calves and, to rub salt in the wound, shot the sixty cows that had borne them.[26] Fergus McMaster became chairman of the Burke and North Gregory Prevention of Cattle Stealing Association, which at one time had eighty horses in its mustering camp as well as a posse of Aboriginal trackers.[27]

McGinness had attended a church service in Cloncurry one Sunday morning and was driving off from the Post Office Hotel that afternoon with a date to a picnic, when he encountered a hot and weary McMaster walking towards him. The grazier had presided over a meeting of the Prevention of Cattle Stealing Association at the Post Office Hotel and had been driving back to Devoncourt when he had hit a large rock in the sandy bed of the Cloncurry River, about five kilometres out of town. His car slewed, and his front wheels became bogged. He had to get the car out quickly before the river rose in the heavy rainstorm that was threatening, so he headed back to town on foot, searching for help.

As the commanding but weary figure of McMaster approached, McGinness stopped his car. The young fighter ace was only too willing to help another returned serviceman in distress. 'I did not know McGinness very well,' McMaster recalled, 'and was surprised when he said he would make arrangements for his friend to attend the picnic with someone else, and that he would give me a hand to fix up my car.'[28] McMaster was always intrigued by the way chance so often plays a part in people's lives.[29]

McGinness drove McMaster back to his stranded car. They found that it had a broken stub axle and would need major repairs before going further.[30] Then McGinness drove him back to Cloncurry, but it being Sunday all the garages were locked. If McMaster wanted to get back to Devoncourt that night, McGinness would have to do something unconventional – and, showing the older man his flair for innovation, he did. The garage was locked tight, but McGinness took some tools from his car, removed a sheet of tin from the side of the garage, walked in and found what he needed to repair the stranded vehicle. McMaster left a note apologising to the garage owner about the urgency of the mission and the address to send the account for the damage to the garage and the cost of the car parts.

During the afternoon as McGinness repaired and regreased the car, he and McMaster talked about the drive from Longreach to the Darwin train. McGinness spoke of his and Fysh's part in establishing landing grounds for the Great Air Race. They also talked about the almost incommunicable experience of war, and the horrors they had both seen.

Before too long McMaster was on the road again, but he couldn't stop thinking about the ingenuity, resourcefulness and generosity of the young airman who had given up a special event to help an old digger. Although McGinness hadn't broached the subject with McMaster of how flying could eliminate the need for long journeys over rough roads between outback towns and stations, he soon would.

THERE HAD BEEN THREE choices for Fysh when it came to making his return journey to meet McGinness in Cloncurry and make plans for their little bush aviation business. He could go south to Adelaide, making part of the journey by camel or horse before catching a train north; he could return to Brisbane by the monthly boat from Darwin; or he could drive. He decided to see more of the country by car.

On the morning of 27 May 1920, Fysh climbed into a rickety old Ford alongside three friends leaving Darwin for Cloncurry. One of his fellow passengers was Tom Lynott,[31] who had discovered zinc, lead and silver while managing the McArthur River Station thirty years before, and was to be dropped off at the Tattersall's Hotel he ran in Borroloola. Since Tom was seventy-two, Fysh had expected their car to be rather comfortable; now he wondered if the battered machine would make the 1750 kilometres to Cloncurry. It was loaded with so many supplies and so much petrol that it looked bow-legged, and Fysh couldn't help but chuckle to himself as they set off into Australia's vast interior, hoping for the best.[32] He had brought a rifle in case they got stuck and needed to hunt for food.

Tyres were in short supply in Darwin, and on their first day they suffered five punctures. Even so, they managed to cover more than four hundred kilometres before camping at Old Elsey Station, a property made famous by the pioneering adventures in the book *We of the Never Never.*[33] After months in the relative comfort of Darwin's finest hotel, Fysh went to sleep to the sounds of clinking hobble chains, crying curlews and howling dingoes, and the chanting songs of Mangarayi people in their nearby camp.[34] The Ford's engine was overhauled, but by the next evening when they reached Newcastle Waters, 290 kilometres further south along a cattle track, one tyre was without a tube and stuffed with cloth from a sack, held in place by strips of greenhide.

The next day, the four men turned east. They had to frequently stop to mend punctures until the front tyres were beyond being patched: both were stuffed with sacking that kept them in shape and allowed them to remain mobile, though there wasn't much shock absorption. Before long, sacking filled three of the tyres.[35]

The hot, enervated travellers were about 120 kilometres from Borroloola when Jim Ross, the driver, collided with an anthill obscured by long grass. The Ford's magneto was dented and dead as a doornail as Ross repeatedly tried to kick the engine into life. Fysh and Ross decided to walk to get help in Borroloola while old Tom Lynott and their friend Hogan were to cool their hot heels in the shade of a lagoon they had just driven past. The food supplies were all but exhausted, so Fysh left his gun with Lynott and Hogan, hoping they could shoot a few birds.

Taking just a small billycan of their remaining water, Fysh and Ross headed off across the dry countryside. Soon, all the water had gone to soothe their parched throats. The creeks and waterholes on the map had dried up, and the pair became fearful that they would be cooked by the broiling sun. They trudged on through the heat until they saw flocks of finches in the distance: water had to be nearby. Finally they reached Shady Camp Creek, about halfway to their destination. The weary, dusty travellers sank their heads into a delicious pool under the branches of a wide tree; then they dived in, letting their bodies soak up the lifesaving refreshment.

On the third day of walking, the pair finally stumbled into Borroloola, where they were given a warm reception and cold beers. Fysh organised a team of horses to retrieve the car and the two men who'd been left behind. Lynott and Hogan had spent six days beside the lagoon and survived by using Fysh's rifle to shoot galahs.

In Borroloola, Fysh helped the others to patch up the old Ford as best they could, and new tyre tubes were made by vulcanising old torn ones together. Then, after taking time to recuperate from their ordeal, Fysh, Ross and Hogan set off for Camooweal. But they had driven just eight kilometres when Fysh needed to start stuffing long spinifex grass inside the ripped tyres to keep them rolling. The men made it to McArthur River Station, but the

Ford would go no further – it had sustained forty-one punctures since leaving Darwin.

As luck would have it, Harry Allman – the man running the property – had a luxurious new Buick and was leaving for Cloncurry the next day. Fysh and the others felt like they were gliding on air as they cruised past Anthony Lagoon, Brunette Downs, Avon Downs and Lake Nash Station on the Queensland–Northern Territory border before they pulled up at another of Allman's properties, Linda Downs, about a hundred kilometres south of Urandangi.

Fysh spent the next few days there before he and Allman continued their journey in the Buick, now through heavy rain. The flooded Wills Creek at full power was a confronting sight, and Allman was forced to back out of the raging water before the car was overcome. Seeking shelter, the men found a little homestead beside the road; it turned out to be Bushy Park, where Alexander and Marion Kennedy were domiciled again. The Kennedys, renowned for their hospitality, took the drenched travellers inside, gave them soft beds and warm meals, and dried their soaked clothes in front of a log fire. Kennedy told Fysh that McGinness had stayed with him in Cloncurry and talked up a plan for an outback aerial service. The old man repeated his promise that he was willing to back the youngsters in their venture.

By the next morning the skies had cleared and Wills Creek had subsided, and the Buick made it across with a little pushing. On 3 July 1920, Fysh was finally reunited with McGinness in Cloncurry, six weeks after leaving Darwin.[36] Eleven months since the pair had set off from Longreach with George Gorham, airstrips were dotted across the outback. McGinness told Fysh of his own adventures and of the friends he had made among the wealthy landholders. He declared that their dream of an aerial service for Queensland and the Northern Territory might soon take off.

Chapter 10

Lieutenant Fysh states that the company is being financially backed by leading Queensland pastoralists, and that the chief aim at present will be to assist in the development of the pastoral industry in Central Queensland.

Press report on the idea for a small Queensland and Northern Territory aerial service[1]

HAVING PUT ASIDE THE differences that had flared on the trek to Darwin, Fysh and McGinness left Cloncurry together by car and headed south for Longreach, hoping to drum up interest in their outback airline along the way. At the same time, the British battlecruiser HMS *Renown* was sailing from Hobart for Sydney carrying Edward, Prince of Wales, and his large entourage.[2] The warship had been converted into a royal yacht, complete with squash court and cinema, as the future king was nearing the completion of a goodwill tour that had taken in Canada and the United States. As the *Renown* sailed into Sydney Harbour on 16 June 1920, a flotilla of warships and submarines was anchored to greet the prince amid a veritable sea of boats, bright flags, waving hands and cheers.

High above, Ross Smith and Jack Butler – Fysh's Tasmanian mate from the No. 1 Squadron in Palestine – were part of an aerial exhibition that thrilled the crowd, especially when one of the de Havilland machines, carrying a pilot and passenger, developed engine trouble and fell into Woolloomooloo Bay.[3]

The aircraft had started to sink when the two soaked men were hauled above a launch called *Dinkum*.

From Sydney, the prince was to take a specially outfitted train to Brisbane, where he would perform the official opening of the city's Royal Queensland Exhibition[4] – or 'Ekka' – on the public holiday of Wednesday the twenty-eighth. The Ekka had been Queensland's major annual social and entertainment event for more than forty years. In 1920 there was an extra reason to celebrate the royal visit to the Brisbane Showgrounds: the previous year, for the first time, the event had been cancelled because of the Spanish flu pandemic that had caused border closures in Australia and left returning soldiers stranded; the Exhibition Ground had been turned into a quarantine station with four hundred beds. The 1920 Ekka would be a celebration of life and a return to a new normal after the pandemic and the deaths of ten thousand Queenslanders in the Great War. Many of the wealthy graziers from western Queensland, including Fergus McMaster and Ainslie Templeton, would be in town, casting their eyes over the prized livestock in the show ring.

Fysh and McGinness were ready to put on a show for them too. They drove down from Cloncurry through Winton, talking up the need for the air service with everyone they met – though they still had no firm commitment from any backers, apart from Alexander Kennedy's promise that came with no specific amount of money. Flying, after all, was still a risky business, and the rate of aircraft crashes at the time was unlikely to start a stampede for shares in their little venture. In Longreach, though, Templeton told them that he and McMaster would be staying at the regal 77-room Gresham Hotel in Brisbane for the Ekka, and that McMaster was the man with the commanding presence and the political contacts who could encourage his wealthy neighbours to open their chequebooks.

In the few months since McGinness had helped McMaster with his car on the Cloncurry River, the grazier had left Devoncourt station near Cloncurry after his brother Hugh had become fit enough to run it again. McMaster had moved to another of the family's properties, Moscow Station near Winton, 350 kilometres south-east. But there had been no change in his admiration for the enterprise and generosity of the young airman, and he may have noticed that McGinness and Fysh were attracting publicity on a national scale for their daring and determination. Glowing newspaper reports had appeared, lauding the great overland drive through the Gulf, and they were illustrated with Fysh's dramatic photographs.[5] The climate was perfect for these likely lads to make their sales pitch.

ON THEIR ARRIVAL IN Brisbane Fysh and McGinness booked into the Gresham, on the corner of Adelaide and Creek streets.[6] The next morning, Templeton told them that McMaster was in the hotel's lounge reading the paper and it was a good time to approach him. Templeton said he would sit on the other side of the room and see how things went; if McMaster demurred, Templeton could enter the discussion offering the young men his emotional support.[7]

McMaster was glad to see McGinness again. He had not met Fysh, but it seemed any friend of McGinness was a friend of his. The pilots told McMaster they planned initially to make money for the business through joy flights and aerial taxi work in western Queensland and the Northern Territory, branching out into carrying freight and mail as opportunities presented themselves. McGinness had money, after coming into the inheritance from his father at the age of twenty-one, and said he could put up £1000 to buy a war surplus aircraft. Fysh could contribute a good amount as well.

McMaster had only ever spent ten minutes in an aeroplane during a joy flight at London's Hendon airfield, which he didn't find that joyful. He fixed the young men with the steady gaze of his deep blue eyes, listened intently, nodded occasionally, and now and then uttered his characteristic drawn-out 'Ye-e-s'.[8] He said later that his appreciation for McGinness had already been established, and he was now 'quite prepared to assist not only personally but to raise sufficient capital to finance the venture'.[9] He saw the aerial service not so much as a money-making venture but as a way of helping the outback community he loved.

The men shook hands, and the deal to start an airline was done.

McGinness smiled and nodded at Templeton to signify they had lift-off. When the young pilots had left the room, McMaster told Templeton the pair had covered themselves with glory on Gallipoli and in Palestine, and that even though an airline service in the outback wasn't a guaranteed winner, aviation should be encouraged.[10] There was no doubt in his mind, though, that the airline would take flight. Templeton told McMaster that he would stump up another £1000, and that if McMaster wanted to raise his stake, he would match it. Thus Templeton became the first Qantas shareholder.

McMaster immediately went looking for others. Propelled by the idea of dragging remote areas of Australia into a new age of innovation, he stepped boldly out of the Gresham and called on some of his well-heeled friends. John Thomson,[11] a popular parishioner at Brisbane's Saint Andrew's Presbyterian Church, ran a bookshop in Queen Street, but once he had fought the Germans alongside McMaster in the AIF in France. Remarkably, Thomson had enlisted four years earlier at the age of fifty-two years and one month, and he had survived gunshot wounds to his legs just two days before the Armistice.[12] He told his old comrade he was happy to put up 'a cool £100'.[13]

McMaster next called on Alan Campbell,[14] the managing director of the Queensland Primary Producers Cooperative Association in Eagle Street. Campbell not only agreed to invest but also to act as temporary secretary of the proposed company – and to provide business advice – until it was up and running. Then, in Queen Street, McMaster ran into T. J. O'Rourke,[15] a prominent Winton grazier and storekeeper, who was also visiting Brisbane for the Ekka. O'Rourke had the reputation of being an extremely shrewd investor, but to McMaster's surprise, after they had walked back to O'Rourke's hotel together, the wealthy storekeeper wrote a cheque for £250 with the promise of another £250 if it was required. He was willing to invest even though he wasn't sure the idea was a goer; when he filled in the cheque butt, he wrote 'Donation'.[16] By the end of the day, McMaster had raised £3000. The principal original backers were Templeton and McGinness with £1000 each, and Fysh and McMaster with

£500 each. McGinness and Fysh soon lifted their investments to £1200 and £600 respectively.[17]

Alexander and Marion Kennedy had moved to a grand home called Loreto, in Chasely Street, Auchenflower, beside the Brisbane River. McMaster came calling, and Kennedy offered £200. When McMaster challenged him to 'make it 250', Kennedy said he would and once again stipulated that he wanted the first ticket to ride when the new service began.[18] Kennedy became a provisional director on the company's interim board.

McGinness wired Arthur Baird in Melbourne to say that the business would soon be operational and they wanted him as chief engineer. Baird readily agreed. McGinness regarded Baird as the best aviation engineer he had met in the war, and the airman credited Baird's skill with saving his life many times when he was flying on missions. The hiring of Baird increased the fulltime staff of the new company to three.

Three months would go by before a certificate of incorporation was granted for the business. During that time, McMaster arranged another meeting around a glass-topped table[19] at the Gresham with Fysh, McGinness, Templeton, Alan Campbell, John Thomson and T. J. O'Rourke. Fysh took notes with a pencil. McMaster was elected chairman, and Campbell was made provisional secretary. The small town of Winton, in central west Queensland, was chosen as the home of the company, which was given the provisional name The Western Queensland Auto Aerial Service Limited.[20] It was agreed at the meeting that the company would buy two aircraft and then raise funds through taxi work and joy flights because there would be very little money for operational costs after the purchases.

Campbell and McMaster briefed the law firm Cannan & Peterson, in Rothwell Chambers, Edward Street, Brisbane, to start preparing the paperwork necessary to register a company in Queensland. Roy Peterson was running the firm's original Longreach office, while John Cannan had established the Brisbane headquarters.[21]

Fysh and McGinness now travelled to Sydney by train. They were planning to buy two aeroplanes from Nigel Love,[22] who had the rights to sell Avro aircraft in Australia. He had taken a three-year lease on the mud flats that were the Kensington Race Club's bullock paddock, between Cooks River and Botany Bay, to use as an aerodrome. The site is now known as Sydney Airport.

NIGEL LOVE HAD SERVED in France with the No. 3 Squadron of the Australian Flying Corps alongside Charles Kingsford Smith, and as an instructor with the Royal Air Force. In France, he had souvenired a piece of fabric from the Fokker triplane of the Red Baron, Manfred von Richthofen, after the German ace was shot down just north of the village of Vaux-sur-Somme and the No. 3 Squadron was given charge of his body.

Back in Australia, Love had been trying to raise money for his new business in Sydney and to create interest in aviation by piloting joy flights and charter operations. He took photographers up high for a bird's-eye view of the harbour. On 14 April 1920 he had set off from Mascot with the first fare-paying passenger – a businessman named John Gibson – on the Sydney-to-Melbourne route; the flight through powerful winds and driving rain took nine flying hours over three days.

The Avro 504, on which Fysh had done his initial flight training, had been around since before the war. It had the dubious distinction of being the first British aircraft to be shot down by the Germans, on 22 August 1914. In the years since then, it had undergone several modifications, and the 'K' variant had become popular as a training aircraft. Made of wood, it was covered with fabric and was 28 feet 11 inches (8.8 metres) long and had a wingspan of 36 feet (11 metres). The machine was a product of the Avro company, founded by Alliott Verdon Roe, who built his machines at Hamble, near Southampton.

Avro had made Bert Hinkler their chief test pilot after he had completed the longest solo flight ever, from London to Turin, on 31 May 1920. Hinkler had planned the 1046-kilometre flight as the first stage of a solo trip all the way to his mother's front door in Gavin Street, Bundaberg – but a war in the Middle East led British authorities to ban him from continuing the journey. Still, he had once again brought Australian aviation to the fore, making the flight in a tiny Avro Baby biplane with a kangaroo painted on the tail of the machine. Hinkler's successes were great promotion for Avro machines, and Love sold twenty Avro 504Ks during 1920. One was bought by John McIntosh, who – after making the flight from England with Ray Parer – rode a motorbike across Australia and was planning a business giving joy flights in the West Australian wheat belt.

Love's company was using a defunct rollerskating rink in Petersham to replace the unreliable rotary engines on the Avro 504Ks with the six-cylinder, water-cooled British Dyak motors built by the Sunbeam company in Wolverhampton. The Dyak had an aluminium sump, block and cylinder head, and an overhead camshaft with two valves per cylinder; it produced 106 horsepower, less than most modern small cars. Love's Avro 504Ks had room for two passengers crowded into a second open cockpit behind the pilot's space, and it could cruise at 105 kilometres an hour, though the great weight of the motor in comparison to its power output gave it a disappointing rate of climb in the thin air of the outback.[23]

Nevertheless, on 19 August 1920, Fysh and McGinness ordered two 504Ks at £1500 each for The Western Queensland Auto Aerial Service. Once the deal was made, though, Love told the crestfallen young pilots that due to strike action in England they would have to wait at least two months for the Dyak engines to be ready.

This wasn't the only bad news that Fysh heard at that time. His father, Wilmot, with whom he had become estranged, died suddenly at his home in Canning Street, Launceston, on 29 August, aged just fifty-four. In the last years of his life he had left his little farm to earn a meagre living as a travelling salesman, and the flag at the city's Commercial Travellers' Club was flown at half-mast as a sign of respect.[24] Mary Fysh had spent her early years as a Christian missionary and now, separated from her husband, she felt a sense of shame among her congregation, and the wider community. In her mind she had fallen from the highest rung on the social ladder. With her funds dwindling, she looked to start a new life elsewhere, far from Launceston.

FYSH AND McGINNESS LEFT Sydney for Melbourne to report to the Defence Department over the landing grounds, and to answer questions from the Melbourne press about their great journey through the bush and the aerial route that had been created.[25] Melbourne was still the seat of Australia's Federal Parliament, and the pair spoke with Prime Minister Billy Hughes and some of his ministers about implementing the *Air Navigation Act of 1920*, which would provide for the rules of flying in Australia, registration of aircraft, licensing of aerodromes and personnel, and mandates for the periodic inspection and maintenance of aircraft.[26] The young pilots also sought government aid for their proposed 'aerial service for Central Queensland which they promised would eventually connect with the Northern Territory rail head at Katherine'.[27]

With his enthusiasm for the business gathering more momentum every day, Fysh told the press that 'The Federal Ministry … received the proposal favourably, and promised to give the whole scheme most careful consideration.'[28] The proposal to the federal government included a request that in addition to the four aerodromes the new company would establish in central Queensland, the government should build or subsidise aerodromes at Camooweal, Avon Downs, Anthony Lagoon and Newcastle Waters, and at Katherine, connecting with Darwin by rail. Fysh said such a scheme would bring Darwin closer to Melbourne by as many as twenty-one days.[29]

McMaster, meanwhile, returned to Winton to drum up more business and backers. On 30 September he told the new company secretary, Alan Campbell, that he would send him a copy of the prospectus – and he ribbed his friend:

> Knowing you to have a contrary kink in your constitution I feel that you will go out of your way to pull this document to pieces … I have dealt with the proposition much more fully than in the draft we drew up in Brisbane. I would like to get the Articles of Association printed and everything, as far as you can, pushed along. The sooner we get the Company registered and in order the better. Also, try and

> get as many good Australians as you can to take shares. I hope you will appreciate the typing. There are a few mistakes but I have been going at a hell of a pace – fully ten words a minute.[30]

Further delays over the Dyak engines had put the fledgling business under enormous pressure, with Fysh and McGinness having to pay all their travel and hotel expenses out of their own pockets. Campbell knew little about aeroplanes, but as he looked for something to replace the Dyaks he obtained a quote for the new Westland Limousine aircraft, of which just eight were built at the company's factory in Yeovil, Somerset. The quote noted that the Limousine carried 'a pilot and five passengers'. Westland's agent told Campbell, 'The touring speed is about 100 to 105 miles per hour, and the petrol capacity is 96 gallons, or about five hours flying'.[31] With the 200 Hispano Suiza engine, the aircraft would be £2700 in England, and with the Rolls-Royce Falcon Engine it would be £3500. The firm was also bringing out a larger machine with a 450 Napier Lion Engine, and this would cost about £5000 in England. Prices would be significantly higher after transportation to Australia.

That was a bit steep for a new enterprise, but Campbell wanted to get the company airborne as quickly as possible. Despite the misgivings of the shrewd McMaster, Campbell ordered one of the new Avro 547 triplanes 'to operate a service between Longreach and Winton, a distance of 108 miles, to a schedule of one hour forty minutes'.[32] Campbell had put his trust in 'the fullest enquiries in Melbourne and London'; in a letter to Campbell, McMaster said, 'I would sooner have let this machine stand over until the Company had got further ahead.'[33]

The triplane came with some wonderful advance publicity. Based on the design of the Avro 504, it had an extra wing, and though the pilot would still sit in the open, braving the vagaries of the weather, the machine had a deeper fuselage and, like the Westland Limousine, could carry four or five small passengers in what was then the revolutionary idea of an enclosed cabin. Newspapers described the passenger compartment as being 'as roomy as a railway carriage ... furnished with four comfortable chairs'.[34] It wasn't, and this triplane had proved to be slow and unstable while flying in tests for the British Air Ministry in August 1920.[35]

Avro agreed to ship one to Sydney at a cost of £2798, though no air service in England would touch the machine. Fysh couldn't have known of the problems surrounding the aircraft, though, when he was interviewed by the *Sydney Morning Herald* after his return from Melbourne with McGinness at the end of October 1920. He told the newspaper that the company had placed an initial order with Love's Australian Aircraft and Engineering Company for two 504Ks and the Avro triplane, and he hoped that the 504Ks, to be built at the engineering works at Mascot, would be ready for delivery in the first week of November when he and McGinness would fly them to Winton. They would be used principally between the scattered homesteads of central Queensland and for 'general business purposes in place of motor cars'.[36] The triplane, Fysh told the paper, would be the first of its kind to be shipped to Australia, and it had already left England. A bigger passenger aircraft than a 504K, it could remain airborne for five hours. The new company planned to fly it on a regular twice-weekly service between Charleville and Cloncurry, connecting with the northern and southern mail train service.

While the company would be based in Winton, Fysh said that depots with fuel and spare parts would be established at Longreach, Charleville and Cloncurry, and the object was to establish a regular aerial service between the main railheads in Queensland and the Northern Territory. Fysh explained:

> At the present time, if a man wants to get from Charleville to Longreach, unless he motors across country, where the roads are mere tracks, he has to travel 430 miles to Brisbane, then 300 to Rockhampton, and then 400 more to Longreach, a journey of 1130 miles, and taking over a week's quick travelling to accomplish. The air journey is only about 250 miles, and can easily be accomplished in less than five hours, allowing ample time for taking off

> and landing ... By train from Melbourne to Charleville is about four days' journey; from thence to Katherine by triplane could be accomplished in about three days, with one extra day for the train journey to Darwin. Thus the journey, Melbourne to Darwin and back, could easily be accomplished inside three weeks, even allowing two or three days for business at Darwin.[37]

AT WINTON'S Bank of New South Wales on 14 October 1920, Fergus McMaster opened an account called 'Aerial' with a deposit of £700, which the bank manager noted would be 'transferred to Qld. Aerial Services when articles of approval come to hand'.[38]

'Qantas,' McMaster later wrote, 'was founded on trust and co-operation, and that is what stuck to it through the first severe years of its pioneering life.'[39] He dated the company's prospectus as 14 October, and having consulted with the partners, it expanded greatly on the original modest plans of McGinness and Fysh for joy-riding and aerial taxi work. McMaster's document contained much more detail than the plans suggested around the glass-topped table at the Gresham. He needed to make the prospectus as wide-ranging and appealing as possible at a time when flying was a risky business and investing in a scheme for outback aviation was a long shot. So far funds had only been raised because of the friendships between McMaster and the other backers, and the company was based largely on the foundation of mateship that McGinness had shown to him on the Cloncurry River, as well as the laudable war service of the two young pilots. McMaster wanted to appeal to a much broader audience and also to seek government backing. He knew that the man on the land was now packing more political clout than ever, with the Country Party in New South Wales, Victoria and Western Australia having won eleven seats in the 1919 federal election.

In the prospectus, McMaster proposed the longest direct air service in the world, 'an aerial mail service from Longreach to Port Darwin, connecting at Winton, Cloncurry, Avon Downs, Anthony Lagoon, Newcastle Waters and The Katherine'.[40] He told prospective investors that federal ministers were considering the proposition, and 'it is expected they will assist in establishing the route ... one that would be required in connection with the aerial defence of Australia'. The air route, McMaster said, would help the federal government in the administration of Darwin, and it would assist in the development of the large pastoral and mineral areas by making them accessible to the providers of capital. A regular mail service along the route, the prospectus promised, 'would do more to develop commercial aviation ... than any other in Australia'.[41] McMaster also proposed that the government could use the company's pilots in time of war.

He was apparently doing a sterling job at raising money. Fysh told reporters in October 1920 that nominal capital was now £100,000: 100,000 shares at £1 each, of which half was 'being called up at present'.[42] This was, in fact, still a pipe dream. He added that the business would be called the Queensland and Northern Territory Aero Services Limited, though in reality the partners were still tossing around names. One suggestion was the Australian Transcontinental Aerial Services Co. Then the name was made more area specific: the first advertising flyers for the business listed it as 'Northern Territory and Queensland Aerial Services, Ltd', or N.T.A.Q.A.S. This name was used in an advertisement placed in the program for the Australian Aerial Derby, to be held in Sydney on 27 November 1920.

But then there was another rethink. Eleven days before the aerial derby took place, a clerk from Cannan & Peterson's legal office walked into the office of the Registrar of Joint Stock Companies in Brisbane's Supreme Court building on the southern side of George Street. He handed over the paperwork and registration fee for the company that Fysh and McGinness had envisaged on their drive through the Queensland bush. Its headquarters were listed as Elderslie Street in the small bush town of Winton. The registrar, Francis Sydney Kennedy, signed the document, and the legal clerk walked out with the certificate of incorporation.[43]

The date was 11 November 1920. Fysh and the others knew that making the business work would be an uphill battle. Even the best aeroplanes of the time were uncomfortable and unreliable. Forced landings and crashes were frequent, and most people were still afraid to travel in what were loud, noisy, unpredictable, unreliable and often deadly contraptions. Fysh admitted that confidence in aviation in 1920 among the general public was 'almost nil'. Anyone suggesting air transport over the vast distances between Australia's biggest cities usually provoked nothing but ridicule, especially among big business and banks. Their only funding was coming as a result of 'postwar sentiment for two young returned servicemen and the hope that in the roadless and bridgeless western plains, where all road transport ceased following heavy rains, perhaps the aeroplane might fulfil a useful purpose'.[44] This bold new enterprise needed the spirit of patriotism and adventure for its launch.

At the outbreak of World War I, the Winton district had a population of 2554. Of that figure, 519 men enlisted, a huge proportion of the eligible males. There were ninety-two killed.[45] In naming their airline, Fysh and the others decided they wanted a name inspired by the Australian and New Zealand Army Corps, one that could be shortened to an acronym like Anzac. The patriotic spirit of the servicemen would be the wind beneath their wings.

The business dreamt up on an outback track was duly registered as Queensland and Northern Territory Aerial Services Ltd. Fysh and the others soon condensed this to Q.A.N.T.A.S.

Chapter 11

It seems imperative that the Commonwealth Government should encourage commercial aviation and the building of machines to the fullest possible extent. It is the encouragement of a new mode of travel, and new industry, which will soon employ many thousands of people besides forming a great means of defence for the country.

HUDSON FYSH ON THE NEED FOR THE FEDERAL GOVERNMENT TO SUPPORT THE EARLY AVIATION INDUSTRY IN AUSTRALIA[1]

FYSH BECAME EVER anxious over the delays for the Dyak engines, and it seemed he might become the first Qantas redundancy before the company even got off the ground. The engines finally arrived in Sydney a week after Q.A.N.T.A.S was registered but stayed on the docks for another week, then the order for the second 504K was eventually cancelled. The triplane, tested by Bert Hinkler in England, was on its way from Southampton but would need assembly work when it arrived. In the meantime, only one pilot would be needed to fly the 504K to Winton, when the aircraft was finally ready. The Dyak engine would still need to be fitted to the machine, and with the *Air Navigation Act*[2] having just been passed, the 504K would still have to be registered, and several test flights would be required before an airworthiness certificate was issued.

All of this would take time, and time was money. Fysh and McGinness were spending all their savings on accommodation and expenses at Sydney's Metropole Hotel at 16 shillings sixpence a day.[3] Nigel Love estimated that he could deliver the 504K with the Dyak

Qantas paid £1500 for its first aircraft, an Avro 504K with a six-cylinder Sunbeam Dyak engine that produced far less power than most modern small cars. It was sold to a private owner in 1926 and deregistered in 1932. *State Library of NSW FL8576702*

engine early in December, though the two pilots reckoned it would be mid-December at the earliest. Alan Campbell, who met them while visiting Sydney, correctly predicted that Christmas would come and go before it would be ready.[4]

The cancellation of the second 504K threatened Fysh's job security. He had only just been awarded his pilot's wings a week before catching the boat home from Egypt, and had only thirty-four hours fifty minutes of flying time to his credit. On his second solo flight on a 504K at Aboukir, Fysh had crashed, confused by the tricky fine-mixture adjustment on its rotary engine. McGinness, by contrast, had more than four hundred hours in the air and was lauded as one of the best pilots in Australia.

McGinness was the senior partner in the operation, and Fysh knew that if Qantas needed just one pilot for a while, McGinness was the man. Still, even with his job in jeopardy, Fysh kept pushing and promoting aviation as vital to Australia's interests. Interviewed by the press in Sydney, he expanded on his new company's plan to launch an air service between Cloncurry and Charleville, and advocated for the federal government subsidising such enterprises 'because serviceable weight-carrying machines would be called up for service in time of war'.

> Machines used for fighting, reconnaissance, photography, artillery work, coastal patrol, and such like work, are of the smaller class of machine, and are not great weight lifters, such as the latest types of commercial aircraft to-day. The most suitable types of bombing machines are those which can carry a large load over a long distance; such as the majority of present-day commercial machines. These machines, in a few hours, could be adapted to carry a cargo of bombs instead of their ordinary load of passengers or goods. It is this type of aircraft which will play the part of bombers in any future war; and every suitable type of commercial machine which comes into use in Australia today means another valuable unit in the defence of our country.[5]

Writing for Brisbane's *Daily Mail* a few days later, Fysh opined:

> In time of war in any country which places reliance on aerial defence, the first step would be the immediate calling up of all commercial aircraft, with probably part of its personnel and equipment. This would immediately put at the disposal of the Defence authorities a bombing fleet which would be escorted by fighting machines and whose service in the defence of the country it would be impossible to overestimate. Owing to her vast coastline … if ever a landing is affected on our shores it will only be the aerial arm that will witness this catastrophe, and attempt to fight off or delay the enemy. To do this work, bombing machines are necessary, and every additional machine is valuable.[6]

For the time being, though, Fysh knew the biggest fight he faced was to keep his job, and to see Qantas take to the air.

THE METROPOLE WAS a grand hotel close to Sydney Harbour, with frontages on Young, Bent and Phillip streets, so

Fysh and McGinness at least enjoyed a degree of luxury while the days ran into weeks and work on their aircraft dragged on. The hotel had 260 guestrooms, several dining rooms, sumptuous furnishings and electric lighting, and – along with lavish stained-glass windows – a roof promenade from which the young entrepreneurs could take in views from The Heads almost to Parramatta,[7] a landscape they hoped to soon see from several thousand feet above terra firma.

One morning in the new year of 1921, a welcome visitor stepped across the mosaic tiled floor of the Metropole's entrance and asked to see the two young airmen. The visitor was the bespectacled John Flynn,[8] a Presbyterian minister who had devoted his life to spreading Christian charity in the outback. He had read in the papers about the aviators staying in Sydney and their outback airline that would cross the same territory as Flynn's work for the Lord.

Flynn was born forty years earlier in the Victorian village of Moliagul, where the largest known gold nugget, the 'Welcome Stranger', was discovered. But wealth didn't follow his schoolteacher father; unable to afford university study, Flynn became a pupil-teacher with the Victorian Education Department in rural areas where he got to know some of McGinness's family.[9] After his ordination, Flynn began ministering in outback South Australia in 1911 before surveying the Northern Territory and being appointed superintendent of the Presbyterian Church's Australian Inland Mission, a charity initiative 'without preference for nationality or creed'. The mission was comprised of him, one nursing sister, a nursing hostel at Oodnadatta and five camels.[10] The camels were good, but Flynn knew that aircraft would be much better in helping people to receive urgent medical assistance in the furthest corners of Australia.

Inside the smoking room of the Metropole, as Fysh and McGinness lit up, Flynn told the young pilots that while they were fighting in Palestine, Australia had been held in thrall by the story of a Kimberley stockman, Jimmy Darcy, who had suffered massive internal injuries on 29 July 1917 when his horse fell in a cattle stampede. An eighty-kilometre ride on a dray over a rough track had taken Darcy to the nearest settlement of Halls Creek in the far north of Western Australia. He needed immediate lifesaving surgery, but the nearest doctor was thousands of kilometres away. For days, newspaper readers around Australia were gripped by the drama of the young stockman's desperate struggle for life as a doctor in Perth, Joe Holland, instructed the Halls Creek postmaster Fred Tuckett on how to perform emergency surgery with the help of Morse code, a penknife and some morphine. Tuckett's hands shook because he was afraid he would make a mistake – that he'd kill the man everyone was trying to save – but Dr Holland told him if he didn't make the incisions, Darcy would die anyway.

Fysh and McGinness listened spellbound while Flynn told them how Tuckett, using his little knife, cut through Darcy's flesh above the pubic bone as the stockman's brothers tried to ease his agony and shoo the flies away from the blood. With the Morse code key rattling away furiously, delivering precise instructions, Tuckett worked feverishly for hours, cutting and stitching, stopping every few minutes to check the doctor's telegrams. The operation on Darcy's ruptured bladder was a success, but the 29-year-old stockman was weakened by malaria. The situation was so grim that Dr Holland made a mercy dash from Perth, boarding a cattle ship that took an excruciating week to reach Derby. The doctor then spent six days in a Model T Ford held together by leather straps, bumping and thumping across the desert to save the stockman's life. Aboriginal men helped to push the car across riverbeds and up sandy banks, but Dr Holland still endured punctures, radiator leaks and engine stutters. At one point he had to use the rubber tubing from his stethoscope to siphon the last drops of petrol from a can. The car finally conked out forty kilometres from Halls Creek. Dr Holland walked for two hours to a nearby cattle station, then rode a horse through the night to reach the town at daybreak. Jimmy Darcy had died a few hours earlier. Fysh and McGinness gasped at the tale.

Then Flynn related how not long after Darcy's death, Lieutenant Clifford Peel[11] of the Australian Flying Corps had written to him[12] suggesting that the new technology of aviation could carry medical aid to remote places, and then detailing how it might work. Peel had not survived the war – he was shot down over France six weeks before the Armistice – but his letters outlining the costs and advantages of aeroplanes compared with ground travel would ultimately help to save thousands of lives.[13]

Flynn heartily shook the hands of Fysh and McGinness. He told them that despite the war impeding his progress, he had expanded his mission from the first nursing hostel in Oodnadatta to bases in Port Hedland and Broome in Western Australia, Pine Creek in the Northern Territory, and Cloncurry in Queensland. 'How wonderful,' Flynn told them, 'if the Inland Mission could get an Aerial Medical Service going.'

THE QANTAS AVRO 547 triplane, fitted with a Beardmore 160-horsepower, six-cylinder, water-cooled engine, arrived in Sydney on 17 January 1921 aboard the Commonwealth government's cargo steamer *Gilgai*. It still needed reassembly and testing that would take weeks.[14] In good news, the final tests of the 504K were completed at last, and on 25 January, McGinness wired McMaster that he was ready for take-off in the first Qantas aircraft. What was Fysh to do? Money was tight and he was in danger of running out of savings before Qantas could pay him a wage.

The problem of his 'temporary redundancy'[15] was solved after they were approached by thirty-year-old Charlie Knight,[16] a leading stock and station agent from Longreach. Knight had bought one of two B.E.2e war surplus machines imported into Australia to promote the Perdriau Rubber Company, a business that would eventually merge with Dunlop. Knight wanted Qantas to deliver the now-famous machine to Longreach.

Six months earlier, Fysh's mate Jack Butler, Perdriau's chief pilot, had made headlines with what was to become Knight's new machine. The jockey-sized Butler[17] took off from Mascot at 7.50 a.m. on 9 July; after dropping off special issues of the *Sydney Sun* newspaper at various towns along the way, he landed at Brisbane's quarantine ground in Lytton at 5.11 p.m., becoming the first pilot to travel between Australian cities in one day.[18]

But aircraft of the time were fickle things, and pilots often came unstuck. Just a few weeks later, Butler was flying the machine over Longreach and carrying a passenger, Herb Avery,[19] who owned the Western Motor Works in Magpie Lane. Butler was flying too low; after almost crashing into the parapet of the Imperial Hotel, he slammed into the telephone line across Duck Street, grazed a blacksmith's shop and crashed into a tree. The aircraft's tail was left pointing to the sky, while the front portion of the plane resembled a concertina and the propeller was splintered. Butler and Avery escaped with their lives, but Avery, who had occupied the front seat, sustained a cut face. The aircraft was eventually rebuilt. Butler limped away from the wreckage with 'a cut nose and the prettiest pair of black eyes imaginable'.[20]

Fysh's belly was understandably full of butterflies on the morning of 31 January, when he and McGinness prepared to take their two aircraft from Mascot to Winton two thousand kilometres away. Fysh was given the B.E.2e to fly, even though he had spent only thirty minutes on the model in Egypt. He and McGinness planned to reach the NSW country town of Moree before nightfall, taking a course north to Newcastle before turning inland. Storm clouds were massing along the coast to the north of Sydney, and Fysh's heart beat faster with the knowledge it would be a rough ride.

McGinness was piloting Qantas's first aircraft, the 504K, and was taking the company's first paying passenger, Hugh John Teahan,[21] a station overseer from Kynuna who sat in the rear cockpit. Alexander Kennedy still had dibs on the first ticket for the airline's regular scheduled services. Fysh had Arthur Baird as his passenger on a machine with the words 'KNIGHT Longreach' emblazoned along the fuselage in large white capitals, an example of the latest hi-tech medium of advertising. Only a few weeks earlier, Butler had piloted the rebuilt B.E.2e to second place in the handicap section of Sydney's Aerial Derby behind a 504K flown by Nigel Love.[22] Now at Mascot, Baird swung the B.E.2's four-bladed propellor and, as Fysh revved the ninety-horsepower V8 engine, he climbed into the passenger seat in the front cockpit. McGinness started the Dyak engine using a crank handle located between his knees. Each pilot then taxied their aircraft to the end of the grass runway, opened the throttles, and lifted off into the great and dangerous adventure. In the event of becoming separated during the flight, Fysh and McGinness had agreed to rendezvous at Singleton.

Fergus McMaster was busy preparing for their arrival. He had informed the press that the two war heroes and their passengers would be in Queensland's central west soon and that the triplane would quickly follow, with low fares on offer between Winton and Longreach.[23] He organised three and a half tons of petrol and three cases of engine oil to be stored at the home of Frank Cory, a Longreach stock and station agent, before they were taken to a storage shed. But McMaster was facing difficulty finding insurance for the pilots and the machines, after Queensland's State Government Insurance Office told him that it couldn't give cover to his pilots under the *Workers' Compensation Acts* because their pay exceeded £400 a year.[24]

Each aircraft had a magnetic compass that swung all too easily with the erratic motion of flight, as well as an inclinometer, a rev counter, and an air speed indicator. The B.E.2 had a top speed of just 116 kilometres per hour, and if it flew slower than 76 kilometres per hour it would stall and drop out of the sky. The spinning propeller drove air backward in a spiral pattern, so Fysh had to push constantly on the left rudder bar to keep the wood and fabric machine in a straight line.

He and McGinness cruised over Sydney Heads and past the bridgeless harbour towards Manly, then followed the coast north. But in their noisy open cockpits, the airmen were assaulted by strong winds and constant turbulence as the sky darkened. McGinness, the far more experienced pilot, flew the Avro through gaps in the thickening clouds and came down to just 500 feet (160 metres) in order to dodge others.[25] Fysh's old feelings of apprehension made his nerves taut and his palms sweat; he feared that with his limited experience as a pilot he wasn't up to blind flying through cloud, but other than turn back there was nothing he could do except plunge hopefully into the thick billowing blackness. The compass began to gyrate wildly. Sitting on the front pew, Baird wondered what the hell Fysh was doing as the B.E.2 was tossed about by the weather. He knew in these sorts of conditions there was the likelihood of hail, which would tear the plane apart.

Fysh, enveloped by the storm cloud, quickly became disoriented. Then the engine cut out. The aircraft began to spin as Fysh grappled for control. As the B.E.2 plummeted, he feared that the cloud was on top of a hill. Spinning wildly, he wasn't sure which way was up.

A ray of light appeared in the darkness. Through their goggles, Fysh and Baird glimpsed a valley below. Fysh's disorientation lifted with a point of reference, but he could see no way through the clouds ahead and didn't want to risk another dizzy spell. There wasn't enough flat ground for them to land safely; there was, though, a bushy hillside near a mining pithead. Fysh turned in towards it and – with McGinness far ahead, looking around and wondering what had happened to his mate – the less experienced pilot jolted along to a bumpy but safe landing as he and Baird tore through the bushes on an uphill slope.

The aircraft stopped next to a miner's cottage. As Fysh and Baird, every nerve trembling, climbed out to inspect themselves and the machine for damage, the startled miner's wife ran from her home to check on these visitors from the heavens. Relieved that they had survived, she said she'd put the kettle on and make them a cuppa while they waited for the weather to lift. She told Fysh and Baird that they were at Redhead near the Lambton coalmine, just south of Newcastle; she later sent Fysh a photograph of the mine as a souvenir.[26]

He and Baird drank the kind lady's tea, thanked her for her hospitality, turned the aircraft around to give it a downhill runway, cleared some of the bushes and, when the clouds started to break, took off again. They followed the road to Singleton, where McGinness and Teahan were waiting anxiously for their arrival. When McGinness constantly teased Fysh about his unexpected tea-break, Baird thought the ribbing was too much and told McGinness to shut up or he would leave them and get the next available transport back to Sydney.[27] The four of them stayed in Singleton for the night. At 7 a.m. the next day they flew on, following the railway line to Narrabri and then on to Moree.[28]

The frightening experience of the forced landing had taught Fysh important lessons: don't 'fly in heavy cloud, always keep a line of retreat open in bad weather, and don't get overtaken by darkness'.[29] Safety was something he would drum into all the pilots who worked for Qantas over the next decades.

After stopping in Moree he and McGinness continued on to St George, flying for two and a half hours against a cold, bracing headwind at about 5000 feet (1600 metres). Fysh was the first of them to land in Wilga Park at 11 a.m. on 2 February, stirring up the dust like the prewar polo games on the historic ground.[30] Only one aircraft had visited the town before, but after a large crowd rushed to the rudimentary airstrip to shake hands with the famous pilots, the local paper declared, 'Now the flying machine has come to take the place of the horse, or car.' Removing his goggles, Fysh smiled 'as if pleased to be on terra firma again'.[31]

After lunch, the pilots took off and circled steadily over St George with the intention of leaving for Mitchell. Owing to the 'warmth of the atmosphere and the peculiar nature of the air currents making the ascent difficult', they came down and waited till the following morning. Baird made some minor repairs, and at 6.30 a.m. the next day the two 'planes rose gracefully and ascending to a great height soon became specks in the far away distance'.[32] Teahan had a ball throughout the journey, admiring the countryside from on high, and at times defying the bumps of air pockets and the blast from the propeller to read a book.[33] The aircraft reached Mitchell, then Charleville, where Fysh and McGinness were the first airmen to land on the newly cleared aerodrome. Fysh thought it rather funny that the townsfolk had an ambulance on standby for their arrival.

Over Australia's inland plains, he noted, 'the air is good for flying up to 11 a.m., after which time an air traveller can expect to meet with "rough air"'.[34] The small air pockets disturbed the even progress of the machine, but the big air currents were of an 'entirely different calibre'. Already possessing an eye for marketing and publicity in the aviation industry, Fysh wrote to the Melbourne

Herald to say these currents were 'apparently formed by columns of ascending or descending hot or cold air, and in negotiating these the aeroplane either goes at a great pace or sinks, despite all efforts to climb'. 'These currents,' he wrote, 'do not form to any extent close to the ground, as the earth forms a cushion for the flowing currents of air. Bumps and air currents are of no danger and their relation to the calmer flying is just that of a calm sea to a stormy one. The Northwest is evidently the "Bay of Biscay of the air", at any rate during the hot summer months of the year.'[35]

The following day the four flew on to Tambo, then Blackall,[36] and finally on to Barcaldine, cruising over the vast Mitchell grass plains of western Queensland that stretched away to the horizon and were occasionally marked with patches of dark timber. McMaster had caught the train from Rockhampton to join them on 5 February in Barcaldine. The summer heat was stifling, but that wasn't the only reason he was sweating: he had agreed to fly the hundred kilometres to Longreach with McGinness and then the 180 kilometres from Longreach to Winton, showing his public support for this bold adventure. He knew that selling joy flights to all those who saw the Qantas men in the air was the best way to make some fast money for the business, so he had to demonstrate to locals that flying was not just the thrill of a lifetime but also perfectly safe. To his dismay, McMaster found the 'so called aerodrome' at Barcaldine 'was a small clay-pan completely surrounded by gidyea[37] stumps'.[38]

Away to the west, towards Longreach, whirling columns of dust rose across the dry mile-wide stock route. The two aircraft came into sight on schedule, passing the Barcaldine train along the way,[39] and landed safely, guided by a large pillar of smoke – 'with only inches to spare'.[40] After something to settle his nerves at the Shakespeare Hotel, McMaster climbed nervously into the Avro behind McGinness, who as usual seemed as cool as ice despite the summer sun. He turned the machine around and, with a cloud of dust and pebbles, taxied to the edge of the claypan. He swung the Avro around again, then opened the throttle. 'If ever anyone realised that he was in the hands of God and the pilot,' McMaster recalled, 'I did.'[41]

The Avro roared across the claypan and cleared the gidyea stumps. McGinness and McMaster were off the ground, rocking from side to side as they circled the airfield waiting for Fysh to take off in the B.E.2. Despite the rising panic that McMaster felt while sitting in the flimsy aircraft, he tried to calm himself so he could enjoy 'the splendid view of the surrounding country'. As McGinness aimed the machine west, McMaster was soon over Kelso, one of his properties between Ilfracombe and Longreach, and he saw five of his dams in a way he had never seen them before.

Soon he was more focused on the approaching lightning as they came into land, amid rough weather in the late afternoon, before a cheering crowd at Longreach. The stormy weather produced gusty air currents that had tossed both aircraft about like corks in a wild ocean, and McMaster was delighted to touch the earth again. The ride had been nailbiting, but the practicality of air transport was reinforced to McMaster and the crowd when the mail train from Barcaldine arrived twenty minutes after the airmen despite a head start of more than two hours.

That night in the Shire Hall, Longreach hosted a 'Smoking Concert' for the aerial party and about fifty guests. The celebration was organised by an Englishman, Dr Archibald Hope Michod,[42] who ran his own private hospital in town. He proposed a toast: 'Success to the Central Queensland & Northern Territory Aerial Services'.[43] He told Fysh and McGinness that the townspeople were proud to see Longreach 'being made a centre for aviation' and that McMaster, Ainslie Templeton and Charlie Knight were also 'to be congratulated on their enterprise'.[44] Knight, the first agent to purchase an aeroplane in Longreach, was heartily cheered. McMaster promised a tremendous future for aeroplanes in the west and that aviation would play a great part in Australia's welfare, 'both commercially and from a defence point of view'.[45] He told the audience that the revolutionary Avro triplane was expected to arrive in Longreach shortly.

Despite the enthusiasm of the audience and the conviviality of the town, McMaster didn't sleep well at Longreach – he knew he would have to follow the rough flight from Barcaldine with another to Winton the next morning. This time Templeton squeezed into the passenger cockpit next to McMaster on the Avro as McGinness led the two aircraft off the ground after breakfast on 6 February for the final leg of the 2000-kilometre odyssey from Sydney. Fysh had Knight as his passenger. Despite the summer heat all the men wore coats because it would become freezing up high. A telegram was sent to Winton to say the aircraft had left, and to ask for a large plume of smoke to be billowing by 11 a.m. and for the landing ground to be clearly marked. The 180-kilometre flight would take just ninety minutes, or so the men thought.

McMaster and Templeton were supposed to know every road and track between Longreach and Winton, and McGinness didn't bother studying the ordnance map. But from ten thousand feet the terrain looked very different, and they couldn't get a bearing. Half an hour out from Longreach, when there should have been open country around them, McMaster saw only the hills and rough terrain of the Opalton Range. It was far too windy and noisy for him to shout to McGinness over the Dyak engine as the Avro was being pushed along by a tailwind and topping 160 kilometres per hour. So McMaster wrote a note in pencil and passed it to the pilot sitting in front of them in his tight-fitting leather helmet and goggles.

McGinness was cool – 'too cool for Templeton and myself,' McMaster recalled. 'He kept pointing to his compass, which showed west.' McMaster tried again with another note: 'One hour from Longreach. Too many mountains on our right.' Still the wiry man in the helmet was unmoved. A third note said: 'Should be open downs country on the right.' But as the Avro droned on, McGinness calmly maintained his course west. McMaster scribbled so hard on the fourth note that he nearly snapped the pencil. 'Winton,' he scrawled, 'is north north-west of Longreach.'[46]

McGinness, who was often strong-willed and cocky, had been steering west for two hours and overshot the town by thirty minutes. He finally took out his map to check his position. As McMaster and Templeton said their prayers, the pilot turned off the engine so he could read the Lands Department map without being distracted by the blast of the propellor.[47] The machine seemed to hover for a while, then McGinness switched to the reserve fuel tank above the wing and started up the Dyak again. He now headed in the right direction. Before long, McMaster sighted a faint discolouration on the ground in the far distance and then a slate-coloured patch 'the size of a dinner plate' that gradually, as they flew closer, resembled a settlement. Templeton passed a note to McMaster: 'Thank blazes we are somewhere!' It would not be the last time that McGinness's personality would grate on the nerves of his Qantas partners.

Three hours after leaving Longreach, McGinness and Fysh, who had been following the lead plane, arrived over Winton. On the hill to the north of the town's artesian bore, a large crowd had assembled, and they were wondering about the fate of the famous airmen. Young Harriet Riley[48] – who as an infant was one of the first people to hear the song 'Waltzing Matilda'[49] – had organised a large white bedsheet to mark the landing field. She was the niece of Sarah Riley, the poet Banjo Paterson's long-time fiancée in the 1890s. He and Sarah had holidayed together on the huge stations around Winton, and he wrote the song there with Sarah's best friend Christina Macpherson – a very close collaboration that reportedly ended the engagement. Harriet's late father, Fred,[50] had once run the 100,000-hectare Vindex station; seven years after his passing, his son Fred Jr[51] had an office for the family's stock and station agency in Elderslie Street, Winton.

Fysh and McGinness circled the town, and McGinness touched down first.[52] With fuel in their tanks for only fifteen minutes more flying, they had made the trip from Mascot with seventeen hours thirty minutes of flying time at an average speed of 68.5 miles per hour (110 kilometres per hour).[53] That night they celebrated their arrival with another Smoking Concert before a sizeable audience at the North Gregory Hotel. The speeches celebrated a new epoch in the history of the district, recalling the days when 'Cobb & Co coaches were the most up-to-date method of rapid transport'.[54]

Charlie Knight might have preferred that slower ride, though.

The next day, 7 February, Fysh and McGinness flew back to the railhead at Longreach, Knight again a passenger on his own machine. Fysh missed the track before deciding to follow the Thomson River into town. It was a bumpy ride, and Knight felt like he'd been on board a sinking ship in a wild ocean. He said to hell with flying and told Fysh that nothing would ever induce him to go up in the air again. He sold the B.E.2 to Qantas for the bargain price of £450, though McMaster insisted that £200 of it was in shares.[55] Qantas now had two aircraft and the triplane on its way.

THE ORIGINAL MEMORANDUM and articles for Qantas declared: 'We, the several persons whose names and addresses are subscribed, are desirous of being formed into a Company in pursuance of this Memorandum of Association, and we respectfully agree to take the number of shares in the capital of the Company set opposite our respective names.' The signatories were:

John Thomson, 311 Queen Street, Brisbane
Alan Campbell, Commercial Union Chambers, Eagle Street, Brisbane, company manager
Hubert Thorne Hill Weedon, Oondooroo Street, Winton, solicitor
Fergus McMaster, Dagworth Street, Winton, grazier
Charles James Anthony Brabazon, Elderslie Station, Winton, grazier
Ainsley Neville Templeton, Acacia Downs, grazier
Paul Joseph McGinness, Winton, grazier and aviator
Hudson Fysh, Winton, aviator.
The provisional directors were: McMaster (chairman), Templeton, McGinness, Alexander Kennedy, and Thomas Brassey McIntosh, a grazier, who listed his address as Wollongorang, Northern Territory.
The secretary was Alan W. Campbell, company manager, Brisbane.[56]

Each shareholder knew that the airline was a risky venture in an industry where most start-ups quickly failed. Of the fifteen or so fledgling airlines that were launched around Australia during the time of Qantas's birth, Fysh and McGinness's company was the only one to remain airborne – and this pattern repeated around the world. In the Netherlands, KLM Royal Dutch Airlines – the world's oldest continuing airline business – had been registered in October 1919, thirteen months before Qantas, but it had taken off with a grant from Queen Wilhelmina as well as major financial backing from Dutch banks and business leaders. SCADTA,[57] which also began operating in 1919, eventually morphed into Avianca, still flying today as the Colombian flag carrier.

On 10 February 1921, Qantas had its first board meeting at the Winton Club. This was the only time the company met there, as the decision was made to eventually use Winton as a secondary hub; the aircraft would be based in Longreach because of its railhead and the ongoing need for spare parts. A working committee of eleven was formed from the men present in the club's dining room. McMaster pointed out that the funds of the company were threadbare, and everyone present was urged to sell more shares and find customers for joy flights. Kennedy stepped down from his position as provisional director on the interim board; he was an £8000 guarantor for the company's first bank overdraft of £30,000 and stood to have lost a considerable fortune if the company had gone into a spin.[58]

McMaster borrowed the back room of Fred Riley's Winton stock and station agency to use as the first Qantas office. Because the company secretary, Alan Campbell, was in Brisbane, all the early typing for Qantas was done by Harriet Riley, free of charge; she was at the birth of an Australian icon, just as she had been at the Banjo Paterson performance. Her friend Kit Tighe was the first Qantas bookkeeper, but the airline's finances were in such a fragile state that she was also an unpaid volunteer.[59]

A lot of other people were willing to chip in, too. The Alba Wool Scouring Company donated a Winton woolstore to Qantas, and the young airline paid £85 to convert it to a hangar in order to protect the fragile machines from wind, rain, hail, souvenir hunters, and straying cattle and horses. The hangar was fitted with heavy canvas hanging doors lined with metal strips to keep them rigid, just like those on the hangars Fysh and McGinness had used in Palestine.[60] While most of the landing fields around Australia were paid for by the Defence Department or owned privately, the Winton Shire Council became the first local authority in Australia to support commercial aviation. It agreed to subsidise half the cost of establishing a proper landing field in town, to the sum of £20.[61]

Chapter 12

Turkey shooting by aeroplane promises to become a popular sport in Queensland. Lieutenant Hudson Fysh, of the Queensland and Northern Territory Aerial Services Ltd, who took part in a 'hunt' recently, describes the sport as most fascinating.

REPORT ON A SIDELIGHT OFFERED BY QANTAS AIRCRAFT IN THE EARLY DAYS[1]

QANTAS NEEDED TO MAKE money immediately after the two aeroplanes arrived back in Longreach, and Fysh was given the task of taking as many Queenslanders as he could on their first flight to help pay off the massive debts the company was incurring. While McGinness ran the Qantas office in Winton, Fysh worked out of the Longreach branch where the Avro 504K and B.E.2 were to be based. He made his home at the Imperial Hotel and took flight bookings in a building in Eagle Street. McGinness and Baird were about to catch a train to Sydney, where they planned to collect the Avro triplane from Mascot.

The two pilots decided on a site at the bottom of the Longreach showground for their landing field as it was suitable for wet weather, and Baird erected another hangar there. Herb Avery, the mechanic from the local motor garage who had crashed with Jack Butler, assisted Baird in his work on the aircraft engines, and before long Avery agreed to fly with Fysh as his mechanic on barnstorming missions across the remote centre.

Rates for passengers were set at two shillings per mile,[2] decreasing the more miles flown, with joy flights starting at £1 one shilling,[3] though Fysh could get three times that in the more remote areas of Queensland that had never seen an aeroplane. No flying exhibitions were given, as the old B.E.2 was incapable of much in the way of aerobatics, though McGinness would occasionally loop the loop in the Avro if the passenger was willing to fork over £5 and risk upending his breakfast.

Under the new *Air Navigation Act*, registration letters had to be painted on the machines, and Fysh and McGinness had to apply for their pilot's licences, though they still flew without them. Fysh was finally licensed on 28 June 1921.

The company also needed more capital from the sale of shares in order to expand. 'Please do your best to absolutely nail any prospective passengers, as well as shareholders,' Fergus McMaster's friend Frank Cory, a Longreach stock and station agent, wrote in a letter to Alan Campbell in Brisbane, urging the lawyer to get any interested shareholder to sign the form of application 'then and there, so you have some hold on him'.[4]

Fysh, McGinness and McMaster were the cornerstones of the little company, and vital for its success. Their personalities were complementary. Fysh was the pragmatic, diligent, determined workhorse, McMaster, the canny pastoralist with political connections, and McGinness the extroverted salesman, full of enthusiasm and bluster, who could walk into any pub, anywhere, bang the drum and sign up prospective shareholders.[5]

Despite the talents of those three, Qantas still needed a government subsidy of some kind to assure its future, enabling it to provide cheap fares and an inexpensive mail run. McGinness had started to correspond with his local member in Victoria, Geelong-born Arthur Rodgers,[6] the Minister for Trade and Customs in the Hughes Government. McGinness formed an instant rapport with Rodgers, who like him had Irish heritage and a farming background. And McMaster had a powerful ally in Donald Cameron,[7] a veteran of the Boxer Rebellion, Boer War and World War I. Cameron had run his family's station near Longreach for many years and was the Federal Member for Brisbane.

As McGinness and Baird headed to Sydney to fly the Avro triplane back to Queensland, McMaster wired McGinness telling him to head to Melbourne and set up meetings with as many powerful politicians and bureaucrats as he could, including Rodgers, Cameron and Richard 'Dicky' Williams. The former commander of No. 1 Squadron in Palestine, Williams was now Director of Air Services at Army Headquarters in Melbourne and about to form the Royal Australian Air Force. 'Let them know we have actually started operations,' McMaster told McGinness, 'keep the position fully before Federal Ministers.'[8]

Fysh had begun these 'operations' in the most rudimentary fashion, as joy rides seemed the only way to keep Qantas solvent. Hopes for charter work as an aerial taxi driver didn't materialise because the western Queensland tracks remained dry, and the mud that had bogged and dogged so many road trips was absent.

Another problem was that much of the appetite for first-time aerial thrills had already been sated in that part of Australia. Ron Adair,[9] from Maryborough, Queensland, had served with McGinness and Fysh in Palestine, and he was barnstorming across Queensland too. Frank Roberts,[10] one of the pilots closely associated with Nigel Love, was also offering joy rides out west. During the war Roberts had flown in Egypt and France, then he had worked alongside Bert Hinkler at Avro bases in Southampton and Manchester before returning to Australia. In the second half of 1920, he had made a long and successful aerial tour into western Queensland. While at Longreach he flew Ranald Munro and A. J. Gilchrist from Dalgety's on an eighty-kilometre flight to Darr River Downs, where they planned to inspect seven thousand ewes for purchase,[11] on the condition that he would only be paid if the sale went through – it did.

Fysh decided to hit smaller towns in the west that hadn't had the Roberts and Adair treatment. Although his natural reticence did not augur well for a salesman of aerial sensations, Fysh was enthusiastic and diligent as he and Avery visited Blackall, Muttaburra, Isisford and Wellshot.

The B.E.2 had been out of its depth in aerial combat during the war, but it was stable, rugged and reliable, and an ideal aeroplane for bush joy-riding, able to operate on short, rough runways. Avery or a brave volunteer was usually called upon to swing the four-bladed propeller, and sometimes Fysh would have to start the machine himself, setting the throttle, lashing the joystick back to keep the tail down, jumping out of the machine to swing the propeller, and then, after the engine started, racing round to climb back into the cockpit. Sometimes on cold mornings it could take half an hour for the engine to turn over, while he made repeated journeys between the cockpit and propeller.

Fysh organised the printing of ticket books and advertising leaflets to be distributed in each of the towns he and Avery visited. He also arranged for a Qantas representative in every town to work without pay, simply for the excitement of being involved with real airmen. The volunteer agent would notify the public that an aeroplane was coming and would organise flat ground for a landing field as well as oil and petrol, hopefully without too much water in it lest the engine cough and die in midair.

Because Fysh had to take so much equipment in the little aircraft, the only room for their two small suitcases was between Avery's legs. Fysh carried with him a petrol funnel and a piece of chamois to filter any water out of the petrol, and if the chamois was damaged Fysh would use his felt hat. He also carried spare cap-and-goggles for the passengers and spare parts for the aircraft, and he always had a good supply of pegs to tie down the B.E.2 in case the unpredictable winds carried it away when it was parked after a day's work.

In each town Fysh would find a nightwatchman, 'usually some old deadbeat,'[12] to stay from dusk till dawn under the aircraft's wing. The volunteer guard would be tasked with chasing away souvenir hunters as well as wandering cows and horses who found wingtips as tasty as apples. Fysh and Avery slept like cats on these remote assignments, keeping one eye open the whole time. Once at Cloncurry, Fysh was lucky to wake at 3 a.m. when a freak wind had whipped up the nightwatchman's fire, its flames licking at the cloth wings. And it seemed that just about everyone in the small towns had to scribble a name or some other graffiti on the wings and fuselage. In one town, when the nightwatchman was having a nap, a local hooligan expanded on the registration lettering of Fysh's machine, 'G-AUBF', with the explanation: 'Go away you bloody fool.'[13]

IN THE LAST WEEK OF FEBRUARY 1921, Fysh had carried seventy-eight paying passengers and told McMaster that he had cleared '£100 off the BE'. At Blackall he had taken up an Aboriginal man free of charge, a passenger who Fysh reckoned was among the most nervous of all his customers, so jittery that he went white as a ghost. Fysh guessed that he was the first Aboriginal man to fly in an aeroplane. After they landed he couldn't wipe the smile off his face, telling Fysh that aircraft were better for station work than horses. 'Plenty quick fellow,' he told Fysh, 'and see all about.'[14]

At Wellshot Station, the vast sheep property near Ilfracombe, Fysh gave joy rides using the road that led up to the homestead as a landing ground. The station manager, Mr Murray, paid Fysh to help him hunt wild turkey from the air. The manager sat in the front passenger seat with his shotgun while Fysh scanned the flat open plains for prey and then manoeuvred the machine into position for the best shot. Murray lived up to the name of the property, and stockmen and their dogs collected the kills. Avery took photographs to commemorate the feat and sent them to various newspapers with some paragraphs penned by Fysh.

> In a B.E. machine Lieutenant Fysh visited Wellshot station, some distance from Longreach, where there are large numbers of wild turkeys. The plain turkey [or bustard], by the way, is a popular food in the North. With the manager of the station, Mr H. [Murray], Lieutenant Fysh decided to use the machine in the sport, and early in the morning when the turkeys were out feeding, they set out, and soon put up four big birds. One was singled out and chased, and after a great deal of twisting and turning, the plane 'took the offensive,' and the passenger opened fire. At first, however, accurate aiming was difficult, owing to the necessity of allowing for speed, but after securing the first bird, four more were put up, and one of them secured. The large tracts of open downs country are perfectly safe for low flying.[15]

The report didn't mention that Wellshot's owner, Jock Inglis, arrived towards the end of the shooting spree none too pleased with the manager's work ethic, and gave him the bullet as well.[16]

Fysh was soon helping McMaster fire some heavy shots on Qantas's behalf. The meeting between McGinness and the Trade and Customs minister Arthur Rodgers in Melbourne had been so encouraging that McGinness wired Fysh to get McMaster down posthaste. Fysh sent McMaster a telegram saying the winds were favourable for flying, before carrying him in the B.E.2 from Longreach to Charleville to catch the Melbourne train.

Running an airline, even one as small as Qantas, was expensive. Fysh assured McMaster that Longreach was the best place for the company's headquarters and that a Qantas bank account should be opened there as soon as possible. He needed money, too. 'As you understand,' he told the Qantas chairman, 'I have all my money in the company and I should like to pay my hotel bill.'[17]

Fysh wasn't only putting his personal finances on the line for Qantas – his life was at risk, too. There were many dangers associated with providing joy rides in remote areas for passengers unfamiliar with aircraft. Seven months after John McIntosh and Ray Parer had arrived in Melbourne following their epic flight from London, McIntosh and the Avro he had bought from Nigel Love were at Pithara, 240 kilometres north-east of Perth. The 29-year-old McIntosh was trying to placate a couple of rowdy wheat contractors who'd been quenching a powerful thirst all day. Alfred Joy and Albert Loughlin each paid £3 15 shillings – half a week's pay – to go for a ride in the 504K with the famous aviator. It was a quarter to six, and the two locals were yahooing and wrestling over a bottle of beer when they climbed into the rear cockpit behind McIntosh. His mechanic delayed starting the Avro because of the state of the passengers, and McIntosh told Loughlin to throw away the cigarette smouldering between his lips as there was petrol all around. Finally, McIntosh started moving his machine down the runway. As he lifted the Avro into the air, Loughlin jumped up in the rear cockpit and started waving a red handkerchief to the onlookers on the ground.[18] At 150 metres the Avro went into a spin and crashed into scrub. McIntosh and

Joy died instantly; Loughlin survived. It was the first fatal air disaster in Western Australia. The scene of the crash was renamed McIntosh Park.

AT THE END OF FEBRUARY 1921, Arthur Rodgers had given McGinness and McMaster every indication that the Hughes Government would back their new aerial venture. But talk was cheap, and the value of Australian commodities had started to tank. The boom of postwar euphoria was waning. There had been a trade balance in Australia of £50 million for the 1920 financial year,[19] but Britain and the United States had curtailed spending on Australian exports. A recession loomed, and a subsidy for a fledgling outback airline wasn't high on the federal government's agenda. Meetings with Billy Hughes and his ministers went nowhere, even though McMaster was still confident that the company had a foot in the door at Parliament House.

McGinness travelled back to Sydney to take delivery of the triplane. The company was counting on the machine's ability to carry four passengers in its enclosed cabin, after heavy rains had hit the west and flooded the waterways,[20] making road transport slow and hazardous. In another newspaper report to promote Qantas flights across the outback, Fysh wrote: 'Only those who know Western Queensland can properly appreciate what the black soil is like when it is in the state of mud. After a good downpour of rain the roads, which till then would present a smooth beaten surface and form veritable speed tracks, became bottomless while the creeks and rivers, which are in most cases unbridged, become for days or weeks on end little better than roaring torrents.'[21]

Fares for the Avro 504K, which carried only two passengers, had been calculated at £10 a head for the Longreach–Winton trip. But Qantas figured that fares for the four-passenger triplane could be sold for just £6 6 shillings each.[22] The advance publicity for the new Qantas machine was gushing in its praise. Sydney's *Daily Telegraph* lauded the enterprise by the 'Queensland and Northern Territory Aerial Service, Ltd' in attempting to link the railheads of central Queensland with the interior, and expanding its equipment 'by the addition of the only commercial triplane of post-armistice design that is in Australia'.[23] The newspaper reported that:

> In charge of Mr. N. R. Love ... [the triplane] was successfully tested in a flight of nearly an hour over the city and suburbs on [February 28]. The cabin which is very comfortably fitted, contains four plush seats, set on springs, is entered by a door similar to that of a railway carriage, and has three windows on each side. The pilot is accommodated in a special cockpit behind the cabin, and has an excellent view for taking off and landing. The triplane, whose passengers included Mr. [Paul] McGinness, a director of the Queensland and Northern Territory Aerial Service, Ltd., responded generously and gracefully to the trials put upon it. A north-east wind prevailed, but it in no way adversely affected the machine's flight. Rising gracefully to 2000 feet, the machine was turned for the city, and flew over it at an altitude of 5000 feet for a considerable time. The passengers, on alighting, were enthusiastic concerning their feeling of security and of comfort during the flight. They experienced very little noise, and they were able to converse without inconvenience or the raising of the voice.[24]

McGinness thought that report was bunkum. He described the triplane as a 'flying monstrosity'[25] and declared that the 'comfortable' passenger cabin was more like a coffin. The exhaust blew oily fumes into the pilot's face.[26]

In his first test of the machine, McGinness aborted three take-off attempts before finally getting airborne over Mascot, but the triplane wobbled about like a drunk and stalled as he came in for a landing. As Baird watched on, horrified, the undercarriage sheared off, and McGinness and the triplane skidded across the Mascot grass together. The aircraft, in which Qantas had invested almost half its capital, was badly smashed. McGinness climbed out of the wreckage unhurt. McMaster was philosophical. 'I am pleased in a way,' he told Campbell, 'that we are up against it for a start, because it will make the business stand on its own feet.'[27]

McGinness now refused to take delivery of what was left of the triplane, telling Love that Qantas had bought the machine under false pretences. He demanded that his company be given a refund. In the end, before he and Baird headed back to Queensland in early March, he insisted that Love make more than £200 in repairs at the expense of his Australian Aircraft and Engineering Co.

Bert Hinkler's relationship with Love and his Avro machines was far cosier than the frosty relationship that had developed between Love and McGinness. With his solo flight from England having been thwarted by war in Syria, Hinkler brought his Avro Baby to Sydney by ship, arriving on 18 March 1921. After Hinkler's world record London–Turin flight, Love wanted to show off his Avro at Sydney's Royal Easter Show, the biggest exhibition of technology, produce, pigs and sideshow performers in Australia, due to start on Monday, 21 March. Hinkler was now planning to fly from Sydney to Bundaberg in one go to break the London–Turin record. While country roads were pretty abysmal in Australia, he told reporters, it dismayed him that his country had yet to embrace aviation in the way Europe had.[28]

Late on 10 April, Hinkler donned a suit and tie, and just before midnight he left the boarding house where he was staying. He had two apples and a piece of apple pie to eat during the flight he had planned. Just before 6 a.m., he straightened his tie and bounced the Baby along the Mascot grass, avoiding the ruts and lifting off over Botany Bay. He aimed for the Hawkesbury River, then headed slightly inland towards the Great Dividing Range and stuck to it as though massaging the spine of Australia. He passed over Armidale at about 9 a.m., followed by the village of Ben Lomond perched on the range 1400 metres above sea level, as he photographed the heavily wooded, uninhabited areas around it. He crossed Glen Innes at about 10 a.m., still snapping away merrily with his pocket camera to give everyone at Bundaberg a rare bird's-eye view of their country. Some cumulus clouds spoiled his sightseeing, but at 10.47 he crossed the state border into Queensland at Wallangarra and the sky finally cleared. He had a lovely view of Toowoomba and the Darling Downs and then, as he recalled, 'nothing but wild forest as far as the eye can reach'. Flying at 1200 metres, he kept slightly inland and passed over Blackbutt and Kilkivan.[29] At about two o'clock, as the autumn sun illuminated the vista through the spinning blade of his propeller, he caught sight of the cane fields and red-and-green quilt of the landscape around Childers. Almost fifty kilometres further ahead, the ribbon of water that was the Burnett River stretched from the azure sea through Bundaberg.

Even though he was a shy man like Fysh, Hinkler knew how to put on a show. People began to swarm into Gavin Street, where the pilot had grown up, and he brought the Baby down on a long stretch of grass in front of the Bundaberg Foundry where he and his father had worked as labourers. Waving at the growing crowd to stand clear, he wheeled the little machine around and brought it off the grass onto the dirt track that was Gavin Street. Amid the dust and the noise, he taxied the two hundred metres along to his old home while hundreds of men, women and children ran or cycled furiously beside him, hollering and whooping as they dodged the gravel being kicked up by the Baby's tail skid.

Hinkler had just made the longest ever solo flight: 1448 kilometres in eight hours and forty-five minutes. The cause of long-distance flying in Australia had been boosted again.

FYSH AND McGINNESS hoped that Hinkler's heroics would prompt the settlers of western Queensland to embrace long-distance flight too, but for the time being the two pilots restarted their barnstorming visits to small towns, offering joy flights. Fysh and Avery set off south in the B.E.2 to cover the Lower Diamantina and Coopers Creek areas but played it safe. 'I am cutting out every possible risk,' Fysh told McMaster, 'and am not taking up any children. As you can see, business in this area is not good. However, I got a few station people interested and one or two promised to take shares when we operate down this way.'[30]

McGinness and Baird flew the 504K north. They based themselves at Oban Station near Urandangi, on the Queensland side of the Northern Territory border. They arrived in time for the week-long celebration of the picnic races, and business for Qantas joy flights was boosted after McGinness climbed out of the cockpit and onto a saddle to ride a winner. After also winning a recommendation to every station owner in the district, McGinness and Baird flew west to Richmond late in March, showing off the Avro and telling everyone who came to marvel at it that Qantas was a good bet as an investment.

Everywhere the Qantas airmen went they were greeted as heroes and celebrities. Crowds rushed the makeshift airfields to greet them with cheers and hurrahs, and there was always a degree of havoc as the pilots waved the starstruck bushies away from the spinning propeller. Some spectators tried to tear pieces of fabric from the aeroplanes as souvenirs; at other times outback women threw themselves at these visitors who had arrived from the heavens, and showered them with kisses.

Though Fysh remained bashful, he forced himself to speak publicly at every opportunity in whatever small town he found himself. He carried a set of glass slides and a lantern for primitive slide shows wherever there was a white wall as a screen; he would extol the great developments in aviation, the convenience of aircraft to defeat distance in the bush, and the exceptional safety of the latest aerial machines – he rarely spoke of crashes or forced landings.

Fysh also kept newspapers around the country up to date with the latest improvements that the business was making in the lives and livelihoods of Australians living in Queensland's remote centre. On 1 April 1921, the *Sydney Stock and Station Journal* reported on a charter flight made by McGinness:

> From Longreach comes news of a sheep deal in which an aeroplane figured well. On account of the heavy rain, the impassability of the roads, and the flooded rivers and creeks, a motor car was out of the question, and so there was left only the aeroplane in which to get about. There were 25,000 wethers on Moscow, Winton, for sale, and the problem was how to get the buyer to them. The prospective buyer was Mr. Nalty, who was prepared to inspect for Mr. Naughton, of Victoria. The New Zealand Loan and Mercantile Agency Co., Ltd. Longreach (local agent, Mr. D. E. McCartney) engaged one of the aeroplanes owned by the

> Queensland and Northern Territory Aerial Services, Ltd., and although the country at Moscow was still too boggy for a car to travel, a successful landing was accomplished, and the sale satisfactorily completed. This deal shows that the aeroplane is going to render good service in the Further Out! The Longreach agents deserve a pat on the back for their up-to-dateness.[31]

While Fysh and McGinness were away promoting flights and the purchase of Qantas shares, Longreach finally became the official headquarters of the business after Fred Riley's office in Winton burnt down. Harriet Riley bravely dodged the flames, running into the burning building to save the papers for Qantas and her family business. She bundled as many documents as she could into a sack and then ran for her life, her hair and clothing being singed in the process.[32]

McGinness was involved in a desperate race of his own soon after, as the importance of Qantas as a life-saving rescue service became known around Australia. A six-month-old baby named Rose Joliffe[33] had become seriously ill on a property that her father Alec[34] was managing fifty kilometres from Longreach. Heavy rain at the time made it impossible to send out medicine or to bring Rose and her parents into town.[35] On 8 April 1921, Rose's condition took a life-threatening turn. Alec and his wife Ava decided the only thing to do was to send a telegram to Longreach to see if the new aeroplanes could save their beloved child.

Like a comic book superhero, McGinness read the wire, ran out into the rain and leapt into the Avro. Knowing the little girl's life was in his hands, he flew as quickly as he could across the black soil plains. Within half an hour he touched down on a flat piece of ground at the Joliffes' station.[36] Bundling Rose and Ava into the rear cockpit and telling Ava to hold on tight to her precious cargo, McGinness roared down the muddy landing strip, ignoring the gooey soil stuck to his wheels, climbed quickly into the ether and headed back to Longreach. Alec waved frantically from in front of his homestead, wondering if he would ever see his little girl again. Within another half an hour, McGinness deposited Rose and her mother at the Longreach private hospital, 'none the worse for their fast trip'.[37] John Flynn wrote about the rescue in his magazine *Inlander*, alerting everyone he could to the importance of aviation in outback Queensland and the Northern Territory – and telling all about how vital Qantas was to the lives of country folk.

As the unofficial publicity director for their little airline, Fysh was delighted not only with the seamless work McGinness had done transporting the sick baby, but also with the column inches that newspapers around Australia devoted to the skill, courage and compassion of the 'Queensland and Northern Territory Aerial Services, Limited'.[38] That was nothing, though, compared to the joy generated when Rose survived her emergency and eventually went home with her mum.

Many outback settlers believed there was something messianic about saviours arriving from the clouds. Once when Fysh was on a distant foray selling shares and raising capital for Qantas, he arrived at a remote property and circled down to land in front of the homestead. The station owner had watched with astonishment as Fysh approached from a great height, and the local man stood slack-jawed as the celestial vehicle taxied towards his front door in a cloud of dust. Fysh was still wearing his helmet and goggles when the old bushy, unacquainted with the man or his machine, approached this caller from the heavens and extended a hand of friendship. 'Good day, God,' the station owner drawled, 'my name's Smith.'[39]

Chapter 13

THE PERSON WHO HAS NOT BEEN IN THE AIR has not yet started to LIVE.

THE FIRST LINE ON FYSH'S ADVERTISING LEAFLET FOR A BARNSTORMING MISSION[1]

ON 12 APRIL 1921, Fysh and Herb Avery set off again from Longreach down the Thomson River for the Coopers Creek districts. The two men were intent on giving people in the south-western corner of Queensland their chance for a first flight, and the chance to see how this new technology could easily make vast distances evaporate. Fysh's advertising leaflet asked, 'Have you ever had a flight in a British war machine?' and it invited customers to fly in the famous B.E.2e, second in the Sydney Aerial Derby.

Their lightning air raid took in towns such as Quilpie, Thargomindah, Hungerford and Charleville. Fysh would announce their imminent arrival at each place a day or two earlier, and offer the volunteer Qantas agent there a complimentary flight if he would select an open level space near town free from stones or potholes. The field was to be at least four hundred metres square with no high trees around the edges. The agent was to mark it with a white sheet in the centre, and he was also to organise an oil smoker on the edge of the landing strip to give Fysh an idea of wind direction.

After flying into Cunnamulla for the races there, Fysh was hired to drop boxes of chocolates at several nearby station homesteads, a task he would later perform after a race meeting at Muttaburra.

Fysh and Avery arrived over Thargomindah to see 'literally hundreds of the local residents frantically working away with the town's brooms sweeping up pebbles from the perfect claypan' that had been chosen as the landing place.[2] At Farnham Plains, on the Paroo River near Eulo in Queensland's far south-west, the Qantas agent had misread the note about the smoke – he had a bonfire raging in the middle of a narrow airstrip surrounded by big trees, making it almost impossible for Fysh to land. Somehow, dodging the wild flames, the pilot managed to bring the B.E.2 down, but he did so amid a cacophony of his own curses, so blue and so loud that he startled even himself and was blushing when he finally climbed out of the rear cockpit before a startled population.

He and Avery made it to Windorah just in time for a three-day race meeting and, when not flying, spent much of their time trying to back a winner or sell shares in Qantas. At McPhie's Hotel, Windorah's only pub, cow dung was burnt continuously throughout the ramshackle building to ward off the mosquitoes coming up from Coopers Creek, but there was no room for the two airmen. So Fysh and Avery bedded down for the night under the pub's billiards table, as fights broke out all around them. And then the dogs, eager to show their masters what it was like to brawl with a bit of mongrel, joined in the fray.

BERT HINKLER TOOK OFF from the street outside his mother's house in Bundaberg on 23 April 1921 bound for an Aero Club dinner in his honour in Sydney four days later, before his planned return to England by ship. He stopped in Gympie to visit his grandparents and on Anzac Day landed in Windsor Park, just north of the Brisbane River, where he was met by his old mate Lieutenant Frank Roberts, whose Avro 504 was based there.

Hinkler left Brisbane at 6.15 a.m. on the twenty-seventh, dressed in a suit donated by a Bundaberg menswear store, for the Aero Club dinner that night. In atrocious weather he was forced to land on the immense length of sand called Stockton Beach, near Newcastle. In the gale-force winds, the frail aircraft with the red kangaroo painted on its tail lifted off the ground, swung on to its nose with a sound of tearing fabric and cracking wood, and then completed the somersault onto its broken back. Hinkler buried his head in his hands and cried.

That night – as the little aviator warmed himself at an isolated farmhouse and waited for someone to collect him and his smashed machine – 150 kilometres south amid the grand opulence of the Hotel Australia, the Aero Club president Colonel Oswald Watt made a stirring speech before the fifty or so wartime pilots gathered to honour the absent Hinkler. Watt said Hinkler's mishap – with no proper landing ground to save his machine – demonstrated the urgent necessity for a thorough organisation of civil aviation in Australia.

Major Norman Brearley, president of the West Australian Aero Club, then told the audience that Hinkler had done a marvellous service to aviation. Brearley had been awarded the Military Cross and Distinguished Service Order in the war. Like Fysh and McGinness, he had started giving joy flights upon his return to Australia, mostly around Perth, using two surplus Avro 504Ks. He had formed his own airline, Western Australian Airways Ltd.

Three weeks after Hinkler's Avro Baby was wrecked,[3] McGinness set off from Longreach on 16 May 1921. He was flying the Avro 504K to Elderslie station, about sixty kilometres west of Winton, where the owner, Charles Brabazon,[4] had about a hundred men working for him. Brabazon, one of the first graziers to take up McMaster's offer to buy Qantas shares, also gave the new airline its first commercial work in the Northern Territory. He had hired McGinness and Baird to fly him six hundred kilometres west to Austral Downs in the Territory so he could visit some unexplored land and assess the possibility of grazing cattle there. In a mission that might have taken days, even weeks, on horse or camel, McGinness flew Brabazon around the desert landscape in a matter of hours, looking for a permanent water supply. There was none to be found, so the cattle grazing idea had to be abandoned.

At Camooweal and Urandangi, though, McGinness gave many residents their first flight – forty of them at Urandangi – as people rode or drove from the stations all around, clamouring to touch the sky. He never let an opportunity go by there, either, and according to one of the station workers during his visit, he was going 'great guns' with a Cloncurry girl staying at Walgra.[5]

While McGinness was away, Qantas held its first annual general meeting in Longreach on 21 May 1921. Qantas moved its headquarters to a wood and galvanised-iron building on Duck Street. The office came rent free, thanks to the stock and station agent Frank Cory, and there was a rail where customers could hitch their horses. Cory became the company's part-time secretary, while Dr Hope Michod was appointed McMaster's deputy chairman.

At its AGM, Qantas announced that there had been applications for 6850 £1 shares for which there were cash receipts of £6075. The company's plant was shown as one Avro Dyak biplane three-seater, one B.E.2e biplane two-seater, and one Avro triplane five-seater (broken). The cost of the plant was set down as £5671 18 shillings and 11 pence, including spares. Qantas was declared clear of debt, apart from £100 on the B.E.2 as well as the cost of repairs to the triplane 'owing to an accident in Sydney during the trial flights'. McMaster told the board that the directors had decided to allow the damaged aircraft to remain in Sydney until August and then have it returned to Longreach to operate in the wet months, when road traffic was problematic. McGinness remained the biggest stakeholder in Qantas, having increased his holding to 1450 shares. Templeton had one thousand, Fysh six hundred, and McMaster five hundred.

McMaster reported that the pilots had 'absolutely put safety first'. McGinness, flying the Avro, had covered 7400 miles and been in the air 111 hours, taking up 285 passengers; the £1500 purchase had made a gross return of £934. Fysh – in the B.E.2 bought for a bargain-basement price – had 'done exceptionally good work', McMaster said, with his machine having already made 'a gross return of double the purchase price'. It had flown 6370 miles, been in the air ninety-eight hours, and carried 296 passengers, bringing in £837 for the company. For all the time in the air and the vast distances covered, there had been no major repair bills for the two operational aircraft, an extraordinary result given the fickle nature of the machines and their engines.

The Qantas office in Duck Street, Longreach came complete with a rail so that prospective passengers could hitch their horses. *National Library of Australia nla.obj-151430031*

Everywhere in Queensland the aircraft were met with such enthusiasm, McMaster said, that opening bases in Charleville and Cloncurry was warranted. That move would, he assured the board, pave the way for 'what must eventually be one of the greatest aerial routes in Australia'. 'Considering the serious financial times that the country has gone through,' he concluded, '… the position of the company must be considered satisfactory.'[6] An almighty crash, though, was just over the horizon.

FYSH AND McGINNESS had been the prime movers behind a proposition for an 'Aerial and Motor Service' to operate between the railheads of Longreach and Winton – a service that would require government assistance. They put the idea to Captain Edgar Johnston,[7] a Gallipoli veteran who had flown Bristol Fighters during the war, and who was now superintendent of aerodromes in the new Civil Aviation Branch for the Commonwealth Department of Defence. Johnston visited Longreach and told Qantas he was 'very interested in the idea'; he advised the company that an application should be lodged with the federal government before 30 June 1921. Around that time, the Department of Defence invited tenders, to close on 30 July, for an air route in Western Australia covering the

1900 kilometres from Geraldton to Derby. Fysh and McGinness resolved to do everything they could to win airmail contracts in Queensland and the Northern Territory.

In the meantime the two pilots continued their barnstorming, trying to find anyone, anywhere, who wanted a ride. The economic outlook was growing grimmer everywhere, and Fysh had to charge £8 an hour for longer flights. After one four-day mission he accrued £49 10 shillings for the airline and considered it 'a very good trip and a profitable one', but told McMaster that he couldn't afford to fly passengers at the usual rates.[8] Although the cost of operating his machine was just 7 pence halfpenny per mile,[9] money was hard to find everywhere. In England, with the cost of the Great War weighing down the government coffers, Winston Churchill was selling off the nation's airships, including 200-metre-plus zeppelins captured from the Germans.

Fysh and McGinness continued to attract national publicity. Newspapers throughout Australia reported that Lieutenant-Colonel Horace Brinsmead,[10] controller of the Civil Aviation Branch, had 'received good reports regarding the prospects of commercial aviation in Queensland'.[11] Brinsmead, a slight, quiet and gentle man, had started the Great War as a private and quickly surged through the ranks to become a senior officer. He could see men of real resolve bringing aviation to Australia's remotest areas, and he told the national press: 'Messrs McGinness and Fysh have made extensive tours of the districts … When aerial centres are opened at Charleville in the south and Cloncurry in the north, and permanent landing grounds marked out at all the intermediate towns and stations, it will be a long step towards opening up what must eventually be one of the greatest routes in Australia, connecting Southern Australia with the Far North and the Northern Territory.'[12]

A gratifying feature of the support given by people in the outback was the gift of landing grounds from property owners, he said. On the Longreach–Winton route, these were to be spaced just sixteen kilometres apart, marked with crosses made from empty petrol tins that were cut into pieces, wired together and painted white. A number of landing grounds also had wind sleeves added,[13] and the strip at McKinlay had to have gravel laid on its black soil.

On 14 July 1921 Arthur Rodgers wrote to Qantas to say that there was no immediate likelihood of his government backing a Charleville–Cloncurry aerial service, though there were prospects of a small grant for a trial service. There had been a sudden expansion of the RAAF, and Edgar Johnston was busy planning aerodromes that would eventually include Mascot in Sydney, Essendon in Melbourne, Archerfield in Brisbane and Parafield in Adelaide. Aerodromes stretching through Charleville and Cloncurry would tally with the RAAF plans as a means of defending the nation's interior.

A week later John Flynn, speaking in Adelaide, continued to talk up the 'unique scheme for medical assistance, to residents in the interior of Australia'.[14] The object of such an organisation, he said, was to institute a band of 'what is called "flying doctors" to provide adequate medical attendance to Australian pioneers'.[15]

WHILE FYSH AND THE OTHERS continued to lobby for government assistance, McGinness and Baird headed for the north of Queensland in July, planning to take passengers on scenic flights over the cane fields of the coast and above the islands of the Coral Sea. McGinness wrote to town councillors in Ingham asking if they could arrange a landing place for the 504K, as he was coming for the annual race days. The local jockey club organised a landing site near its racetrack, and McGinness advertised taking passengers on his machine for fifteen minutes at £2 10 shillings a turn.[16] He charmed everyone he met with his extroverted personality and cool daring, taking up 138 passengers. But on 31 July, the visit ended disastrously when two passengers were sent to hospital after the Avro fell out of the sky. A propellor blade was smashed, and the fuselage and framework were badly broken up.[17] The press reported:

> An aeroplane smash occurred last Sunday, something quite sensational for Ingham. The Sunbeam Avro machine under control of Lieut. McGinness had the misfortune to fall with

> two passengers, Miss D. Neilsen and Mr. George Hubbard. Neither passengers were seriously hurt, but Miss Neilsen is still in the General Hospital, suffering from shock and a nasty cut on her face. Mr. Hubbard escaped, with a cut on his chin, which necessitated the insertion of one stitch. Lieut. McGinness escaped without a scratch. It appears that as the machine was soaring some 1000 feet over the town some engine trouble set in, and Lieut. McGinness had to make a forced landing … and the machine crashed down into Mr. F. Fraser's cane paddock. The wrecked machine has been dismantled and sent to Sydney for repairs. Much sympathy is felt for Lieut. McGinness who had become very popular during his short stay in Ingham, and it is stated that it was only due to his coolness and courage that an accident of a more serious nature did not happen.[18]

With the triplane still in pieces in Sydney, Qantas was left with only Fysh's B.E.2 and a maximum passenger carrying capacity of one. Fysh oversaw the purchase of a spare RAF engine for the lone surviving machine, just in case.

Joy rides had been making Qantas pocket money, but everyone involved in the start-up knew their future 'hung by a hair'[19] without a government subsidy for a regular route. On 2 August 1921, the federal government accepted Norman Brearley's tender for the Geraldton–Derby route, a service that would make the mail delivery so much quicker in that remote stretch of Western Australia. Brearley was about to hire five pilots, including Charles Kingsford Smith, to partner him in flying a fleet of Bristol Tourers.[20] Fysh and his colleagues implored the government to realise the importance of a similar service through Queensland's outback.

Qantas now had most of its capital tied up in broken aircraft. At a board meeting at Longreach on 13 August 1921, plans were made to mount a vigorous campaign to secure more shareholders; five days later, a new prospectus was drawn up to raise £15,000. Lieutenant-Colonel Brinsmead stipulated that all moneys subscribed would be refunded if Qantas failed to secure a government contract. McMaster again headed for Melbourne to ask for the same support that Brearley's airline had been given with a £25,000 grant to cover expenses.[21]

On 20 August, Fysh took off in the B.E.2 for a six-week tour, heading south for Charleville and Cunnamulla before going north along the proposed mail route to Cloncurry. He was now targeting potential investors, rather than joy riders. McGinness was to cover the northern towns in a car until the Avro 504K was repaired.

The barman at the Longreach Club chipped in £10 for shares to help the cause. W. J. Bell, a railway fettler, wrote to McMaster to say: '[I have] read your advocacy … and I beg to state that I, a common worker, am willing to take up 10, 20 or 30 pounds' worth … so you can fire away.' Touched, but aware of the financial risk for a battler, McMaster wrote back warning Bell not to risk more than he could afford. Bell cut his application by half, telling Qantas that their letter bore 'the imprint of honesty'.[22]

McMaster addressed meetings in Rockhampton on 16 September, then he took a steamer to Townsville and joined McGinness there for more public rallies. They drove inland to Charters Towers, meeting and greeting, and winning support from the mayor. McGinness then went by car to the north-east coast, while McMaster took the train to Hughenden, where the mayor chaired a meeting of town and country people. Then it was on to Cloncurry where 'the project had a splendid reception'.[23]

At Tambo, Fysh heard that Tom Stirton,[24] the owner of properties in Queensland and New South Wales, had just left by car for one of them nearby, Minnie Downs. Stirton had a reputation as 'one of the most lovable and generous citizens of Australia'[25] – and Fysh knew he was loaded. The pilot jumped into a car and chased the grazier down the narrow bush track, rattling along at speed for an hour until he saw a dust cloud ahead: he had his man. Four years in McGinness's company had helped draw Fysh out of his reticent shell. As Fysh tailgated the grazier's car, he signalled for Stirton to pull over and he then gave the 61-year-old grey-haired grazier his well-rehearsed spiel. Stirton bought two hundred Qantas shares beside that bush road.[26]

By October 1921, more than a dozen shire and municipal councils plus the Charleville, Charters Towers, Townsville, and

Rockhampton Chambers of Commerce and the New Settlers' League at Cloncurry had sent telegrams to Prime Minister Billy Hughes backing Qantas for the outback mail route. McGinness travelled by train to Melbourne and, boosted by his support from Arthur Rodgers, wired Fysh from Geelong Railway Station on 29 October: 'Absolutely vital our subsidy McMaster come Melbourne immediately Air Board Meeting Wednesday Fysh fly him Mungindi tomorrow can do it that way stop If we put up strong case will get subsidy altered from one of other routes under consideration DO NOT FAIL.'[27]

McGinness was still the boss, and Fysh followed orders immediately. He flew McMaster to the Mungindi rail line on the border of Queensland and New South Wales. McMaster then met with Brinsmead in Melbourne to argue the Qantas case, putting together an impressive deputation of all Queensland members of the House of Representatives and Senate. They were led by McMaster's friend James 'A. J.' Hunter,[28] a small, slight Glaswegian accountant with pince-nez spectacles. He had just won the federal seat of Maranoa for the Country Party from his base in Dalby.

With Canberra still a sheep station, the deputation marched on Prime Minister Hughes in Melbourne on 10 November. They surrounded the deaf and cantankerous 'Little Digger' in his office in the basement of what is now, once again, the Victorian Parliament House on Spring Street. Tenders for a government-subsidised air service between Sydney and Brisbane and between Sydney and Adelaide were to close the following day.

Hughes had a very narrow majority in Parliament, and the Country Party had been voting almost solidly with the Labor opposition, so when Hunter introduced the deputation, Hughes was, in McMaster's words, 'bristling like an old war horse'.[29] With the salty language he had learnt as a bush cook after arriving from England almost forty years earlier, Hughes abused Hunter and the other Country Party members for coming to him cap in hand after being a thorn in his side so often. McMaster and McGinness weren't sure where to look.

Suddenly Hughes stopped his outburst and picked up a black box-like device he used to assist his hearing. He placed it against a large but inoperative ear, then looked at McMaster and, with the impatience of a man who might have been late for a train, told the otherwise imperious grazier, 'Well, well, out with it. What do you want? Everyone who comes here has an axe to grind. What is yours? Out with it!'[30]

'I am reputed,' wrote McMaster, 'to have a very loud, resounding voice. Possibly on that occasion it was not absolutely under control.' He bellowed about the dire need for an aerial service in western Queensland, how it was the way of the future and could dramatically improve the lives of everyone in the outback as well as being an important cog in the wheels of Australia's defence.

After a short time of McMaster's voice booming around the small basement office, Hughes interrupted him. 'I might be deaf, McMaster,' he growled, 'but I'm not as deaf as you think.'[31] When McMaster finally lowered his voice and finished his plea, some of the Queensland parliamentary delegates offered supporting speeches. Hughes's eyes glazed over as he looked at his watch. Then the crotchety prime minister turned back to McMaster. 'When are you going back to Queensland?'

'As soon as we secure your reply?' McMaster said.

'You have it now. The Government has no money. My reply is no!'[32]

As McMaster, McGinness and their backers filed out of Billy Hughes's office, the plucky little prime minster was left all alone, still with the black box on the table, but with the demeanour of 'a conquering bull'.[33]

BACK IN LONGREACH Fysh was disheartened. He couldn't see Qantas lasting another year without government funding, and the business was faced with refunding all the new shareholders if the mail run wasn't approved. After one meeting of directors, the men adjourned to the Longreach Club next door. On leaving the club Dr Hope Michod, the chairman in McMaster's absence, said good night, and as he closed a metal gate with a clang, he said consolingly, 'Well, Fysh, hard luck – it looks as if we'll have to close down.'[34] The sound of that gate closing made Fysh think that all of their hard work had been a waste of time.

Still in Melbourne, McGinness had put an alternative proposal to the prime minister, asking Hughes to give Qantas some of Britain's war surplus DH.9 aircraft. But that idea was grounded, too. 'Two shots had been fired,' McMaster wrote, 'and both had missed the target. The position was anything but encouraging. However, this did not disturb McGinness, for he was still out for a fight, and Fysh and the Longreach Committee urged continuing action.'[35] Hughes was encouraging his government to take up Winston Churchill's offer of their surplus airships to begin England-to-Australia flights, as Australia could 'in one bound be loosened from the fetters imposed by our remoteness and brought in close touch with the western world'.[36]

When McGinness heard that 'a certain member of the Government'[37] would be travelling on the Melbourne–Sydney train, he suggested he book a ticket too – and do some lobbying. McMaster said McGinness was always 'adventurous, quick and daring'; that he had a restless personality and 'something out of the ordinary as an occupation'. On the train to Sydney, McGinness quickly won the politician's ear as the vista of Australia passed by the windows of the carriage. During the early hours of a Saturday morning, McGinness called McMaster from a railway station somewhere along the train line to tell his chairman that if it were possible to get the Country Party to ease up its opposition to the Estimates in Parliament, sufficient money would probably be made available for a service between Charleville and Cloncurry.[38]

On 22 November 1921, James Hunter told the House that if certain unexpended funds could be used for subsidising commercial air services throughout Australia, 'it would probably induce Honourable Members to pass these Estimates as they stand'. 'Since February last,' Hunter told the House, 'a Company [Qantas] has been operating in the interior of Queensland with headquarters at Longreach ... It has been found impossible to carry on if dependent on passengers only.'[39] Within a week McMaster was advised that tenders for a Charleville–Cloncurry service would be called at 'an early date'. The subsidy would be £12,000.

EVEN WITH GOVERNMENT backing, airlines faced great difficulties and dangers flying in remote areas of the country. Norman Brearley's new air route in Western Australia began disastrously after he led three Bristol aircraft into Geraldton on 4 December following a three-hour flight from Perth. The next morning the aircraft took off on the first leg of the mail run to Carnarvon – but forty-five minutes into the journey, one of the machines developed engine trouble and made a forced landing. The young lieutenant Bob Fawcett was circling the stricken machine in his plane when suddenly it fell to the ground from about fifteen metres and was smashed to pieces. Fawcett and his mechanic Edward Broad died with broken necks. They were buried in bush graves near Murchison House Station homestead, about a hundred metres from the crash site.[40] Brearley blamed a lack of suitable emergency landing grounds for the disaster.[41] Just three weeks later, Brearley, Kingsford Smith and their mechanic were forced down on a beach near Broome because of an oil leak, and their Bristol Tourer had to be dismantled and returned to Perth on a steamer.[42] But Fysh and the Qantas board, with government money in the wind, pressed on.

Fysh was still heading the 'propaganda committee' for Qantas, churning out newspaper articles that highlighted the safety of flying – despite the frequent crashes of the time – and the absolute necessity of an aerial service to link the western railheads of Queensland and the Northern Territory.[43] He told readers that Qantas pilots – all two of them – had flown over the Charleville–Cloncurry aerial route many times, and were 'thoroughly conversant with the country and conditions … The fact that the whole 580 miles between the two terminal air ports is situated over stretches of open country where a landing can be safely affected in case of engine trouble, lends an aspect of safety to the route, which is probably not enjoyed by any other aerial route of the same length in the world.'[44] The vast areas of country that stretched westward from the outback Queensland railheads, Fysh wrote, were dotted with different station homesteads, 'some of them hundreds of miles from the nearest railway line'.

> All these way back homes and townships are without adequate communications, especially in the wet season, when isolation is often enforced for periods of weeks, the country being entirely without made roads, and few bridges … aviation must not only prove itself a good commercial asset, but must supply a long felt want to the pioneer populations, and must prove a great and strong factor in the opening up in further peopling of our empty interior. The advent of the Rev. John Flynn's scheme of 'flying doctors' should not be far off, and forms another phase of the utility of aircraft, which will help to make our back country a safer and more pleasant place in which to live.[45]

Fysh also oversaw the printing of an eight-page booklet, *Flying's the Thing*, which highlighted the link between the government subsidy for commercial aviation and national defence. He planned on using the booklets to canvass for more investment in both Brisbane and Sydney throughout the early months of 1922. 'Success in any undertaking,' Fysh's leaflet explained, 'comes not to the man who idly waits for his great opportunity, but to him who seizes whatever opportunity comes and makes it great.'[46]

In their campaign to draw new money, Qantas hired a curly-haired New Zealander named John Clarkson, a man Fysh called a 'professional share-getter',[47] to partner McGinness on flying visits across northern Queensland at the end of 1921. In the weeks before Christmas, armed with Fysh's booklet, Clarkson sold two thousand shares at £1 each, promising the money would be refunded if Qantas did not win the tender.

On Sunday, 18 December during a four-day trip to Barcaldine in the Avro, McGinness and Clarkson flew out to Auteuil, a station near Aramac. It was owned by Carlyon Foy,[48] from the famous family of Sydney merchants. The next day, while sweet-talking as many graziers in the district as they could to invest in Qantas, McGinness and Clarkson picked up Foy and took him to Barcaldine, where he was to catch the train home to Sydney. Foy had misplaced his satchel containing valuable documents and 'amidst much agitation and annoyance'[49] had to leave town without the bag. Shortly after his train had departed, the bag was found. McGinness and Clarkson leapt back into the Avro to chase down the locomotive and deliver the documents.

The mail train had three-quarters of an hour's head start, but McGinness caught it halfway between Barcaldine and Jericho. Clarkson attached a long rope of calico to the bag and swung it from the Avro as McGinness had the machine 'ducking and diving and encircling the mail train much to the excitement … of the passengers'.[50] Eventually McGinness brought the Avro down to within ten metres of the train, and the bag was dropped a few yards from the engine, which duly stopped. The bag was retrieved, and the aviators were loudly cheered by all on the train.[51] Foy sent a telegram of thanks to McGinness and bought 750 Qantas shares.[52]

THE BRISBANE WAR HERO Jimmy Larkin[53] had received the Distinguished Flying Cross after bringing down eleven enemy aircraft over France, and he quickly made it his mission back in Australia to shoot down Fysh's baby. Though Qantas had put capital and effort into pioneering aviation routes in western Queensland, and had mustered widespread financial and government support, the tender for the Charleville–Cloncurry mail service was very much an open one, and Larkin was a fierce competitor. He had been the signals clerk for Generals Monash and Chauvel in Egypt and on Gallipoli before becoming a fighter ace on the Western Front. After the war, Larkin, with his younger brother Reg[54] and other members of his 87 Squadron, formed a Sopwith agency, and then when Sopwith went into liquidation, the Larkin Aircraft Supply Company, better known as Lasco. It was based at Melbourne's Coode Island.

The Larkins had won the government airmail contract for the Sydney–Adelaide run, but they had difficulties raising both the money and the aircraft to honour the tender. So they formed a partnership with Frank Roberts, the successful bidder for the Sydney–Brisbane service. The resulting Australian Aerial Services Ltd took until 2 June 1924 to commence the Adelaide–Sydney run, but their Sydney–Brisbane operation failed to take off.

The Larkins also held the Australian agency for the Handasyde H.2 monoplane, the big experimental British aircraft. The cabin had a forward space for luggage and a toilet at the rear, which Fysh thought would save much embarrassment – especially on cold winter mornings. Many Qantas passengers had to sprint for backyard dunnies or bushes upon landing, or, having forgotten to bring an empty jam tin, found other ways to relieve themselves midair.

The Larkins had bought three H.2s to serve their airmail routes and were hoping to sell more to Qantas. Jimmy Larkin had made an enemy of the Civil Aviation Branch with his aggressive ways and regular complaints, and was notorious for a tactless, impatient temperament.[55] Fysh was much more obliging, and Larkin, who intimidated many of his business opponents, hoped to pressure the quiet, modest man from Qantas into a sale.

Larkin had a hard, straight mouth, deep eyes set close together, and a way of glaring hard at the person he was addressing.[56] Looking at Fysh as though he was trying to bore a hole through his head, Larkin was insistent that Qantas order one of his machines. He even offered to withdraw his competition for the Charleville–Cloncurry tender if the company did as he wanted.

Larkin figured it was an offer Fysh could not refuse. The Civil Aviation Branch was also insistent that at least some of the various experimental machines being planned in Britain would be used on the Queensland outback mail run, and that a British pilot experienced in handling big aircraft would also be hired as part of the deal. But Fysh and McGinness had been burnt by the purchase of the experimental Avro triplane, which was still in Sydney, and they had promised themselves they were done with untried aircraft. For the outback, Fysh favoured smaller, lighter machines that were durable, easier to maintain and less complicated to fly.

Qantas had set down a brief financial forecast that outlined proposals to acquire three Bristol Tourers or – Fysh's preferred machine – the Airco DH.9C for the Charleville–Cloncurry run, at a cost of £9000. Spares, workshops and hangars were costed at £5600, and the aircraft and equipment in hand was valued at £5719, to 'bring total capital to £20,319'. Running costs were estimated at £15,849 for one year, including £1500 for three pilots. Total revenue was estimated at £18,472.[57]

Fysh told Alan Campbell on 19 December that 'the capital is coming in well', and on the same day wrote to McMaster, imploring him to support the use of light aircraft for the tender.

> I find it hard to believe that the Civil Aviation people would want us to put big machines on a service such as this … It is going to be a very bad business risk to start this service with anything approaching a big machine, and we may not get out of it without a very big loss of capital. DH9s are still running on the mail routes in America and England, so they should be quite good enough for us to start with. We know that we can run the service with DH9s, but with these new experimental machines, we know that it will be a gamble. I hope you will get in touch with Colonel Brinsmead and find out what he thinks about the DH9. Colonel Williams should carry lots of weight, and he absolutely recommends the DH9.[58]

Brinsmead actually favoured the eight-seat Vickers Vulcan biplane, nicknamed 'the Flying Pig'. It was still in the planning stages but promised a twelve-cylinder, water-cooled Napier Lion piston engine producing 450 horsepower (340 kilowatts), a huge fifteen metres wingspan, and a hefty price tag of £3700.

Four days into 1922, McMaster cabled the Westland Aircraft of Yeovil, Somerset, asking for prices on three aeroplanes suitable for the thousand-kilometre mail route and able 'to carry three passengers and luggage, one hundredweight [forty-five kilograms] of mails and pilot'. McMaster told the company that the service would soon be extended northward from Cloncurry to the Katherine, and southward from Charleville to Mungindi in New South Wales, 'a total distance of 1600 miles [2500 kilometres] and the longest direct air service in the world'.[59]

Two days later the airline received the form required to tender for the Charleville–Cloncurry run. The following day, 7 January 1922, a meeting of the board decided to request that the Bank of New South Wales increase the company's overdraft limit to £3000.

It was agreed that Fysh, the quiet, thoughtful diplomat, rather than the extroverted McGinness should go to Melbourne to 'make enquiries as to suitable machines and report by telegraph, complete the tender in all respects and lodge same on behalf of the company with the Defence Department'.[60]

Fysh asked Brinsmead for permission to tender with two aircraft instead of the three specified, as two could more than cope with Qantas's weekly services and would be much more cost effective. The request was refused, and it seemed that Qantas would have to tender with three eight-seater aircraft for a service on which they estimated they would carry fewer than ten passengers a week. Brinsmead continued to talk up the Vickers Vulcan.

On 23 January, Fysh had lunch with Jimmy Larkin and Frank Roberts. Larkin continued with the hard sell for the Handasyde, and three days later, in Larkin's Melbourne office, Fysh ran his eye over the Handasyde plans and specifications. He wired McMaster on 29 January 1922 to say that one of their competitors was proposing his company would fly four of the big Vulcans, and that Larkin was planning to undercut everyone by offering to extend the proposed service to Cunnamulla without subsidy. At the end of January, Fysh submitted his tender specifying that Qantas would use two Vulcans, even though the aeroplanes were much bigger than the company needed. He added an Airco DH.4 to the list at the last minute and said Qantas would fly the route at four shillings per mile. Two days later, Fysh walked into the General Post Office in Melbourne and wrote out a telegram for McMaster, dated 2 February 1922.

> Tender accepted this morning against heavy opposition. Have guaranteed experienced pilots England and DH4 for auxiliary. This absolutely necessary. Congratulations. Fysh.[61]

He then sent another telegram to praise the work of McGinness and McMaster in getting Qantas over the line for the government contract, saying their fine work and the 'splendid record' of the infant company had made all the difference.[62]

Chapter 14

I left Longreach at 5.30 this morning, and am here at 11.20, less than six hours. When I first made the trip [53] years ago it took me between two and three months.

Alexander Kennedy, 84-year-old pastoralist, after becoming Passenger No. 1 on the first scheduled Qantas service[1]

HIS ADRENALIN PUMPING at a smart clip, Fysh got cracking to make sure Qantas got off to a flying start with their new assignment. On the afternoon that the tender was approved, he was at Harry Shaw's[2] aerodrome in Fishermans Bend, Port Melbourne, inspecting a DH.4 aircraft owned by Ray Parer, who was in West Australia with a cousin. The two Parers were four months into a flight around Australia in an old B.E.2b to raise funds for purchasing a machine capable of being the first to fly across the Pacific. Shaw was Parer's agent, and he was also the Australian distributor for Bristol and Farman aircraft. He told Fysh that Parer was keen to make a sale. Five days later, Parer became even keener when he continued with his long run of mishaps, hitting a telegraph pole after failing to gain enough height on a take-off from the Boulder Racecourse. That aircraft was wrecked, and the Parers were hospitalised for weeks.[3]

Fysh told McMaster that the big DH.4 – a much larger machine than Qantas had been using, but still smaller than the Vulcan – was 'fitted up well with a cabin, and is in perfect order'. He urged Qantas to buy it at once. 'The Rolls-Royce engine alone is worth £1,200,' Fysh wrote, 'and I think we can purchase the machine complete for £1,500.'[4] The rest of the board needed time to mull over the purchase, but there had been a notable shift at Qantas. While Fysh was now entrusted with winning the airmail tender and buying new aircraft, McGinness was still in North Queensland flogging shares with John Clarkson, who had been given his notice and was working out his temporary contract.[5]

McGinness had clashed with Fysh on their great drive across the Gulf, and he was starting to rile McMaster too. The older man did not approve of drinking, particularly for pilots, and Fysh, along with the deputy chairman Dr Hope Michod, backed the chairman to the hilt. A board memo was produced requiring Qantas pilots to sign an abstinence pledge. This did not sit easily with McGinness, but when he sought to have it revoked, proposing instead that pilots be permitted to drink lager beer, no one supported him.[6]

Only a few years earlier in Palestine, Fysh and McGinness had depended on each other to stay alive in their Bristol Fighters, but their close bonds had frayed. Fysh wrote that McGinness was increasingly 'on the outer' with the other Qantas directors, and that while there was no denying his skills as a pilot, he showed no administrative ability and was simply not cut out for 'organised business life'. Fysh and McMaster had 'a tremendous regard' for McGinness's adventurous spirit and hailed him as 'the bravest of the brave'.[7] Yet when it came to running a business, the quiet, dour Fysh – meticulous with his records, temperate in his habits, moderate in his ways and counting every penny – was now calling the shots.

FYSH TRAVELLED to the new RAAF base at Point Cook, Victoria, for flying instruction on the Airco DH.9. He wanted to make sure his skill level was up to handling heavier machines than the B.E.2e he'd been flying in Queensland. His mother Mary and sister Geraldine had now moved to Melbourne, sharing a house on Normanby Road, Kew, where Mary would spend the rest of her life. It became family folklore that with inflation eating into the money her father had left her, she could no longer keep up the appearances expected in Launceston of Henry Reed's daughter. Though he found his mother's religious zeal increasingly confronting, Fysh continued to take a keen interest in her welfare and that of his siblings.

From Melbourne, Fysh headed to Sydney to discuss contracts and delivery dates for the Vulcans with William Adams & Co., the Australian agents for Vickers. The aircraft were to cost £3700 each, though the sale was dependent on the big machines meeting contracted performance requirements. Vickers guaranteed to Fysh that a fully loaded Vulcan could climb to ten thousand feet in thirteen and a half minutes, a rate he regarded as the minimum acceptable for the job. On 12 February 1922, Qantas cabled £2000 to Vickers in England as a deposit. Fysh busied himself with ensuring all the requirements were covered for the Longreach, Charleville and Cloncurry landing grounds, and he arranged with the Air Council for their representative in London to oversee the contracted tests on the Vulcans.

With the approval of the Qantas board, Fysh bought Parer's DH.4 on 7 March for £500 cash and then £1000 to be paid three months later, telling McMaster they were the best terms he could get.[8] Parer had just checked out of the Boulder Hospital after almost a month there. Baird arranged with a young Scottish-born mechanic, Frank McNally, to ship the machine to Rockhampton where it could then be freighted by train to Longreach. A week later, on 15 March, the company hired builders Stewart & Lloyds in Brisbane to erect a twenty metre by forty metre hangar at Longreach for £1637, a building that still stands today. The old hangar at the local showground was moved to the new aerodrome as an office and store. Qantas asked the Department of Defence for permission to spend £100 of government money on clearing and improving the Longreach aerodrome.

Frank Cory, who had provided Qantas with its free office in Duck Street, resigned as the unpaid temporary secretary. Sixty years later he told John Gunn, writing the history of Qantas, that the company needed someone with bookkeeping skills and he had none.

In the week following Cory's departure, Qantas appointed a manager: a big sun-tanned accountant, Captain Marcus Griffin,[9] who was 188 centimetres tall and weighed 95 kilograms, and who had been awarded a Military Cross while serving in France. He had returned from the war to find that the manufacturing business his brother had run in his absence had failed, and he was deeply in debt; though he had no legal obligation to do so, he personally met all liabilities before he started looking around for a job.[10] He came highly recommended for his integrity as well as his abilities as an accountant and business organiser. Fysh was impressed when he interviewed the imposing man in Sydney[11] and was well pleased with his selection, calling Griffin 'a first-class type of man'.[12] The company also had two first-class woodworkers, George Boehm and George Dousha; the latter was a sixteen-year-old American of Swedish descent who slept in the hangar and cooked his own meals.

Fysh was far less confident, though, in the ability of the Vickers company to deliver on their promises with the new Qantas aircraft. He knew that engine power on machines built for English conditions could be compromised in Australia's interior, because the hot air was less dense and less effective for combustion, making the engines work harder and often causing them to overheat. On 22 March 1922 he wired McMaster from Sydney to say that despite Qantas being committed to the Vulcans he had his concerns that the big biplane – shaped like a plump piglet – might not fly as well as Vickers promised.

Just two days later, Fysh's younger brother Frith died in the General Hospital, Launceston, five years after falling ill in France. Since being repatriated home, he had spent almost all his time in hospital beds in Launceston and Hobart.[13]

EXPERIMENTAL AIRCRAFT were fraught with dangers. As Fysh continued to express his concerns to McMaster over the Vickers Vulcan, the knighted brothers Ross and Keith Smith and their mechanic, the newly commissioned Lieutenant Jim Bennett, were busy testing another of Vickers' experimental machines, the amphibious Viking. The trio had based themselves at the

Brooklands motor-racing track, which doubled as an aerodrome on the southern outskirts of London, as they planned to top the England-to-Australia run with the first-ever flight around the world, starting on Anzac Day 1922.

Six months after crossing the Atlantic in a Vickers Vimy, Sir John Alcock had died in a Viking prototype on his way to Paris, and the Smith brothers were supervising every stage of the preparations for their new model Viking IV. They were checking and rechecking everything about the construction of the aircraft and the testing of its Napier Lion engine, which powered the pusher propellor sitting behind the cockpit.

On 13 April 1922, Ross Smith and Bennett decided to give their machine another thorough examination at Brooklands. They were waiting for Keith Smith to join them on their test flight, but he was running late because the Easter holidays made train timetables redundant. They decided to take off without him at 12.15 p.m.

The Viking performed splendidly for the first fifteen minutes, climbing gracefully to nine hundred metres. Keith arrived at the airport, breathless after rushing from the station, just in time to see the Viking's fifteen-metre wingspan begin to shake and the machine tilt vertically. The huge plane started to spin, then it plummeted. With hearts in their mouths, Keith and the other spectators hoped Ross was stunting while he tested the Viking to its absolute maximum. But the spin became faster, and the Viking nosedived into the earth.

Keith and the others bolted to the scene. Amid the wreckage Ross Smith, the war hero who had flown with Lawrence of Arabia and crossed the world, was trapped in his seat, a gash all the way down his face. Jim Bennett was moaning, but by the time the ambulance men extracted him from the twisted metal, he had died. The bodies were loaded into an ambulance and taken to the local mortuary.[14] Keith, whose life had been spared by a late train, had them embalmed and placed on the steamer *Largs Bay* for transport home to Australia. 'No one knew, or ever will know the cause of the catastrophe,' he told reporters. 'Everybody who takes up a new bus knows the risk and simply conjures with Fate.'[15]

The embalmed bodies were still on their way to Australia when Fysh arrived at Mascot to try to wrangle the repaired Avro triplane, which had been sidelined for a year, and which had now been overhauled by Arthur Baird. Staring at the 'strange-looking machine with its high cockpit and towering tiers of wings, its stove-pipe exhaust topping all',[16] Fysh told himself that if ever there was an aerial malformation, this was it. He couldn't believe the company that had turned out the 504K and Hinkler's Baby could have allowed this thing out of their Southampton factory. But Qantas had paid almost £3000 for the triplane and its repairs, enough to buy five houses in Sydney[17] – which, in hindsight, would have been a much better investment. The machine represented more than half of the airline's initial budget, and they still wanted to use what they believed to be the only aircraft in the country with a cabin.

Fysh felt a dull sense of dread when he saw the triplane waiting for him like a big clumsy dog that appeared benign but could bare its fangs at any moment. It had been built as much as possible out of Avro 504K parts, including the wings. The addition of a deeper fuselage provided room for the cabin, and the third wing and powerful, heavy Beardmore engine allowed the machine to gain more lift than the 504K. But all the extra weight was plonked on the standard 504K undercarriage, and the aircraft was equipped with the standard rudder from that much lighter plane, causing steering problems.

On his first day with the triplane, Fysh was pleasantly surprised: he completed five landings and spent eighty-five minutes circling Sydney. He became so confident that on 6 May, before a crowd of about nine thousand, he took off in the triplane from the Victoria Park racecourse in Sydney's 1922 Aerial Derby, racing against Nigel Love and five other leading airmen. They flew eleven times around a five-mile aerial track, heading out over the Sir Joseph Banks Hotel at Botany and then returning to Victoria Park via Maroubra Bay.[18] Fysh finished second to Love's 504K, covering the fifty-five miles (ninety kilometres) in a shade under fifty minutes,[19] but he reckoned he would have won had Arthur Baird's overhaul of the engine not left the bearings too tight and the revs too low.

No sooner had Fysh landed than he was turning his success in the race into money for Qantas by taking out newspaper display advertisements to coincide with Sydney's Royal Easter Show.

FLYING
ONE OF THE SIGHTS OF SHOW WEEK
SEE SYDNEY HARBOR FROM THE AVRO TRIPLANE
AERIAL LIMOUSINE – AUSTRALIA'S ONLY TRIPLANE.
AT MASCOT AERODROME ALL THIS WEEK.
Parties of up to 4 conducted on AERIAL TOURS OF THE CITY, THE HARBOR.
BOTANY BAY, SHOW GROUNDS. Etc
FARE £2/10/ FROM EACH FOR 20 MINUTES TRIP.
The Avro Commercial Triplane is fitted with a modern comfortable cabin, holding 4 passengers. No wind, noise, or oil. Every comfort, while underneath one of the world's most beautiful scenes.
DON'T MISS IT.
Flights by arrangement only, and before 11 in mornings and after 3.30 in afternoons.
Pilot, Hudson Fysh. Book at Mascot Aerodrome, or Metropole Hotel.
The Queensland and N.T. Aerial Services.[20]

Fysh thought the triplane was safe to fly in fine weather and to land carefully on perfect airstrips, and he made money on the sightseeing ventures over Sydney Harbour.[21]

Two passengers who rode for free were Fysh's new sweetheart, Nell Dove,[22] a Rockhampton girl who'd grown up in New Zealand, and her Aunt Letitia. Nell was a petite, dark-haired beauty with a round face and a dazzling, wide smile. She was stylish with a flair for clothing, and in later years her family remarked that she would have become a fashion designer if more opportunities had existed for women in the 1920s.[23] Her family thought it was funny that a Dove had fallen for a flying Fysh.[24] Like Fysh, Nell was from Australian pioneering stock. In 1853 her grandfather, the Anglican Reverend William Whitman Dove, came out from England to New South Wales as a young man and was first stationed at Newcastle before taking over the parish of Jerrys Plains on the Hunter River. His parish stretched all the way to the Darling Downs, in the new colony of Queensland. The Reverend Dove rode all about his pastoral territory visiting many remote settlements to carry out christenings, marriages and church services; it was a hard life, and his heart gave out when he was just thirty-five. He left behind a 'sorrowing widow'[25] and four young children, including Nell's father, John, who became the manager of the Rockhampton branch of the Bank of North Queensland,[26] and an officer in the local militia, before moving to Auckland.[27]

Nell met Fysh when she was staying at his base in Sydney, the Metropole Hotel, while visiting friends and relatives, and she was twenty-six when she flew in the triplane with him. She was still mourning the death of her younger brother John Jr, who had been killed in France just two weeks before the Armistice. Nell was impressed by the flight over Sydney Harbour – but even more impressed by the pilot. Fergus McMaster had also found love again, almost a decade after the death of his first wife, Edith. The 43-year-old widower was preparing to marry 28-year-old Edna May Faulkner in Brisbane and she would become the stepmother of McMaster's daughter Jean.

The revolutionary triplane and Fysh's cool competence at its controls aided his love life and finances, but he still held no affection for the machine because it clearly needed perfect conditions to be considered safe. He eventually complained to the Civil Aviation Branch about the dangers of its overtaxed undercarriage and underwhelming steering. As a result, the triplane's certificate of airworthiness was withdrawn. Qantas appealed to Nigel Love and Avro for compensation but came away empty-handed.

At the next Qantas board meeting, a resolution was passed: 'Proposed by Mr McGinness, seconded by Mr Templeton that in the event of Messrs A.V. Roe and Company declining to offer any compensation in the matter of their Avro Triplane purchased by this Coy. that this company do no more business whatsoever with the firm of A.V. Roe & Co.' The ban lasted only until

Qantas needed spare parts for the 504K. The Beardmore engine was stripped from the triplane and sent to Longreach as a spare, and the fuselage and its cabin became a chicken coop in a Sydney backyard.

WHILE FYSH WAS LIVING IN SYDNEY, Lieutenant-Colonel Brinsmead visited Longreach to inspect the company's new landing ground and oversee preparations for the start of the mail service. At a dinner in Brinsmead's honour at the Imperial Hotel, McGinness assured the Controller of Civil Aviation that in civil flying the motto was always 'safety first' and the policy of Qantas was 'absolute safety'.[28]

At the time, crashes were common on mail runs in England and especially in the United States, where pilots flew night and day in all weather, often in unsuitable, unstable machines. The American airmail pilots became known as 'the suicide club' – thirty-five pilots hired by America's Post Office Department between 1918 and 1926 were killed while flying the mail.[29] Young Charles Lindbergh[30] was one who survived, twice parachuting out of machines that had become known as 'Flaming Coffins'.

Brinsmead told the gathering he had no doubt Qantas would be an outstanding success because the company had 'jolly good pilots', 'a wonderfully good country to fly over' and 'magnificent machines'; the new Vulcans on order, he said, 'were absolutely the last word in aeroplane construction'.[31] But the fears that Fysh held over the Vulcans were well grounded. The machine had first flown at Brooklands Aerodrome in April 1922, with the company's chief test pilot, Stan Cockerell, at the controls. The weight was well up on expectations and the performance well down, despite repeated tests using the Rolls-Royce Eagle engines that had powered the Vimy on its record-breaking flights. But Vickers continued to assure potential buyers that the machines would prove themselves eventually.

On 24 May, Qantas moved out of Frank Cory's Longreach stock and station agency and into its new offices in the town's Graziers' Building. From there, on 26 June, the company's new manager Marcus Griffin prepared a telegram for McMaster, who was in Sydney on his honeymoon, telling him that the delivery of the Vulcans would be delayed by another three months. Vickers had cabled Qantas that no British pilot would agree to the proposed employment contract clause banning alcohol.

Qantas sent the RAAF a request to buy two DH.9 machines, similar to the one Parer and McIntosh had flown from England. Instead, the company was presented with an opportunity to buy two cheap war surplus British-built Armstrong Whitworth F.K.8s. One belonged to the tall bronzed pioneering Australian aviator Horrie Miller,[32] who had recently dissolved his partnership in a small airline in country Victoria; the other was owned by the Simpson Tregilles Aircraft and Transport company in Perth, and for almost a year it had been taken up on joy-riding flights in Western Australia.

Qantas needed to save every penny it could, as it had lost £4400 for the 1922 financial year despite the joy flights and capital raising. The F.K.8s were significantly bigger than the Avro 504Ks, could carry 450 kilograms more, and had space for three passengers. They were powered by the same six-cylinder Beardmore engine as the ill-fated triplane; an engine that could produce 160 horsepower compared with 100 from the 504K.

Brinsmead wired Qantas on 6 July, telling the company that, while he still favoured the Vulcans, in order for it to begin the airmail contract in early September, Defence Minister Walter Massy-Greene had approved the purchase of the F.K.8s. Baird went to Melbourne to inspect them. While Qantas waited for the machines to be delivered, the DH.4 that Fysh had bought from Ray Parer arrived in Longreach by train on 12 August. It had been so badly damaged in transit that Baird had to do a complete overhaul. This included extensive work on the Rolls-Royce Eagle VIII engine to ensure it could deliver its full force of 360 horsepower, more than double the power that the F.K.8s promised.

Qantas bought a Talbot truck for £275 to cart stores and spare parts between the aerodromes and emergency landing grounds. Construction of the hangars at Charleville and Cloncurry was started, and petrol and oil depots were established to provide about a thousand litres of fuel at Charleville, Blackall, Winton and McKinlay. The Charleville–Longreach fare was set at £11 or 9.9 pence per mile, and the fare all the way to Cloncurry would be £21 or 8.7 pence per mile. For 'the small surcharge of' three pennies per half an ounce, a letter posted along the route early on Monday morning would reach Brisbane at dinnertime on Tuesday by airmail, cutting at least two days from the delivery time by train.[33]

Brinsmead had suggested that the almost thousand-kilometre Charleville–Cloncurry flight could be made in one day. But with McGinness now having much less of a voice at Qantas, Fysh and Griffin successfully argued against flying the British machines in the hot, thin afternoon air of the Queensland summer; they opted instead for an overnight stop at Longreach. Each stage of the trip would start at dawn to avoid the searing heat – and to avoid the maddening midday air pockets and bumps that posed a threat to breakfasts staying down.

AS MUCH AS THE FIRST SCHEDULED government-backed flights of Qantas were important to Fysh, he had other weighty matters on his mind. He had popped the question to Nell Dove, she had said 'yes', and his impending marriage made him rethink his life in Longreach. He was earning good money: more than double the average Australian wage with a salary of £500 a year, plus nine shillings a day travelling allowance. But his frugality had always made him an asset for Qantas, and he now took the bottom-line approach to his living expenses, moving from the Imperial Hotel to a small galvanised-iron hut used as emergency occupation by the mechanics at the Longreach Motor Co. It was a sweatbox in summer and a freezer in winter. 'The shack was an ordeal,' Fysh wrote, 'but it saved me money.'[34]

On 6 September, Fysh tested one of the Armstrong Whitworth aircraft, and he took the second one up nine days later. The Armstrong Whitworth company promised that the F.K.8s could reach thirteen thousand feet and cruise for three hours at ninety-five miles per hour, but those figures had been attained in the leaden skies and cold air of Britain. Fysh found that in the hot air over Longreach the radiators quickly boiled as the engines worked overtime. Baird overcame the problem by fitting bigger radiators as well as header tanks that condensed the steam and kept water in the system, though even then it was common to see the F.K.8s flying with steam billowing like smoke from their engines.[35]

Vickers insisted that despite negative reports emerging from England, their machines were now up to speed, and they offered to send the Vulcans to Longreach at their own expense. The first would be shipped from England on 19 September, with the second coming at the end of the month. The company said it had found an experienced British pilot for Qantas, Captain Godfrey Wigglesworth,[36] late of the Royal Flying Corps. Many people mistakenly called him Geoffrey. He was an eccentric Yorkshireman, but Vickers said he was willing to abstain from alcohol and abide by all of Qantas's safety regulations.

On 11 September, McMaster told Brinsmead that the DH.4 and the F.K.8s would be ready to start the Charleville-Cloncurry route on 26 September and would operate for six weeks until the Vulcans arrived. But McMaster had seriously underestimated the work Baird had to do on the big DH.4. On 26 September McMaster wired Brinsmead to say the whole service would be delayed until 5 October and to ask the Civil Aviation Controller if Qantas could use the Avro instead of the DH.4. Brinsmead was ropeable. 'The delay in the commencement of the service,' he thundered, 'is deplored … two Armstrong Whitworths and an Avro of considerable age are not sufficient to ensure a reasonable guarantee of reliability of service, even for six weeks.'

A new starting date of 2 November was set, but then the Department of Defence's liaison officer in London told Qantas that in further tests the Vulcans had fallen well short of Vickers' claims, and the machines were much heavier than initial calculations had estimated. The Vulcans required about twenty minutes – instead of the promised thirteen and a half – to climb to ten thousand feet. Qantas refused to take delivery until the aircraft came within five per cent of the promised rate of climb,[37] stressing to the manufacturers 'the absolute importance of full climbing capabilities' and insisting that all further correspondence about the Vulcans should 'be passed through the hands of the company's legal advisers'.

Late in September, Fysh and McGinness flew the Armstrong Whitworth machines on inspection tours of the landing grounds along the Charleville–Cloncurry run. Fysh brought Griffin and Baird's new mechanic Frank McNally.[38] They had a forced landing in the open countryside around Kynuna because of a sticking valve, but quick repairs were made and they were airborne again before long. By 21 October, the F.K.8s had been tested, and they proved satisfactory in all areas. Four days later, on 25 October, the DH.4 was tested and approved.

Fysh was put in charge of flying operations, and he would also oversee the company's aerodromes and hangars. McGinness was no longer the enthusiastic driving force of the company. He and Dr Hope Michod did not get along, and McGinness had become disillusioned and disenchanted with the way the airline he had conceived on a bush track was being run. Trying to make the airline turn a profit was difficult, and the glamour of flying that had intrigued him in Palestine had waned. Fysh and McMaster had insisted he stop drinking. McGinness wasn't a man of routine. Getting out of bed at 4 a.m., especially on cold winter's mornings, then flying a route he had traversed dozens of times before, isolated in his cockpit, while feeling isolated in the business given Fysh's developing reputation, pushed McGinness to a personal precipice. On 25 October, as his long-nurtured dream of a regular air service in the outback was just a week from realisation, McGinness sent a telegram to the board, tendering his resignation. He was just twenty-six.

Fysh was saddened but not surprised by this turn of events. He was also angry. With the business still in its infancy and desperate for pilots to fly the mail run, McGinness had left Fysh in the lurch.[39] Yet, Fysh remained forever grateful to his friend, whom he called 'the spark'[40] of Qantas – a daring if impulsive young man, full of ideas, who set the example for courage in war and business.

McMaster implored McGinness to at least fly the first leg of the new mail service he had envisioned. McGinness agreed.

'We desire to place on record,' McMaster wrote to McGinness on behalf of the Qantas board, 'our appreciation of your services to the company, both in regard to the initial inauguration and the subsequent work in obtaining the subsidised contract and the raising of capital, and the directors view with regret the step you have taken.'[41]

THE ARMSTRONG WHITWORTH F.K.8 that had been bought from Perth held the Australian Certificate of Airworthiness No. 36 and was boldly identified with the registration letters G-AUDE. McGinness flew Baird in it to Charleville, where they prepared for Qantas's first scheduled service. On 2 November 1922, the mayor and leading citizens of the town rose at dawn, and in the silent stillness of that vast countryside, they assembled at the aerodrome where Baird had readied the machine and filled its fifty-gallon petrol tank.

In his flying suit and leather helmet, McGinness addressed the hushed crowd. Though he was baling out of the business, he realised the importance of what he had started. Australia, he said, was keeping pace with Europe and America. This service, whose start they were to witness, was destined to link Australia to Asia, Africa, Europe and Great Britain.[42] A cheer went up when at 5.30 a.m. McGinness climbed into the cockpit, Baird swung the propeller, and the engine roared into life. With 108 letters on board, McGinness rolled the aircraft forward to begin the first Charleville–Cloncurry Aerial Mail Service.

Fysh and McMaster were waiting at Longreach with the chairman of the local shire council and a crowd mostly composed of townspeople. In this crowd was 84-year-old Alexander Kennedy, who had caught the train up from Brisbane. He was about to have his long-held wish granted: he would be the airline's first passenger on its first scheduled service.

After stops at Tambo and Blackall, where McGinness and Baird received hearty welcomes, the pilot landed in Longreach ahead of schedule at 10.15 a.m., having covered 266.5 miles (430 kilometres) in a flying time of three hours fifteen minutes. McGinness climbed out of the cockpit and got ready to stride out of Qantas's future. He may have already had second thoughts about quitting as he told the crowd that the trip had been the most enjoyable he'd ever made – and he 'had done a lot of flying'. He could assure prospective passengers that flying was like sitting in an armchair when they were at five thousand feet.[43]

McMaster prophetically told the cheering crowd that he looked upon McGinness's flight as 'a small beginning, which would develop into one of the greatest services in the world'.[44] Qantas would be neglecting their duty, he said, if they did not grow the service and the aerial routes, eventually bringing Sydney within two days' journey of these 'far-back places'.[45]

At dawn the next day, 3 November, as the heat haze prepared for its emergence over the western plains, Fysh took advantage of the cooler air to ready the second F.K.8, registered as G-AUCF. The aircraft was wheeled out of the hangar by many willing hands onto the uneven surface of the stony Longreach tarmac. Baird spun the propeller a few times, until flickering flames burst from the exhaust stubs. Fysh revved up the six-cylinder Beardmore engine. Kennedy climbed in as quickly as his eighty-four years would allow, brushing away helping hands when he stumbled towards the foot notches in the side of the fuselage. After he settled into his seat, his safety belt was adjusted. Then Baird climbed in next to him, and the wheel chocks were pulled away.

Fysh taxied the machine to the far corner of the aerodrome. A light wind was coming in warm puffs from the north-east, and the day was going to be blazing hot. When Fysh opened the throttle, the Beardmore roared with apparent delight as the machine raced across the bumpy tarmac towards the far fence. But Fysh couldn't get enough revs to lift into the air, so he aborted the take-off and taxied back to the far corner for another run. He tried twice more, but the revs were still down; each time he had to pull out before it was too late.

The three men eventually climbed out of their aircraft and into the F.K.8 McGinness had flown the day before. Within half an hour, Fysh had the aircraft buzzing above the little town. Kennedy's flying cap had slipped around, and his long white

The first official Qantas passenger, the almost 85-year-old Alexander Kennedy, waits for Fysh to finish his smoke before the take-off from McKinlay to Cloncurry, the last leg of the first scheduled Qantas passenger flight on 3 November 1922. *State Library of NSW FL7561337*

beard was flowing in the wind like a streamer. Fysh looked over his shoulder to see the old man's face dissected by a huge smile. 'Be damned to the doubters!' Kennedy shouted above the noise.[46] At no stage did he complain that he might be too old for this sort of thing, but plenty of his friends had – they had warned him that it was too dangerous for a man his age to try this flying caper.

After 180 kilometres of flying, Fysh brought the machine down at Winton to drop off some mail and drink some tea with McMaster, who had motored through the night along bad roads to be at the airline's first home for the arrival. There was another large crowd there to cheer and there were more speeches about aviation being the way of the future. Kennedy later wrote to a friend that it was 'more than pleasing to note the great interest all the waybacks are taking in this coming service'.[47] After a quick top-up of the petrol tank, Fysh took off into headwinds, aiming this time for the gravel runway at McKinlay, 132 miles on. The men spent a few minutes there, then flew the short run of seventy-one miles to the Cloncurry aerodrome that Fysh and McGinness had surveyed three years earlier. Fysh kept the F.K.8 cruising at four thousand feet, just high enough to stay within gliding distance of the emergency landing strips cleared throughout the bush. This was the countryside where Kennedy had spent the best part of half a century. The old grazier looked down contentedly on the rugged hills and rough tracks that had once been so difficult to traverse but were now flying by.

Fysh landed in Cloncurry at 11.20 a.m. The first Qantas service from Charleville had taken seven and three-quarter hours in flying time for the 577 miles (930 kilometres).

As Fysh taxied towards the new hangar, Kennedy saw many of his old friends among a crowd of several thousand people there to welcome him on his historic journey. They gave three cheers for the success of the service and three more for its exhilarated first passenger. The beaming Kennedy began handing out the mail to recipients in the crowd. 'I feel splendid,' he told them, 'and 20 years younger since I started on the trip. The aeroplane is a speedy, safe and comfortable mode of travelling.'[48] Air travel, Kennedy said, would revolutionise life in Australia.

Chapter 15

Mr Fysh said it would be cold up top so he put an overcoat on me. [But] it was [not as exciting as] when I did the loop-the-loop on a joy flight over Cloncurry.

YOUNG IVY MCLAIN, THE FIRST FEMALE PASSENGER ON A SCHEDULED QANTAS FLIGHT[1]

THE TINY, VERY PRETTY and very brave Miss Ivy McLain[2] became the first female passenger on a scheduled Qantas service when she flew with Fysh the day after Alexander Kennedy's arrival in Cloncurry. She joined Fysh in the F.K.8. back to Longreach alongside another new Qantas mechanic, the tall and rangy Jack Hazlitt.[3] The 23-year-old Ivy could claim to be the first paying passenger, since Kennedy's ticket had been part of his guarantee to buy shares. Her father Archie McLain had made his money on bullock dray transport in the pioneering days, delivering – among other things – the first load of copper from the mine in Cloncurry to Normanton, 320 kilometres north. He was an old friend of Kennedy's and on the grazier's recommendation had bought a ticket for Fysh's return journey. Ivy, the youngest of his seven children, made the flight after her father became ill.

Draped in a heavy coat on a broiling November morning, the young woman climbed into the machine at Cloncurry for the first leg of the journey. It was a long flight, and she drank tea from a thermos when the wind began to bite at cruising altitude. With the ever-cautious Fysh at the controls, she found the outing rather passé compared to her previous experience in the air. She stayed overnight in Longreach after the first leg. The following morning, 5 November, McGinness and Hazlitt flew her on to Charleville.

Twenty-five-year-old Hazlitt was a remarkable character: one of the rugged, hungry young veterans trying to piece together a life from the wreckage after the war. The son of a theatrical manager who had abandoned his family, Hazlitt had put up his age to enlist in 1915, and he had fought on Gallipoli and in France before being repatriated back to Australia in 1917 with severe 'shell shock'. He learnt to fly in Melbourne but couldn't gain a pilot's licence because of a heart condition caused by the war. Instead, he wrote to Qantas offering his services as a mechanic in Longreach and took the job despite the 'long hours and very poor pay'.[4]

On the day McGinness and Hazlitt arrived in Charleville with Ivy McLain, Qantas hired McGinness's replacement pilot, Fred Huxley,[5] a man with a similar reputation for daring. He had been the first member of the Australian Flying Corps to shoot down a German aircraft in France,[6] and in 1921 he flew his pregnant wife and two-year-old daughter across Bass Strait on one of the first passenger flights to make the crossing to Tasmania.[7]

Accompanied by Hazlitt, McGinness made two more airmail trips for Qantas between Charleville and Longreach before the company presented him with a formal letter of thanks and a cigarette lighter as its parting gift. There was no fanfare. Baird had turned McGinness's little car into a Qantas truck, so McGinness left Longreach on the train, first heading for Brisbane and then on to Warrnambool, where he was met by one of his brothers and taken to Riverview to plot his next move.[8] John Clarkson was now a sharebroker in Western Australia, and he told McGinness it was the land of opportunity. Soon McGinness was in Perth, based at the Palace Hotel on the corner of St George's Terrace and William Street, and selling shares across the state for the Primary Producers' Bank.[9] The venture failed and was taken over by the state government. Locals who invested lost their money and Clarkson left for the eastern states.

FYSH WOULD NOT DEVIATE from the path before him: trying to make Qantas a profitable business. For the next four months he kept flying the Longreach–Cloncurry run, mostly in the fast, powerful and reliable DH.4. His day generally began at 4 a.m. when he had a cup of tea with Baird and any passengers outside the Imperial Hotel. They then climbed into the Talbot truck and drove out to the aerodrome, where they all helped to push the DH.4 from the hangar. Fysh would roar the big Rolls-Royce engine into life and then – in caps-and-goggles and thick coats regardless of the heat – they would all hold on tight as the DH.4 raced down the runway and over Longreach by the dawn's faint light, heading for the first stop at Winton. In dry seasons the landscape was like a bare desert, but after rains the rich green Mitchell grass would appear to be waving at the aircraft as it flew by, while mobs of feasting kangaroos stood to attention.

After the heady excitement of those first scheduled mail runs, months of hard work lay ahead, as well as staff tensions, resignations, forced landings, crashes, major damage and frustrations – and some frankly bizarre events. The government insisted the mail be carried in thick asbestos-lined bags in case the aircraft caught fire, and insurance premiums were so steep that for the first years Qantas planes flew without accident cover, meaning that each passenger had to sign their ticket to say they flew at their own risk.[10]

It was hard to maintain a regular schedule, given the unsuitability of the British aircraft in the most extreme of Australian conditions. The Armstrong Whitworth F.K.8s had seen better days, and their overheating engines made them 'the boiling Beardmores'.[11] Wind direction also played havoc with the timetable Fysh now oversaw. Once when racing a storm into Longreach in the DH.4, he was carried helplessly upward for thousands of feet until he hit a black cloud mass. On other days the tailwind was like a jet stream, and Fysh and Hazlitt once averaged 185 kilometres per hour in the DH.4 between Longreach and Cloncurry.[12] Flying an F.K.8, Huxley once covered the Charleville–Longreach run in just two hours forty-five minutes at an average groundspeed of 156 kilometres per hour. At another time, with the hot winds playing havoc, he took six hours at an average of just seventy-one kilometres per hour. But some passengers didn't mind how long their trip took;

one female customer was annoyed she had to get off the machine at Charleville and take the train to Melbourne, telling a reporter she wished for the day when an aeroplane could take her the whole way.[13]

Qantas hadn't been flying the mail run a month when ten-week-old Peg Glasson[14] became the first baby to fly on the scheduled service, first with Huxley and then Fysh. Peg's father Rupert Glasson was the manager of Oak Park Station on the Nive River, and on 30 November he put Peg and her mother Hilda[15] on the Qantas plane from Charleville so they could visit Hilda's ailing father, who ran a property near Winton.[16] There were no bassinettes or baby carriers in Fysh's DH.4 from Longreach, so Peg travelled in what had been the observer's ammunition box when it was a bomber. Hilda wrapped an orange in a nappy and threw it down to her husband as they flew over Charleville airport.[17] At six-thirty the following morning, Hilda and Peg arrived in Winton. The alternative would have been a meandering week-long train journey via Townsville.

Fysh told reporters that aeroplane travel was becoming 'most popular' in Queensland's west, and that Qantas was booked solidly for the next month, taking in the Christmas holidays. The Qantas agents, he said, could fill big planes if they had them to put in service.[18] Finding pilots willing to fly the long, lonely routes and base themselves in the wild west was tough, though.

QANTAS HIRED A THIRD PILOT, Tom Back,[19] on 4 December 1922, bringing its total staff to nine. Back was from a prosperous family of wine and spirit merchants in Norwich. In 1909 he had come to Australia aged seventeen to work as a jackaroo and overseer on stations in western New South Wales. During the war he became a British air force major, and after the Armistice he returned to Australia, buying a small sheep property near Mudgee in New South Wales. Then he heard that Fysh was looking for pilots. After a quick refresher course at Point Cook in Victoria, he was soon in Longreach. Fysh regarded Back as 'an excellent pilot, resourceful and a very firm individualist'[20] – sometimes too much so.

On 22 December 1922, Back flew Dr Hope Michod from Longreach to Winton, where he attended a patient in what may have been the first emergency treatment administered by a flying doctor. During that same week Qantas carried fifteen cattlemen between stations during a particularly wet period. That Christmas Eve, two schoolchildren were flown to their homes over impassable roads.[21]

Conditions were tough for all of the Qantas employees, especially those who had come to Longreach from the city. They were underwhelmed by the primitive living conditions, the constraints and isolation of a small rural town of three thousand people, and the very real possibility that their jobs would fizzle out if Qantas was permanently grounded by its financial woes.

Although Marcus Griffin had done impressive work organising Qantas and sorting out some of the company's money problems, he quit as manager on 5 January 1923. He was engaged to be married in Sydney but had been enduring a lonely existence in a little room at a Longreach pub. There had been repeated clashes resulting from his lack of tact with some of the directors.

Fysh had been granted a £50 salary increase that took him to £550 a year, and he took over from Griffin as the new Qantas manager on 5 February.[22] In his dependable, no-nonsense way, he assured the board of his 'earnest endeavour to do the best possible for the company and to make a success of [his] new position'.[23] He continued to fly the Longreach–Cloncurry mail run while overseeing the manager's duties. McMaster wrote to him, praising his determination to see Qantas succeed despite the tough start, and telling him, 'I feel there is a big future ahead of you.'[24]

The increasing number of passenger enquiries kept Fysh's confidence high about the airline's future, but he was dismayed when the new pilots Huxley and Back quarrelled over who was more senior. When Fysh supported Huxley, Back began treating his boss with outright contempt.[25] Then another British pilot arrived to give Fysh hell.

When the Commonwealth steamship *Cooee* berthed at Fremantle on 23 January, onlookers were fascinated by a tent-like construction on the afterdeck – and the mystery it contained. The canvas covered an immense packing case, which turned out to hold the fuselage and some of the parts of what the press called a 'monster eight-seater passenger aeroplane': the Vickers Vulcan, aka the 'Flying Pig'. It was on its way to Longreach whether Qantas wanted it or not. Down in the hold, other large cases contained portions of the dismantled plane, with the one on the deck too large to go below.

'Quick and nervy in his actions,'[26] the tall and lean 37-year-old wartime pilot Godfrey Wigglesworth was overseeing the delivery of the aircraft, and he was wary of reporters, dodging their 'questions, entreaties, and verbal traps'.[27] The pioneering aviator Horrie Miller remembered Wigglesworth as a strange character, 'afraid of nothing and nobody. He neither boasted nor bluffed. He was simply indifferent to the feelings of his fellow men.'[28] He had a battered, weather-beaten face, and his hair dropped carelessly down a broad forehead over his right eye. Though Vickers had promised that Wigglesworth would abstain from alcohol when flying the Vulcan for Qantas, Miller saw him drink 'until all others had either retired or fallen in a stupor', then walk from the bar 'as steady as a rock'.[29]

Wigglesworth had qualified as a pilot at the Military School in Shoreham, on the English south coast, on 29 May 1915, training in an ancient Maurice Farman biplane.[30] 'Aye, I fought with the Royal Flying Corps during the war,' the big Yorkshireman told a chap from Perth's *Daily News*. 'But bless you, the war's been over for years, and everybody is trying to forget it.' No amount of prying could get Wigglesworth to open up about his wartime experiences. What he wanted to talk about, though – at length – was the Vulcan. The British pilot spoke of how Qantas passengers would be accommodated in an enclosed saloon with more than six feet (183 centimetres) of head room in what was 'probably the biggest passenger aeroplane that has come to Australia'.[31] Each of the eight passengers would have a window that they could keep open during the flight, providing natural air-conditioning that would be a little icy at cruising altitude. The cabin, trimmed with satin braid, resembled a luxurious railway carriage, with six dove-grey leather-upholstered collapsible armchairs and a fixed double seat.[32] Passengers had individual hat racks, luggage racks, and cupboards.

Wigglesworth told the press that three Vulcans had been operating between London, Paris and Brussels, and had done 'exceedingly satisfactory work during the past year'. He said that with the Rolls-Royce Eagle VIII engine, the 'Pig' could fly fully loaded with passengers and luggage at 105 miles per hour (170 kilometres per hour) for three and a half hours, and it would come into land at a slow and safe forty miles per hour (sixty-four kilometres per hour), almost half the speed of most aircraft.[33]

But Fysh maintained grave doubts about its performance and felt that bureaucracy was tying a heavy anchor to Qantas's tail. Buying the Vulcans was Lieutenant-Colonel Brinsmead's idea, and Fysh had written to the controller on 31 January 1923, during a Longreach heatwave, to say that with the thermometer about to bubble over it would be 'positively unsafe to fly either the Avro Dyak or the [Armstrong Whitworth] between 11 a.m. and 5 p.m. Could there be any hope for the Vulcans on a hot day?'[34] Even if the Vulcans were loaded to only half their specified capacity, he said, they had no better performance than an Armstrong Whitworth. 'The first aim of this company is absolute reliability of machines and safety to passengers,' Fysh wrote. 'It is only on those conditions that commercial aviation can be built.'[35]

By 13 February 1923, the Vulcan was being reassembled at Point Cook by RAAF personnel under Wigglesworth's supervision.[36] Brinsmead finally conceded there could be power problems in the thin air of the outback, and he suggested that the machine could be used as a four-seater in summer and a six-seater at other times. But Fysh feared the aircraft would still be unsuitable and that its failure would leave Qantas with only the DH.4 and the unreliable 'Boiling Beardmores' to fulfil the all-important government contract.

On 25 February, Fysh responded to an emergency phone call to the Longreach Qantas office from a shearing contractor, William Ballinger, who owned the Hulton sheep property, about thirty kilometres east of Barcaldine. Ballinger's child was in

Rockhampton, 550 kilometres away, seriously ill, and he wanted Fysh to fly him there immediately. Fysh was reluctant – it was a long way, and the aircraft was unreliable – but Qantas needed every penny it could raise. Ballinger was in desperate straits and offered good money.

It was already late in the day when Fysh left Longreach in the Armstrong Whitworth G-AUCF for Hulton, 140 kilometres away. He hoped to spend the night at the Ballinger property[37] because the sun was falling and the plane couldn't travel in the darkness. But the desperate Ballinger implored Fysh to get going, and together they flew on to Jericho, fifty kilometres east, sideslipping to land the big aircraft in fading light on a short runway at the showground, completely surrounded by tall gum trees. Fysh tied the machine down and had a fretful night trying to sleep at the local pub, knowing he couldn't get over the trees on take-off the next morning without a strong breeze.

When he climbed fearfully out of bed at sunrise, there was a slight puff of wind. He waited and waited for any kind of breeze to build as the understandably impatient Ballinger badgered him to hurry up. Together they dragged the aircraft back to the far corner of the showground, with the plane's tail brushing up against the trees, and a local garage man juggled the throttle while Fysh swung the propeller to spark up the engine. Still there was no wind of any strength. Fysh rounded up some helpers to hang on to the tail as he flayed full revs from the screaming Beardmore. Above the noise and wind, he shouted, 'Let go', and the F.K.8 was released to scamper across the grass.

Fysh zoomed up fifty feet, heading straight for the top of the wall of gum trees. He decided on the lesser of two disasters. He took a sharp left, pulled the nose up and stalled the machine. It came down with a massive crash. He and Ballinger crawled from the wreckage, but their only injuries were superficial cuts. Fysh returned Ballinger's cheque and put him on the next train to Rockhampton.[38] The F.K.8 was dismantled and sent by train to Longreach for repairs, though it was already worn out.

Fysh called the crash 'the only blot on my copybook in my whole civil flying career, though on a number of occasions I admit it was Dame Fortune who got me through'.[39] More than ever, Fysh and Qantas were dependent on the outcome of the Vulcan tests – and they weren't looking good.

WHEN WIGGLESWORTH TESTED the Vulcan at Point Cook, the results were dreadful. But Vickers decided to send him and the plane to Longreach for final testing at their own expense. In dire need of another aircraft, Fysh hoped that the Vulcan would surprise him.

On the morning of 14 March 1923, Wigglesworth took off in the Vulcan from Point Cook with two newspaper reporters on board for a ten-day journey to Longreach. Ernest Jones, formerly of No. 3 Squadron, was flying two inspectors from the Civil Aviation Department in a Bristol Tourer as an escort. Horrie Miller recalled that Wigglesworth made his passengers clean and brush their boots before entering and 'brusquely reprimanded them for fingering the upholstery'.[40] On the first stage of the flight from Victoria, the passengers were violently sick in rough weather; horrified by the vomit stains on the plush carpet and upholstery, Wigglesworth ejected the offenders at the first stop, before reconsidering.[41]

Fysh told McMaster that their little airline was fast approaching the end of its tether with their present equipment, and they might have to suspend the mail service temporarily if the Vulcan failed its tests. He complained that Brinsmead was now silent about the problem and seemed 'to be leaving us to fight Vickers rather on our own'.[42]

Brinsmead had other problems. At Bourke, the engine on the Bristol Tourer escorting Wigglesworth's Vulcan failed, and the machine crashed into a tree. An Aboriginal man came to the rescue of the three occupants, as the reporters on the Vulcan rushed to file reports rather than help the accident victims. Wigglesworth abused them for their callousness. All three victims were taken to Bourke Hospital with shock. Jones also had a broken thigh, and one of the inspectors, Rupert Hoddinot, a dislocated shoulder; the other, Bob Buchanan, was badly cut about the face. The wrecked plane had to be dismantled.[43]

Wigglesworth lumbered on and brought the Vulcan into Longreach on the morning of 24 March, fighting off crowds who wanted to graffiti their names on the silver wings.[44] Nothing like this huge luxurious flying machine had been seen in Australia before the Vulcan's arrival, and here it was in a small outback town. But its pilot brushed many of the spectators away and was in a fighting mood.

After some hours at work on the Rolls-Royce engine, a grease-covered and dirty Wigglesworth entered the Imperial Hotel's crowded bar. The cattlemen and shearers gritted their suntanned jaws at this odd-looking Englishman, and he surveyed them in return 'with open contempt'. 'What,' roared Wigglesworth, 'are all you convict bastards drinking?'[45] With the ice broken, they all got down to some serious thirst quenching.

For the next two days Wigglesworth worked on the engine, and on 27 March, Fysh, McMaster and the other Qantas directors assembled at the landing ground to see him put the big machine through its paces. The fact that he was wearing old carpet slippers and looked like he'd been on a bender did not fill Fysh with confidence. Still, Fysh, McMaster and some others, including Jack Hazlitt, climbed into the cabin, already as hot as a sauna. The Vulcan was carrying only half its payload, but doubts weighed heavily. 'We were for it,' wrote McMaster, 'in an aeroplane we all knew to be unairworthy for Western Queensland conditions.'[46]

Without any preliminary checks, Wigglesworth jumped into the high cockpit above the passenger cabin and worked the throttle as the Vulcan shuddered along the runway. Wigglesworth barely cleared the fence, and the machine struggled to reach five hundred feet. The noise and heat in the cabin was almost overpowering, and Fysh noted that McMaster's eyes darted from the hospital to the cemetery and back. Wigglesworth kept trying to climb for forty-five minutes, but he couldn't get the machine above 5750 feet (1750 metres). He finally admitted defeat.

Two days later at a meeting of directors at which Wigglesworth was present, a resolution was carried that 'as the local test of the Vickers Vulcan had proven unsatisfactory, and that as both local and English tests had demonstrated that the machine was not in compliance with the contract, acceptance of the Vulcan be refused'. Qantas received £1000 in damages, but they were still in a dire position with only the DH.4 and one F.K.8 operational for the mail run. A leak in the DH.4's fuel tank was patched with chewing gum and adhesive tape.[47]

FYSH TRAVELLED TO SYDNEY to see his fiancée, Nell, and sort out details of the Vulcan cancellation with Vickers' agents. Then he headed on to Melbourne, where Brinsmead approved his emergency plan to buy a single-engine Bristol Fighter F.2.B from Horrie Miller. The machine was similar to Samuel McCaughey's fighter plane that Fysh and McGinness had flown in Palestine. 'Huddy,' Miller said to Fysh, 'you're stuck with an air service and no aircraft, and I'm stuck with an aircraft and no service. Maybe we can help one another out.'[48]

The Bristol's 300-horsepower, water-cooled V-8 Hispano-Suiza engine gave it a cruising speed of 90 miles per hour (145 kilometres per hour). Even in the hottest conditions, the Bristol was ever reliable, but it could carry just one passenger who invariably choked on fumes from the castor oil that lubricated the engine and made clothes and skin stink. Almost all the passengers on the Bristol became queasy, something that wasn't in the Qantas advertisements. Fysh oversaw the production of small cardboard boxes to be handed to all passengers in case of air sickness.

The price agreed for Miller's Bristol was the Vickers' cheque for £1000, on the condition that Miller deliver the aeroplane to Longreach at his own risk – and quickly. He was down to just £50 in his bank account, so he readily accepted.[49]

On 21 April, Miller took off in the Bristol from the Port Melbourne aerodrome run by his manager Harry Shaw. He landed at Longreach seven days later. Fysh asked him to show off the aircraft to a couple of the Qantas directors in a flight around town, but Miller – remembering several close scrapes he'd had in the machine – refused to start the engine up again until Fysh handed him the cheque. The £1000 was paid, and Baird and some of the other workmen soon built a snug cover for the Bristol's passenger seat.

Miller stayed only a few days in Longreach. On 3 May he hitched a ride for Melbourne with Wigglesworth in the rejected Vulcan. While putting oil into the engine at Charleville, the Yorkshireman swayed on a ladder and fell off, much to the amusement of spectators. At Bourke, he and Miller stopped to see how Ernest Jones and the two Civil Aviation inspectors were progressing in the local hospital. Wigglesworth met one of the nurses, Margaret Dundas, and eventually married her in Rockhampton before he returned to Longreach to start a motor engineering business.

SOON AFTER THE DEPARTURE of the Vulcan, the Qantas board approved the purchase of two slow but steady Airco DH.9Cs. Each of them could carry three passengers, their luggage and a sack of mail.

But the company's problems were far from over. Fred Huxley tendered his resignation on 19 May to take a better-paying job with Shell. On 6 June, while waiting to work out his time, he went off in the DH.4 to rescue Fysh. Huxley needed to ferry Arthur Baird to Guildford Park station, where Fysh had landed the Avro 504K in bad weather when returning with the Longreach bookmaker Possum Patterson from the Windorah races. Fysh had forgotten to warn Huxley of the telephone line strung between rough bush poles near his landing place. The DH.4 – the company's most reliable workhorse – hit the line and crashed hard, sustaining serious damage. Luckily, no one was badly hurt.

To add to the woes of the overworked Fysh, he complained that the increasingly hostile pilot Tom Back was getting 'quite out of hand'. Fysh told McMaster that the volatile Englishman had been flying the Bristol, which Fysh labelled 'the best machine on the service', but when Hazlitt declared her unfit for flying and Back was asked to take the Avro, 'he made quite a scene … and a lot of silly accusations about our staff, particularly Baird'. Fysh demanded his resignation, and on his last flight Back looped the loop with the mail over Charleville. He became a grazier near McKinlay.

Fysh wouldn't tolerate anyone criticising Baird. Sometimes, he said, Baird and McNally would work all night. Often one of them would have to accompany a pilot on an early morning run with minimal sleep. At 4 a.m. the sleeping engineer would be summoned, and the little woodworker George Dousha would put the frying pan on and make a cup of billy tea. Day would break; the F.K.8 would be wheeled out and started up. The mail would be collected at the post office, then the Talbot truck would arrive at the hangar with the pilot and perhaps a passenger. Usually the pilot would ride in the back of the truck, his legs dangling over the edge so the passenger could have the more comfortable ride next to the driver.[50]

One of Fysh's comrades from No. 1 Squadron in Palestine, the tall and willowy Fred Haig,[51] was hired to replace Huxley and arrived the day before Back's exit. This meant that Fysh wouldn't have to do all the flying himself – Haig would certainly not be put off by tough conditions in Longreach. In 1918, when he was landing behind enemy lines in an attempt to fly out comrades who had been shot down near Amman, the overloaded undercarriage on Haig's Bristol Fighter collapsed. He was captured by the Turks, then made a prisoner of the Germans for the last six months of the war. But although Haig had the heart and stamina for flying in western Queensland, the equipment was wearing thin.

Newspapers alerted Australians to the problems Qantas was facing:

> … everybody who knows anything at all about aeronautics realises that in the dead air of the tropics a high-powered machine is essential; first, to get off the ground; secondly, to attain the cooler altitudes before the water in the radiator boils away … when the mail plane took off at Charleville for the northern run to Longreach on January 18, the early morning 'shade' temperature in the hangar stood at 108 degrees [42 Celsius]. As an almost inevitable result (owing, of course, to overheating and lack of climb) it arrived several hours late after a prolonged wayside halt for the 'cool of the evening.' On the following day, by way of contrast, a cool southerly sprang up, and blew the same machine from Longreach to Cloncurry at a speed of 108mph, the 320 miles being covered in 2 hours 55 minutes.[52]

In July 1923, George Matthews – whose attempt on the England-to-Australia race had ended in a Balinese banana plantation – became a Qantas pilot, relieving Fysh of his flying duties so he could concentrate more on the business. Matthews was reunited in Longreach with the mechanic Frank McNally, who had served with different squadrons of the Royal Flying Corps in England and France and was a specialist on the Rolls-Royce engine in the DH.4.[53] While McNally was with the Royal Air Force in Persia after the war, Matthews and Tom Kay had arrived in their Sopwith Wallaby while trying for the £10,000 prize. McNally replaced a broken valve spring in the Wallaby's Rolls-Royce Eagle VIII engine. Kay gave him 'a bright picture of Australia', and McNally decided to check the place out for himself.[54] McNally and Matthews had worked for Jimmy Larkin at Essendon, and the repaired Sopwith Wallaby later flew for Lasco on its Sydney–Adelaide route.

When Matthews had to make a forced landing after an engine failure in one of the Qantas F.K.8s,[55] near Blackall on 13 September, the machine suffered major damage to the undercarriage, wing and propeller. With other machines out of action, the repaired Bristol Fighter was the only aircraft in the main fleet able to fly from Charleville. In 46-degree heat, Fysh was forced once again to fly the Longreach–Cloncurry run in the old Avro, with his only passengers being two four-gallon (eighteen-litre) tins of petrol. For most of the flight he kept the aircraft at just 500 feet (160 metres) because of the thin air higher up. He wouldn't ask Haig or Matthews to fly the old plane, although he logged more than forty hours in it over the next two months.[56]

FYSH'S WEDDING TO NELL DOVE was set for the historic St James' Church in Sydney on 5 December 1923. But while romance was in the air, his career faced a rocky, uncertain future. There was optimistic talk of extending the Qantas route to Camooweal and to the new mining settlement of Mount Isa, but by 29 September 1923, at the third annual general meeting of the company, the vice-chairman, Dr Hope Michod, formally referred to the possible liquidation of the company, citing three years of losses and what Fysh admitted had become 'a struggle for existence'.

Fysh told McMaster, 'We have been generally running ourselves into a hole because of second-hand plant and lack of spares.'[57] Longreach was in the middle of an often desolate landscape, and the pilots were concerned about the poor quality of the ageing machines – and, far from their extended families, they longed for less stressful and draining jobs in a more pleasant climate.

Fred Haig was a hardy type, but his wife Phyllis hated the outback. Haig told Fysh he would be taking a better-paying job with the Vacuum Oil Company, and would be gone by January 1924. Fysh noted that Matthews was 'ageing and [would] not stand up to another year of flying'.[58] Qantas hired a new pilot, Captain Arthur Vigers,[59] a British fighter ace credited with shooting down fourteen enemy aircraft when he and his Sopwith Dolphin had formed an unbeatable team. Vigers had served in France with Jimmy Larkin and then worked for him in Melbourne, before surveying an aerial route from Adelaide to Sydney via Melbourne for the Defence Department. Fysh thought Vigers was from the Wigglesworth school of eccentricity, and later recalled that the pilot would bristle whenever he had to take passengers on the mail run because it prevented him flying close to the ground to chase kangaroos.[60]

McMaster stayed on the board but relinquished his position as chairman, blaming 'added pastoral interests and unfavourable seasonal conditions'. He was exploring other business ventures that included a factory in Sydney producing dried vegetables – a business that failed. Dr Hope Michod was elected the new Qantas chairman on 17 November, just after the two Airco DH.9Cs arrived in Longreach and passed Fysh's flight tests, though he noted that the laminated three-ply wood used on the aircraft 'would require renewing at an early date'.[61]

After a year of subsidised operation and constant obstacles, Fysh could report that despite using mostly temporary machines 'of 1915–16 vintage' when they started their regular services,[62] Qantas had carried 156 passengers and just over 1000 pounds (450 kilograms) of freight. While there had been a few accidents and delays, in 71,108 miles of flying there had been no serious injuries to passengers or crew. Brinsmead assured Fysh that the

Charleville–Cloncurry contract would be renewed for a further year at the same subsidy of four shillings a mile – but the situation remained precarious. In the first full year of Qantas operations, the company had received £12,000 from the federal government and carried only 12,985 letters. That meant each letter cost Australia nearly £1 in return for an airmail stamp that cost the sender three pennies. Those sort of sums raised questions in Parliament, despite Fysh maintaining that the service enhanced Australia's capacity for defence and opened air routes for the future.[63]

From 29 November 1923, Fysh was granted six weeks' vacation. He was due to be married a week later, and would spend the rest of the time honeymooning and setting up a new home in Longreach with his wife. The Qantas board increased his salary to £600 a year and gave him a £100 Christmas bonus. But he was confident that marriage to Nell would take him even higher than Qantas ever had.

Chapter 16

She was true grit. She made a great contribution towards Qantas because she helped my father provide some of the early help and moral support to the staff out there who were finding it difficult and lonely.

JOHN FYSH, HUDSON'S SON, TALKING ABOUT HIS MOTHER NELL[1]

THE WEDDING of Miss Elizabeth Eleanor 'Nell' Dove, only daughter of Mr and Mrs J. C. Dove, of Auckland, New Zealand, and Mr Hudson Fysh, eldest son of the late Mr F. W. Fysh, of Ketteringham, Tasmania, and Mrs Fysh, of Melbourne, was celebrated at Sydney's St James' Church on Wednesday, 5 December 1923.[2] The Reverend P. A. Micklem officiated. Jack Butler, Fysh's pilot pal from Tasmania, was the best man though he was in poor health,[3] while Nell was glowing in a gown of white georgette and silver brocade, with a bouquet of lilies and pink and white carnations in her hand. The reception was at Fysh's usual Sydney digs, the Hotel Metropole.[4]

The happy couple had a brief honeymoon motoring around Tasmania. Fysh spent a large part of the time behind the wheel warning Nell about the hardships ahead in Longreach, where they would soon be living.

It took three days of rattling on the train from Sydney for the couple to get to their new home. They made a brief stopover on 3 January 1924 in Brisbane, where Fysh was interviewed about the prospects of his airline after a year of government support. Qantas was 'giving every satisfaction', he said, 'and working most efficiently'. The two de Havilland 9C machines recently arrived 'were a great improvement on the earlier planes', he claimed, and demand outstripped availability on the flights. Qantas hoped to extend its service to Urandangi and Camooweal before the end of the year, while customers sending gifts over Christmas had been happy to pay a little extra for parcels to go by airmail.[5] The DH.9s were quieter and more comfortable than the other Qantas machines, and also much more mechanically reliable, Fysh explained – though he didn't tell the press that his airline had been on the verge of collapse before their arrival in November.[6]

Fysh and Nell had a sleeper cabin on the run from Rockhampton to Longreach. In their narrow bunks they woke out of a restless sleep covered in thick black coaldust from the cheap low-grade fuel used in the Queensland trains.

Nell's introduction to the Qantas headquarters was when the company's new secretary Doug Miller, sweating profusely in high collar and white starched shirt cuffs, met them at the station in his old Model T on a broiling Saturday, 5 January 1924. He loaded their luggage into the machine, crank-started it and chugged away. All around the car, whirlwinds of dust were sucking up dirt, stones and twigs as though whistling a warning to Nell, who had grown up among the lush green hills and azure bays of Auckland. Fysh had told his new bride that she wouldn't have to worry about the cold and rain of New Zealand in Longreach, but he hadn't told her the full story. Nell's family recalled that she probably would have been most at home at a garden party or ball in Victorian England rather than the dry heat of Queensland's outback, and her new surrounds were a culture shock.[7]

Fysh had moved out of his bachelor pad in the tin shed behind the Longreach Motor Works, and he and Nell set up their marital home on Kingfisher Street, in a small rented house with a galvanised-iron roof, bare wooden walls and a tin lean-to kitchen at the back. There was a galvanised-iron outhouse surrounded by flies in the bare yard. An iceman called daily to replenish the hessian cooler that was hanging in a breezy position to keep food fresh through evaporation.

There was also a waterbag swinging on the verandah. The bore water in Longreach erupted to the surface from 1100 metres down, but it was too salty to drink or use on vegetation, and because it came out of the ground at a scalding 70 degrees, it had to be cooled even for showering. Drinking water came from tanks or was carted from the muddy Thomson River and strained. While there was a small dairy in West Longreach, almost all the fruit and vegetables for the townsfolk had to be railed in from Rockhampton, seven hundred kilometres away on the coast. There were few gardens in the town, and the sparse trees that survived the heat invariably fell victim to goats that jumped fences to eat washing off lines, and paper hoardings in town.

The iron roof usually generated soul-destroying heat inside the house, but on the afternoon after the Fyshs arrived there, a violent thunderstorm struck. Nell ran around to the back of the house where rainwater was gushing from the roof into a galvanised iron tank, and she jumped under the torrent as though it was a tropical waterfall. She spent much of her first night in Longreach in the little home's water cooling tank. She told Fysh she would never forgive him for taking her there, though she soon did.[8]

Nell never complained about the privations, though, and became the rock on which her husband could build a future. He was still something of an introvert who avoided social gatherings, but the couple would watch silent movies in an open-air theatre that employed a town crier to march down Eagle Street announcing each night's offering. Dinners at the Imperial Hotel were delicious, and picnics on the Thomson were a delight, as was the swimming – even if the freshwater crustaceans nipped now and then. There were regular tennis parties at the local stations, and then there were the annual race balls; these were highlighted not by the charging thoroughbreds but by the belly-butting competition between two well-fed locals who charged at each other, stomach first, from opposite ends of the Shire Hall floor. There were also sightseeing trips for Nell as a passenger on the DH.9 to Winton and Cloncurry. She improved Fysh's dress sense almost immediately, buying him well cut suits and making sure he was always turned out smartly.

Compared to most people in Longreach, Fysh was well off; he could even hire a housekeeper to come in during the day.[9] However, at times he had to wonder how long it would be before Qantas shut up shop and its hangar was sold off for a woolshed. A recent incident had increased his anxiety. While the Fyshs were on their honeymoon, Captain Vigers had crashed the Bristol near the Longreach rail line right after take-off. He had just finished telling Fred Haig that he didn't need his advice on how to fly the damn thing. Vigers was sacked not long after, when he looped the loop in a DH.9 much to the horror of two elderly ladies riding with him.[10]

Qantas remained in business, though, and soon, with Fysh now on £600 a year, he and Nell saved enough money to buy their own house in Emu Street. It cost £1000, with Qantas lending Fysh three-quarters of the money at seven per cent interest. But Fysh was thinking beyond providing for his family. When it came to rebuilding the Bristol Fighter after its crash, he oversaw Baird converting it into an air ambulance by adapting the passenger cockpit to hold a stretcher and painting red crosses on the side. The machine, however, wasn't suitable for rough bush landings, and before long Fysh was eyeing a replacement.[11]

HOPING TO EXPAND THEIR OPERATIONS, the Qantas board voted on 2 February 1924 to order one of the new de Havilland DH.50s, the first aircraft the company had purchased that was specifically built for civil work. The large biplane could carry four passengers in an enclosed cabin, and its 230-horsepower Siddeley Puma engine could push it along at 109 miles per hour (175 kilometres per hour).

While drought during 1923 had caused heavy losses for local graziers, the mines at Mount Isa promised a huge windfall for investors and thousands of jobs for workers who all needed transport. In mid-February 1924, the drought finally broke with a deafening deluge that turned much of the area around the Thomson River into a lake, covered the long bridge and turned many of the western roads into quagmires. Fysh had never been busier doing taxi work in the B.E.2e. He took owners on stock inspections, flew in supplies to homesteads and took train passengers from distant properties to make their rail connections. He flew in thirty-six shearers to a station marooned by floodwaters, then flew in bottles of beer and spirits for the West Longreach pub, after those shearers had drunk it dry.[12] One of his passengers was perched on a small keg of rum.[13]

Fysh often ferried Dr Hope Michod on medical trips to isolated stations, as a forerunner to John Flynn's idea of 'flying doctors'. On 21 February 1924, Fysh made an urgent flight alone to Corona Station, ninety-eight kilometres from Longreach, in order to collect the heavily pregnant Lorna Armstrong, the wife of the station overseer Tom Armstrong. His pregnant first wife had died in 1922 when floodwaters prevented medical help from reaching her. Fysh landed on the roadway at Corona, kept the engine running while Lorna was lifted into the front passenger compartment, and took off for the Longreach hospital. He was back there within an hour, and Hope Michod took over. John Cay Armstrong was born at the Longreach hospital on 13 March. As a token of his gratitude, Tom Armstrong gave Fysh's brother, Graham, a job on the property. Graham had trained as a bookkeeper but had endured a difficult relationship with his parents. He had already left Tasmania for work on Queensland's Darling Downs.[14]

Two months later, Fysh flew Hope Michod and Arthur Baird to Winton in the DH.4 so the Qantas chairman could perform emergency surgery on a lorry driver after an accident.[15]

But flying still wasn't for everyone. And some passengers refused to get on the Qantas flights unless their favourite pilot was at the controls. Once at Camooweal, a passenger 'in a very alarmed state of mind' made out a last-minute will before climbing into the machine with Fysh, who could only wryly admit: 'Apparently he did not like the look of me.'[16]

Whenever Fysh had to fill in on the Longreach–Cloncurry mail run, Nell would be up at 4 a.m. making him breakfast, cutting sandwiches for his lunch and giving him a flask of tea. Then she would nervously await his return, especially if storms were about. Even on the ground there were hazards. Fysh was at

Winton aerodrome once, trying to peg down the DH.9 as a storm approached, when the wind turned the aircraft almost vertical. Its pilot had to hang on to one of the wings. His hat was blown off and carried so far that he never found it, although the earth was as 'bare as a billiard ball' for miles around.[17] The same unpredictable winds occasionally deposited fish and shells, sucked up by wind vortexes from the river, into their neighbourhood.

PAUL McGINNESS WAS OF a mind to marry and settle down, too. On his travels selling shares, he had fallen in love with the rolling hills of the wheat and sheep country north-east of Perth – and with Dorothy Baxter, the daughter of a prominent West Australian Country Party politician. They were married in St Mary's Catholic Cathedral in Perth on 24 September 1924. McGinness immediately transported his bride into a world of hardship at Morawa, 380 kilometres to the north-east, where McGinness had qualified for 2500 acres (1000 hectares) at 2 shillings 6 pence an acre under a scheme to help returned servicemen.

The couple drove from Perth in McGinness's Armstrong Siddeley car, staying at Perenjori on the way in a hotel with blood-spattered walls. Dorothy lived at a Morawa boarding house for some time and saw McGinness only on weekends, while he stayed in a tent on the thousand hectares he was clearing and fencing, and which he named Mount Morawa farm. He also erected windmills and made bricks for the home he had planned. He bought six draught horses, an old Chevy truck and two thousand sheep. Dorothy later helped him with the horrors of farm life, pouring petrol down snake holes to burn them alive, culling kangaroos and poisoning emus. When McGinness needed to take out a mortgage on the property, he used his 1450 Qantas shares as collateral.[18]

Qantas had long recovered from McGinness's departure. In 1924 the company hired two new pilots, and these ones were keepers who became foundation stones for the airline. P. H. 'Skipper' Moody,[19] another Gallipoli veteran, joined Qantas on 20 April 1924 after George Matthews went back to Lasco to fly its new Sydney–Adelaide route. Moody was to stay with the airline for six years, and for a time he made his home at the Commercial Hotel in Longreach. When he left Qantas, it was to run a sheep station outside Charleville, and then to purchase slices of paradise in North Queensland – he bought two islands, Hamilton and Daydream.[20] Lester Brain,[21] who succeeded Arthur Vigers, impressed the Qantas directors as soon as he arrived on 15 May, and Fysh immediately thought him a 'sensible sort of fellow'. The twenty-one year old had just graduated top of his class from the RAAF flying school at Point Cook; he was the first pilot employed by Qantas who hadn't honed his skills in the war. He would spend more than two decades with the company.

Now that there were new airmen on the payroll, Fysh agreed to the Civil Aviation Branch's request for Qantas to fly the Charleville-Cloncurry run in one day, eliminating the overnight stop in Longreach. At the same time, Brinsmead responded favourably to their request for an extension of the airmail route to Camooweal. With a marked jump in passenger numbers, Fysh organised to have the dirt floor of the Longreach hangar concreted, and Baird cast a concrete roller for the gravel runway.

In August 1924, Brinsmead arrived in Longreach as part of an around-Australia flight on a DH.50 that also carried Ernest Jones and Bob Buchanan,[22] now recovered from their crash in Bourke. Brinsmead was inspecting Fysh's proposed route for the Mount Isa–Camooweal extension, and the extension of Brearley's West Australian route between Broome and Wyndham.[23] Fysh was also keeping an eye on the extension to routes in Europe, envisioning the time when Qantas would cross countries in the manner of the newly formed British enterprise Imperial Airways Limited, a merger of Handley Page, Instone, Daimler, and British Marine Air Navigation. In September 1924, Imperial opened another link in their London–Continental air services, enabling travellers from the English capital to reach Zurich in a day. The service would soon be extended to Rome.[24]

The arrival in October 1924 of the first Qantas DH.50, registered as G-AUER, marked the end of the era of cap-and-goggles for passengers. While the pilot still flew in an open cockpit, four passengers could enjoy the comfort of an enclosed

Soldiers line up at a delousing parade on Gallipoli. Fysh is at the far left, with the water bottle. *State Library of NSW FL8576694*

Fysh (right) with fellow Tasmanian Captain Sydney Addison, standing in front of a Bristol F2B Fighter during their time with the Australian Flying Corp. Captain Addison commanded No 1 Squadron from 28 June 1918. *Australian War Memorial B02182*

Fysh and Paul McGinness after returning from the Great War. *Qantas Founders Museum*

George Gorham (left) and Paul McGinness with dinner during the great trek with Fysh from Longreach to Katherine in 1919. *State Library of NSW FL8648695*

Certificate of Incorporation

OF

The Queensland and Northern Territory Aerial Services Limited.

I, FRANCIS SYDNEY KENNEDY, Registrar of Joint Stock Companies, HEREBY CERTIFY that "THE QUEENSLAND AND NORTHERN TERRITORY AERIAL SERVICES LIMITED" was on the sixteenth day of November, one thousand nine hundred and twenty duly registered and incorporated under the provisions of "The Companies Acts 1863 to 1913" as a Company limited by Shares and Numbered 146 of 1920.

Given under my hand and Seal at Brisbane, this sixteenth day of November, one thousand nine hundred and twenty.

F. S. KENNEDY,
Registrar of Joint Stock Companies.

CANNAN & PETERSON,
Solicitors,
Queen Street, Brisbane.
And at Longreach.

Qantas's Certificate of Incorporation was signed on 16 November 1920.
Qantas Founders Museum

Fysh (left) with engineer Arthur Baird, passenger Alexander Kennedy, the first Qantas chairman Fergus McMaster and a crowd of onlookers after the first mail flight from Longreach to Winton on 3 November 1922. The flight continued to McKinlay and Cloncurry. *State Library of NSW FL8576694.*

The de Havilland DH.4 was Fysh's favourite aircraft in the early days at Longreach. *State Library of NSW FL8581236.*

Fysh and his new wife Nell beside the Qantas hangar in Longreach not long after their wedding in Sydney in 1923. Raised in New Zealand, Nell found the western Queensland heat oppressive but spent 10 years in the town as Fysh built Qantas into an international business. *State Library of NSW FL8633851*

Flooded roads between Winton and Longreach forced Prime Minister and Gallipoli veteran Stanley Bruce to fly with Fysh on a DH.50 on 5 October 1925. The Prime Minister's party, from left, Donald Cameron, Skipper Moody, who flew the luggage on a DH.9, Fysh, the new Qantas chairman Dr Hope Michod, and Ethel and Stanley Bruce. *State Library of NSW FL8585842*

Nell and the children, John and Wendy, along with Qantas manager George Harman welcome Fysh home in November 1933 after his historic flight aboard the *Astraea* and his visit to the British and American aircraft factories. *State Library of NSW FL8576709*

By his mid-30s Fysh was no longer a shy and awkward young man but an astute businessman who was charting a profitable course for his airline. *State Library of NSW FL8634110*

John and Wendy grew up in a happy home in Brisbane and often accompanied Fysh to work at his office and at the Archerfield Aerodrome. *State Library of NSW FL8633999, FL8633978*

The great silent film star Charlie Chaplin with his wife, actor Paulette Goddard, her mother Alta Mae and Qantas captain Russell Tapp in Batavia. Captain Tapp's grandson, James, became a Qantas Boeing 737 First Officer. *Qantas Heritage Collection*

The Qantas Empire flying boats offered passengers new levels of comfort and space on their travels to London via Singapore, with a smoking lounge and promenade deck that offered spectacular scenery. *Qantas Heritage Collection*

Fysh snapped this photo after the Singapore III flying boat moored on a wide expanse of the Roper River in the Northern Territory in 1936. Some of the braver occupants made it ashore across crocodile-infested waters. *State Library of NSW FL7552794*

Qantas House became a landmark building in Hunter Street, Sydney after it was completed in 1957. It was a far cry from the first Qantas offices in Winton and Longreach. *State Library of NSW FL8644409*

The Boeing 707 brought a new age of air travel to Australia. Qantas Heritage Collection

Fysh posed beside a replica of the original Qantas Avro 504K and the airline's newest acquisition, the massive Boeing 747 Jumbo jet, demonstrating the great strides Qantas and aviation had made in half a century. *State Library of NSW FL8577444*

cabin, and the aircraft had better all-round performance than the DH.9. Fysh tested it in early October and declared it 'a very great advance on anything we have from almost any point of view'.[25] The DH.50 was so easy to fly, Fysh admitted, that it could be dangerous. Like other pilots, he sometimes had trouble staying awake on long flights over a route he flew constantly. On some days the drone of the engine, the rush of wind around the pilot's screen, and the warm sun on his face could act like a lullaby. Once he had to keep slapping his face to stay awake on the mail run, and another time he stuck a pin hard into his arm to keep his eyes open.

The introduction of the DH.50 was timely. Prime Minister Stanley Bruce, who had succeeded Billy Hughes for the office, was in Winton on 31 October during a tour of western Queensland. Bruce had arrived by train from Townsville and was planning to drive with his wife Ethel and entourage, including McMaster's old friend Donald Cameron, to Longreach, when a thunderstorm and torrential rain cut the road. Fysh was called on to fly the nation's leader and his main party in the DH.50, while Moody flew the luggage and the rest of the group in the DH.9. Bruce was reluctant at first, but Ethel and Cameron convinced him to climb into the machine. Fysh and Bruce had much to talk about, as Bruce had been wounded twice on Gallipoli and received the Military Cross.

On his arrival back at Parliament House in Melbourne, Bruce wrote to E. J. Hart, the editor of *Aircraft* magazine, to say: 'The flight I have just completed from Winton to Longreach has further impressed upon me the great importance of aerial transport to a country such as Australia with its enormous areas and great distances ... In the past, progress in aviation has been retarded to a great extent by the apprehension of danger with regard to it but such advances have been recently made that today to travel in a modern aeroplane is practically as safe as travelling in a motor car.'[26] With the success of the DH.50, Qantas now entered a ten-year association with de Havilland, and the British company helped Fysh and Qantas turn a corner. After three years of losses, Qantas returned a profit each year for decades to come.

In October 1924, Fysh's salary was raised to £750 per year. But he wasn't afraid to get his hands dirty. When the extension of the mail route to Camooweal was agreed, money from the government was still tight; rather than build a hangar at the end of the line, Fysh organised a cheap lean-to instead. Qantas had to meet the Federal Treasurer Earle Page halfway when it came to building aerodromes at Duchess, Mount Isa and Camooweal, and Fysh went on another overland trek, in another Model T, to survey the new route solo. He marked places for emergency landing grounds, and cut down stunted trees here and there to clear the way for aeroplanes. Somewhere in the inky darkness between Mount Isa and Duchess, he decided to bed down for the night in his swag. The lights on the car were still on, and they illuminated a huge dingo nearby. Fysh grabbed his rifle, but the dingo disappeared, only to reappear with its eyes glistening in the light when he'd put the gun down. The frightening dance repeated again and again as sweat pooled on Fysh's brow. He stood beside his swag with his gun, looking all about and behind him to see where the animal had gone. A curlew cried as though warning Fysh to get away. He took the hint and packed up as quickly as he could, then raced the fifty kilometres to Duchess, banging on a hotel door to wake the publican.

Arthur Baird's annual salary had risen to £500, and he was awarded a £50 Christmas bonus to end 1924 as a reflection of the sterling work he and his team were doing. Forced landings had become much less frequent, but Baird and his men in the Talbot truck were still always at the ready to drive out to distant claypans where they overhauled and sometimes replaced engines. Baird's fox terrier, Kelly, rode with him on the bonnet of the truck, and his black cat slept in the fuselage of Fysh's B.E.2e, near the elderly nightwatchman Jim Eason,[27] nicknamed 'Wobbly Tooth' by Nell Fysh on account of his loose front fang. Eason made his bed in an aircraft packing case at the rear of the hangar, and after a few drinks he would recall his role in the British naval bombardment of Alexandria in 1882. The luck of Baird's black cat ran out when he overslept in the aircraft's cockpit and awoke with a start while it was being flown over

Longreach. The cat leapt out of the side hoping to get back to familiar surroundings.

Baird's work was of such a high quality that in January 1925, after discussions with Brinsmead and the de Havilland company, Qantas was given permission to build its own DH.50s at Longreach, under licence. They would pay a royalty of £100 per aircraft, later reduced to £50. In having his own machines built, Fysh was now emulating the grandfather he so admired, Henry Reed, whose ships had been constructed in Tasmania to support his trading business.

The decision allowed Qantas to keep its mechanical staff in constant work – men such as Baird, McNally and Hazlitt – and to have the aircraft exactly to their specifications. Modifications for conditions in western Queensland would include bigger radiators, header tanks to conserve the steam and save water, and double wheels on extended axles for operating on muddy runways.[28] And so, one of the country's biggest aircraft factories was established in the remote centre of Queensland.

LESTER BRAIN AND McNALLY set off from Cloncurry for Camooweal in the DH.4 at 6 a.m. on 7 February 1925. This was the first official flight of their 400-kilometre route extension. They had one passenger, Charles Johnston,[29] who was the manager of the vast Alexandria cattle station, and a cargo of ice and butter, with the ice proving a big hit when they touched down in a town without refrigeration; ice would become a regular freight item. The route extension helped Qantas turn a profit of £2847 for the year to 30 June 1925, and their deal with the government was extended to a three-year contract rather than a year-to-year deal.

From Cloncurry, Brain carried out many aerial taxi jobs between mail assignments,[30] including a pair of long flights across northern Australia for the American mining engineer L. J. Stark. On one of these trips, they flew to Banka Banka, in the Northern Territory, where they met Indigenous people living as they had done for thousands of years. The landing was rough, and many of the local people helped Brain clear a runway so he could take off.

Brinsmead warned Fysh that the federal government did not agree to extending the route further south from Charleville to

Brisbane or north from Camooweal to Darwin. But Fysh was certain this would come, eventually. His opinion was based partly on what was happening in European aviation circles, and the way Imperial Airways was conquering distances across Europe.

In Italy, Marquis Francesco de Pinedo, a 35-year-old nobleman, was paving the way for more global highways and had embarked on the greatest aerial journey ever attempted. On 20 April 1925, he and his mechanic Ernesto Campanelli took off in a Savoia S.16 flying boat with a monstrous 450-horsepower engine and pusher propeller from Sesto Calende in northern Italy. The two men were bound for the Great Southern Land in a round trip that would cover almost fifty-five thousand kilometres. When they touched down in Broome on 31 May 1925, having covered the first 16,900 kilometres of their adventure, they were the first aviators to reach Australia from Europe since Parer and McIntosh five years before.[31] In July and August 1925 de Pinedo and Campanelli visited Sydney, Brisbane, Rockhampton, Townsville, Innisfail, Cooktown and Thursday Island. Then they flew to New Guinea, the Philippines and Tokyo. On 7 November 1925, to a fanfare worthy of Caesar's homecoming, they landed on the Tiber in Rome.

Fysh found all of this very exciting – he was even more excited, though, by the fact that Nell was pregnant with their first child. He and Nell were determined to stay in Longreach, regardless of the isolation and hardships, so that she could have the baby at the local hospital, where the Qantas chairman, Dr Hope Michod, just happened to be in charge. In late 1925, Fysh became a father when Nell, ably assisted by Hope Michod, delivered their son John Hudson Fysh at the Longreach hospital.

Soon after Fysh, Nell and baby John visited Melbourne to see Fysh's mother and sisters Peggie and Geraldine. Nell and Mary Reed did not gel, though. Nell found the old woman grim, foreboding, and overly pious.[32]

Back in Longreach, Fysh's wife and son were his bulwarks against the relentless pressures of trying to make his airline grow, and life was tough for the young family in their remote home. Hot bore-water showers remained the main way to wash, but even though fresh water was scarce, in the hottest weather Nell would watch on as John and some of the local children paddled about in big open tanks at the back of the house. Rain was so rare that when John was about three he came running into the house crying when water began falling from the sky and made a clanging sound on the iron roof – he couldn't remember seeing such a phenomenon before. Fysh maintained his love for photography, documenting all the happy days he enjoyed with his wife and son despite their hardships and isolation.

McGinness was also embracing fatherhood. On 29 December 1925, at Nurse Harvey's Hospital in Bulwer Street, Perth, he and Dorothy welcomed their first child, a daughter Dorothy Veronica, they called 'little Dot'.[33] There was friction under the happy facade, though, and their family life would soon start to fracture. McGinness had great ideas, but maintaining relationships was not a strong suit.

In contrast, with his quiet, modest ways and methodical approach to life, Fysh had become a master at winning friends and influencing people – especially those in high places.

FYSH TRAVELLED TO Melbourne in March 1926, as Donald Cameron brought together all the Queensland members of the Federal Parliament in a deputation to the Minister for Defence, Sir Neville Howse.[34] As a doctor for the NSW Medical Corps in the Boer War, Howse had earned a Victoria Cross for rushing out under a heavy crossfire to pick up a wounded man and carry him to shelter.[35] Fysh regarded the war hero as 'a delightful man' but thought it extraordinary that Australia's defence minister wanted to spend nothing on aerial defence. Fysh told Howse that not only would an extension of the aerial route aid in protecting Australia's interior, but that it would also cut thirty hours off the 42-hour train journey from Brisbane to Longreach and reduce the government's Qantas subsidy. The airline would make Brisbane its base and provide a workshop for the overhaul of aeroplanes. But Howse 'openly expressed his opposition to, and disbelief in all forms of aviation, both military and civil', Fysh recalled, and in case the airman still didn't get the point, the minister told him that subsidies for aerial routes were 'a complete waste of money'.[36]

Howse thwarted Brinsmead in his plans too, and the lieutenant-colonel later wrote to Fysh, 'If it were not for Qantas and the Aero Clubs I would resign immediately. It has been a continual and nerve-racking fight to get anything done ... I can't thank you sufficiently for your unswerving support and straight dealing, even when we do not quite see eye to eye on certain matters.'[37]

They certainly did not see eye to eye on the extension of the Qantas route. Fysh maintained that it should go to Brisbane, but Brinsmead was constantly pushing for the airline to meet the NSW railhead at Bourke. Fysh retained his own clear vision of Qantas's future, and he found an important ally in Australia's governor-general, John Baird, Lord Stonehaven,[38] who in December 1916 had been appointed Britain's Parliamentary Secretary to the Air Board in Prime Minister David Lloyd George's coalition government.

On 28 July 1926, Fysh flew Lord and Lady Stonehaven and their hosts Fergus and Edna McMaster from Winton to Longreach during a vice-regal tour of the west, passing over Evesham station to watch sheep feeding on hay bales during drought. Stonehaven described the eighty-minute journey as 'very pleasant'.[39] With that sort of vice-regal endorsement, and after Qantas returned a bumper profit of £6370 for the 1926 financial year, Fysh pushed even harder for expansion – not just for the company but for his own mind as well.

Embarrassed by his lack of formal education, Fysh began studying at night. He spent his days managing the peculiarities of his pilots: both the hardened veterans and the eager youngsters such as Lester Brain. Fysh insisted they observe his timetables, and be diligent with trip reports and logbook compliance, and ordered them to be especially careful with passenger comfort by doing their best to cruise in the altitudes of least turbulence. At night after dinner with Nell and baby John, Fysh would read voraciously, and he took correspondence courses in economics and Pelmanism, a system of mental training to develop his memory and intellect. He began to infuse his staff with the power of positive thinking, sending them notes with messages such as 'Are you known as a man who has practical ideas?' and advice on how 'pep, progressiveness, loyalty, and a willingness to carry out instructions were the secrets to secure advancement in life'.[40]

Fysh's long-range vision for Qantas received more positive reinforcement a year after de Pinedo's visit to Australia. On 10 August, Alan Cobham, de Havilland's chief test pilot, landed in Longreach at 2.05 p.m. Forty-two days had passed since he had set off on floats with his mechanic Arthur Elliott in a modified DH.50 from the River Medway, at Rochester near London. Cobham, who was Bert Hinkler's constant rival in all the British air races, had replaced his machine's Siddeley Puma engine with a more powerful fourteen-cylinder Siddeley Jaguar that upped the output to 385 horsepower.

Cobham's DH.50 was fitted with floats because he believed that the facilities for land planes were totally inadequate between Calcutta and Darwin. However, a lack of suitable landing fields was really the least of his concerns while flying in stifling heat across the Persian Gulf from Baghdad to Bushire (now Bushehr) on 6 July. Cobham, in the open cockpit, took off from the Tigris River with Elliott inside the four-berth passenger cabin. They were about 190 kilometres from Basra, flying low because of a dust storm, when over the roar of the engine Cobham heard an explosion. He looked back through the cabin porthole and saw Elliott's face ashen and blood spurting from his arm and chest. A bullet from a Bedouin rifle had torn through the fabric of the fuselage, gone through a parcel of photographs – bound for Australian newspapers – of the drawn Second Ashes cricket Test from Lord's, burst a petrol pipe, ripped through Elliott's arm, broken one of his ribs, blown apart his left lung and lodged in his back.

Cobham was aghast. With the dust storm raging, the DH.50 leaking petrol and Elliott gushing blood, Cobham flogged every one of the 177 kilometres per hour he could get out of the aircraft until they reached the British air base at Basra. He beached his machine on the Shatt al-Arab waterway. As he screamed for help, some locals waded into the water to put his wounded mate onto an improvised stretcher. Elliott underwent surgery but died the following night.[41] The shattered Cobham wanted to fly home immediately, but Sergeant Alan Ward, an RAF engineer stationed at Basra, said he could handle the DH.50's big engine.

On the beach in Darwin, Cobham had switched his aircraft's floats with wheels for the flight through Australia. He would use the route surveyed by Fysh and McGinness. From midday on 10 August, much of Longreach's population was at the Qantas aerodrome waiting with breathless anticipation after a telegram arrived from Cloncurry stating that the English aviators had left there at 10.35 a.m. Fysh ensured there was a large supply of Shell fuel and oil for Cobham's machine, which came in so low to the aerodrome that Fysh said it could have been mistaken for a speeding car. The big crowd surged around the plane, cameras clicked, and as the beaming aviator stepped from the cockpit, three ringing cheers broke out.

Fysh ushered Cobham into quiet seclusion, where together they studied a series of maps, Cobham's focus was so much on the journey and conditions ahead that he paid little attention to the wad of congratulatory telegrams and letters that Fysh handed him.[42] Cobham and Ward were given a hearty lunch followed by pineapple slices and ginger ale, as Fysh discussed with them the merits of various aircraft and engines. Cobham told Fysh he had made a great decision buying the DH.50s for Qantas. 'That Jaguar engine of mine is a wonder,' he said. 'It stood the test when we were travelling through the Sudan at 118 degrees in the shade.'[43]

After just fifty-five minutes on the ground in Longreach, and with his mechanic and machine both replenished, Cobham fired up the Jaguar engine and headed towards Charleville. It was his first stop on the way to his final destination in Australia, Essendon, where 150,000 people – said to be the biggest crowd Melbourne had ever seen – knocked over the forty policemen hoping to maintain order and charged onto the landing ground to greet the DH.50. Many barely avoided the whizzing propeller.[44]

Eight days after Cobham had left Longreach, Lord and Lady Stonehaven were back in town as part of a two-month journey through the remotest parts of Australia. Much of it was spent with

'Skipper' Moody, who flew them the two thousand kilometres from Morven, Queensland, to Newcastle Waters.[45] In Longreach, Lady Stonehaven formally christened the first Qantas-built DH.50A[46] with a bottle of Australian champagne. Baird and his team had constructed the machine from Queensland maple and three-ply wood, using a chart and blueprints supplied by the de Havilland company. Hope Michod, who was about to hand back the chairmanship of Qantas to McMaster, told the crowd at the aerodrome that the DH.50A was the first large plane to be built in Australia, and that it would bear the name '*Iris* – a Messenger of the Gods, and the Personification of the Rainbow'. He said the confidence of the vice-regal party in Qantas planes was such that they were about to continue their onward journey on *Iris* to the Northern Territory.[47]

On 1 October 1926, Cobham, dressed in a suit and tie, brought his DH.50 down on the Thames near the Houses of Parliament, completing the first aerial return trip between England and Australia. Fysh was already thinking about how Qantas planes could eventually fly to London. He saw a Qantas link with Brisbane as crucial to his expansion plans, and in October 1926 the airline established an advisory board there consisting of Hope Michod, Alan Campbell and Fred Loxton, a director of the Burns Philp trading company.

Such was the success of Cobham's machine and its high-powered engine that Fysh and Baird decided to significantly up the power of the DH.50s they were building. Qantas turned out four with the standard Puma engines, while the last three, DH.50J models, were loaded with the new 450-horsepower Siddeley Jupiter engine, which delighted pilots with the rush they produced on take-off.

THROUGHOUT 1926, Fysh had been urging Qantas of the need to introduce newer and more economical light aircraft for the company's taxi work. At the same time, he talked up the potential Qantas had to teach flying to a new generation of aviators who could access the recently released two-seat DH.60 de Havilland Moths that were as affordable as motorcars. At the time there were no flying schools in the whole of Queensland, but with the support of the Queensland Aero Club, Fysh gained permission from Brinsmead to open teaching facilities at both Longreach and Brisbane. While in Melbourne negotiating with Brinsmead, Fysh had hired several new pilots who could also give flying lessons: two recruits from Point Cook, Alan Davidson[48] and Arthur Affleck,[49] as well as a graduate of aerial combat in France, Charles Matheson,[50] and the enigmatic Englishman C. W. A. (Charles) Scott.[51]

A big, glamorous, charismatic swashbuckling type, Scott had been born on Friday, 13 February 1903, and had been riding his luck ever since – working on a sugar plantation in South America, twice winning the RAF's heavyweight boxing title, and earning himself a devilish reputation as a ladies' man. A lover and a fighter, he was also an aerial acrobat. Urbane, charming, highly educated and erudite, he was the son of the musical genius who had founded what became known as the London Philharmonic Choir. Scott was a talented violinist and poet. He often visited the Fyshs and would bury his nose in the treasures of their small library. He baked extraordinary cakes for his landlady at Longreach and was always lending a hand when it came time for washing up at the Riley family's Camooweal Hotel – especially when the comely Riley daughters were around.[52] Fysh regarded Scott as a unique character, an enthusiastic hard worker, and 'a brilliant but over-volatile pilot … too brilliant to be stable'.[53]

Once on a wet runway at Longreach, Nell had to trot about in high heels to test out the firmness of the surface so Scott could take off. She often joked that Qantas used her as a guinea pig after she had flown with Lester Brain from Longreach to Winton in a tiny compartment with a lid over her head.

Charles Matheson became the main instructor at the Longreach aviation school when it opened on 27 December 1926, but the hot, gusty conditions made flying tricky, and frequent accidents – combined with a drought that crashed the local economy – forced it to close quickly. The Brisbane school opened at Eagle Farm in March 1927, with Lester Brain as the main instructor. It thrived using the first private hangar in the Queensland capital, though Brain was under instructions to look for another landing site because Eagle Farm became unusable in heavy rain.[54] Before long Brain had made a landing on Franklin Greiner's Farm in Brisbane's south-west, on a much drier area that would become Archerfield Aerodrome.[55]

Fysh had great faith in Brain despite him crashing just a few weeks before the school opened, his third crash in two years. Flying a DH.50,[56] he turned over when taking off from McKinlay on boggy ground. His female passenger was a popular employee at Cloncurry's Post Office Hotel where the Qantas pilots stayed, and she was carrying her pet cockatoo in a cage. The cage broke open in the crash, but while the bird flew away it eventually made its way back to Cloncurry, 110 kilometres away.[57] The aircraft was rebuilt and no one was seriously injured but Fysh told Brain he had 'made a serious error of judgement … in not taking sufficient care on soft ground', but he also told the Qantas board that while Brain was 'worthy of censure in this case',[58] he had for three years successfully flown on the most difficult sections of the route and done the most hazardous taxi trips.

Brain was assisted at the Brisbane flying school by Charles Scott and another new Qantas pilot, Russell Tapp.[59] Once on an assignment taking a tourist to capture photographs over Moreton Bay, Tapp had to make an emergency landing when the passenger found a snake in his cockpit and almost jumped out of the machine into the water hundreds of feet below.[60]

There was nothing light-hearted, though, in what happened when Alan Davidson came down in his DH.9[61] at Tambo on 24 March 1927. Only a month earlier, Fysh had written in the company's bulletin that Qantas had flown 449,008 miles without injury to personnel or passengers. He regarded Davidson, the son of a prominent Adelaide journalist, as a 'highly promising young man'.[62] He had been an RAAF flight instructor at Point Cook and had logged more than five hundred hours in the air. Moody flew him over the Charleville–Longreach route before Davidson was put in charge of the regular mail run. At 6 a.m. on 24 March he left Charleville with two passengers, 55-year-old Archibald Bell of Winton and Bill Donaldson, the forty-year-old manager of Rocklands station, near Camooweal. Donaldson was returning home full of life after having sold thirty thousand sheep.[63]

Somewhere between Charleville and Tambo, two hundred kilometres north, Davidson brought the machine down and spent thirty-five minutes on the ground. When he reached Tambo he appeared to be flying very low. A hundred metres short of the landing strip, the aircraft suddenly dived into the ground, 'smashed to matchwood'.[64] Davidson and his two passengers were killed. Fysh lamented that the pilot had apparently let his airspeed fall below the 'minimum necessary to keep his aircraft airborne',[65] with the cause of the crash almost certainly 'stalling on approach … the most common cause of fatal aircraft accidents …'[66]

The tragedy played havoc with Fysh's emotions for a long time, casting a pall over his family home. He blamed himself for the deaths and was sorrowful with grief about failures in pilot training. He insisted that the initiation of new pilots into western Queensland should immediately be made much more thorough. Still, he remained adamant that flying was as safe as riding in a motor car, and he had plenty of supporters in that opinion.

Two months after the Tambo crash, Charles Lindbergh – having honed his skills on the US Mail runs still known as 'the suicide club' for the danger of the work – completed the first solo flight from New York to Paris. He arrived to a $25,000 prize and the kind of celebrity fanfare rarely seen on the world stage. Civil aviation had never before enjoyed such a pedestal, and few endeavours by an individual had created such excitement around the world. Lindbergh's remarkable feat made the possibilities of civil aviation seem boundless.

On 1 July 1927, Fysh inaugurated a service from Cloncurry four hundred kilometres north to the remote cattle town of Normanton in the Gulf Country. Before long, Arthur Affleck was flying there with Lord Stonehaven, now a Qantas frequent flyer, on a hunting and fishing trip. The prime minister and his wife also flew with Fysh from Longreach to Hughenden, after Ethel Bruce had christened a Qantas DH.50 *Pegasus*. Soon after, in September 1927, Bruce announced that Federal Cabinet would increase funding of Australian civil aviation from £115,000 to

£200,000 because of 'the urgent necessity' of bringing it into line with developments overseas.[67]

Fysh saw the announcement as fuel for Qantas's further expansion. But even though he was the boss of operations, he was still flying taxi work to remote areas of Queensland, and he wasn't too proud to admit his limitations. He was not alone in having trouble starting up the Siddeley Puma engines, especially on cold winter mornings in towns such as Charleville and Mitchell, which were often covered in frost at dawn midyear. Once Fysh flew a group of punters to Blackall for the races, but at the aerodrome, ready for the return flight, he couldn't start his aircraft. He took the plugs out and ignited them with petrol to warm them up, then sat down for half an hour and tried again. After two hours, he couldn't get a single kick out of all that horsepower. Finally he had to send a telegram to Arthur Baird, asking him to fly over from Longreach, two hundred kilometres away. Baird arrived and started the engine first go – and Fysh felt a proper fool.[68]

Chapter 17

The call for help was carried 80 miles by an Aboriginal stockman splashing through flooded country, then relayed by station telephones. The pilot flew the 'Victory' to Normanton through low clouds and rain … Dr Welch decided to take the mother and baby back to Cloncurry and the baby recovered in hospital.

REPORT ON AN EARLY FLYING DOCTOR MISSION FLOWN BY QANTAS[1]

THE VISION FYSH HAD for aviation to shrink the entire globe for Australians was shared by many flyers of the time. They saw the coupling of courage with the latest scientific advances as the solution. Bert Hinkler was among the most intrepid and since the end of The Great War he had planned a solo flight from England to Australia. So it was that on 22 February 1928, after a breakfast of two bananas – the only food Hinkler trusted in Bima, on the Indonesian island of Sumbawa – the pint-sized Queenslander with the huge heart set off in his tiny Avro Avian biplane just before daybreak for a 1450-kilometre flight to Darwin.

It was the last leg of a titanic solo journey that had started at the Croydon aerodrome in South London fifteen days earlier, and that had seen Hinkler fly over European alps, through volcanic smoke, and across jungles and seas. Twice he had bedded down in the Sahara Desert because it was getting dark and he had no way of flying through the night. Now, with his exhaust pipe red-hot and gleaming, Hinkler headed out over the Indonesian rice paddies and then the Savu Sea, crossing Kupang on Timor at about 10.30 a.m. Then came five hours across the Timor Sea, the most desolate stretch of water Hinkler had seen – but his spirits were soaring because he was almost home.

Just as they had done for Ross Smith and his crew nine years before, huge crowds of Territorians gathered at the aerodrome at Fannie Bay,[2] sweating together. Hinkler saw Darwin in the distance at 5.40 p.m. and, as he neared the city, he came down to six hundred metres before dropping sharply to glide in. He had sunburn across his face where the leather helmet couldn't protect him, but despite the pain he couldn't stop smiling. No one had ever flown from England to Australia alone before, and no aircraft had ever done it so quickly; he had almost halved the Smith brothers' record as aviation entered a new age.

Hinkler was exhausted from the gruelling flight, his head kept wobbling from the vibrations of the aircraft, and he was partially deaf from the noise of the engine. As the crowd surged around the little aviator, Darwin's mayor pushed a bottle of Melbourne Bitter into Hinkler's hand and he guzzled down the amber fluid. *The Times of London* said Hinkler had opened the way for bigger, faster planes with passengers, cargo and mail to fly between Britain and Australia.

Less than forty-eight hours later, Hinkler took off from Darwin, aiming for Cloncurry 1600 kilometres away on his journey to his home town of Bundaberg. He was following the route through the Northern Territory and central Queensland laid down by Fysh and McGinness. Hinkler passed over Katherine at 9.18 a.m. The heat was so intense that he felt he was flying into the open door of a blast furnace.

He was expected at Brunette Downs, 320 kilometres north-west of Camooweal, at 4 p.m. But he didn't arrive. Concern about his whereabouts turned to rumour, then to fear. At 5.40 p.m. when there was still no news, the alarm was raised. In Cloncurry, another 580 kilometres further on, several aeroplanes had been arranged to escort Hinkler into the landing field in triumph that evening; now they sat abandoned as more than twenty cars were arranged with their headlights on in case he somehow came in for a night landing. The drivers waited in vain for the buzz of a small aircraft engine to break the pitch-dark, eerie silence of the outback.

There was still no word from Hinkler the next morning as newspapers around Australia tried to assuage the fears of readers by telling them that a safe landing was possible in the 'Never-Never … as Q.A.N.T.A.S. planes often come down in isolated places'.[3] It was estimated that Hinkler had enough food for four days, but everyone knew he wouldn't be able to survive long in the desert without water.

At 10.30 a.m. on Saturday, 25 February, the Qantas pilot Arthur Affleck took off from Cloncurry bound for Camooweal with his eyes peeled, flying low as he scoured the inhospitable terrain between Cloncurry and Mount Isa. Just after midday, Big Charles Scott also took off from Cloncurry – carrying an executive from British Imperial Oil, one of Hinkler's sponsors – and started his own search.

When Affleck arrived at Camooweal he received word that a day earlier Hinkler had landed at Brunette Downs, halfway between Katherine and Cloncurry. The pilot had eaten lunch and taken off in plenty of time to reach Camooweal by nightfall. No one knew where he'd disappeared, but at least the search area had been slashed in half. Affleck refuelled and went into the town of Camooweal, about two and a half kilometres from the landing ground, to gather supplies. He had just returned to the airfield when he heard a faint buzzing noise off to the west. Squinting into the sky, he made out a speck that was getting bigger and louder as it approached. He soon realised it was the missing world-beater.

Hinkler touched down in Camooweal at 12.12 p.m. While he was delighted to meet Affleck, he wondered what all the drama was about.

Affleck was startled to find the famous pilot dressed absurdly in a leather helmet, a suit coat, dress shirt, tie, tennis shoes, and a pair of lady's shorts – the only shorts that friends in Darwin had been able to find for the slightly built Hinkler to cope with the desert heat in his rush to start the cross-country flight.

Hinkler told Affleck that he had simply gotten off track and slept the night in the desert, just as he had done twice in the Sahara on his way to Australia. He and Affleck headed across the sunburnt, treeless plain for Reilly's Hotel and a party in Hinkler's honour.

Hinkler was up again before sunrise on Sunday, 26 February. He left Camooweal for Cloncurry accompanied by the Qantas aircraft piloted by Affleck and Scott. Fysh lauded the little pilot's achievement, even though he despaired that Hinkler's record flight inspired others to risk their lives in topping his benchmark, sometimes with fatal consequences. Another epic 1928 flight by Bill Lancaster and his lover Jessie Miller, the first woman to fly from England to Australia, was soon followed by the crossing of the Pacific by a team led by the Australians Charles Kingsford Smith and Charles Ulm.[4]

THE SEARCH FOR HINKLER was just the start of Affleck's career flying rescue missions, as Fysh and John Flynn began to enact their plan for flying doctors. Flynn promised a 'mantle of safety'[5] for people in the outback, with what was initially called the Aerial Medical Service.

Since outlining his scheme to Fysh and McGinness at the Metropole Hotel in 1921, Flynn had sought out Fysh at every opportunity to press his case for Qantas to support the expansive, expensive plan. Flynn kept weird hours, Fysh recalled, and would often badger him until well into the early hours of the morning with maps and diagrams spread over the floor wherever they met. The DH.50, with its enclosed cabin, would be the ideal air ambulance, but lack of finances and the inadequacies of radio technology at the time meant it wasn't until 2 November 1927 that Fysh could outline plans to his board for this 'flying doctor' idea to take off the following Easter.[6]

Initially the service was to have a radius of 480 kilometres from Cloncurry, bringing medical aid to isolated people within an area of more than 650,000 square kilometres, or three times the area of Victoria.[7] On 27 March 1928, an agreement was signed under which Qantas, in return for the provision of an ambulance aircraft and pilot, was guaranteed twenty-five thousand flying miles at two shillings a mile, with the Civil Aviation Authority paying half and the rest coming from donations to Flynn's charity.

Flynn had told Fysh that if the first-year experiment proved a success, the mission proposed to make a national appeal for funds that would support six flying doctors to cover all of inland Australia. Recent experiments by Flynn's technical expert, Alf Traeger,[8] had proved that cheap and reliable wireless sets could be produced, and there were plans for a durable transceiver to send Morse code with a system powered using bicycle pedals to drive the generator. Augustus Downs was the first property to install Traeger's typewriter transmission system, in which a Morse code message could be sent simply by typing it onto a keyboard rather than remembering all the dots and dashes.[9]

On 17 May 1928 – more than a decade after the death of stockman Jimmy Darcy – Affleck and Dr Kenyon St Vincent Welch[10] took off from Cloncurry for Julia Creek, 140 kilometres east, in Flynn's first flying ambulance. Qantas had received the DH.50 G-AUER from Britain in 1924. It was rebuilt after Lester Brain crashed it at McKinlay and renamed *Victory*. Welch had been selected from twenty-two applicants who had responded to an advertisement in the *Australian Medical Journal* for a fulltime doctor, to be paid £1000 a year. His first assignment saw him attend to two patients at the Julia Creek Bush Nursing Home, one of whom had attempted suicide by trying to cut his own throat. Affleck brought the DH.50 down nearby, and Welch successfully performed two minor operations.[11]

The doctor spent a year in Cloncurry before he returned to Sydney to continue his practice. During his time in the west he made fifty flights, covered twenty-eight thousand kilometres and treated 225 patients suffering a range of medical complaints, from typhoid fever to gunshot wounds.[12] It was not uncommon for Welch to 'make a 300-kilometre flight before breakfast, perform an operation then be back to base by lunchtime, flying at night only on urgent cases, landing in paddocks and rough clearings'.[13]

The Flying Doctor Service missions were the riskiest of all the Qantas operations, and Fysh sent out instructions to station owners throughout the west on 'How to Make Your Own Landing Ground, and How to Receive an Aeroplane'. There was frequently a delicate balancing act for Affleck and, from 1931, Eric Donaldson, the Flying Doctor pilot who replaced him, when it came to responding to urgent medical needs in dangerous flying conditions. Donaldson had grown up on cattle stations, so he knew the west and its fickle weather like the back of his sun-bronzed hand. Sometimes a storm warning meant he had to refuse to fly to an emergency, and there was often tension between him and the flying doctor.

Accidents and breakdowns still happened far too often. Dudley Wright, who went on to become one of the most revered engineers at Qantas, joined the airline in 1927; like Arthur Baird, he was often called out to repair aircraft in the most inhospitable, inaccessible, godforsaken places. In those early days the mechanics had to rely on their resourcefulness to get things done, particularly if an aircraft had been forced down in the middle of nowhere. Once when Affleck was flying a DH.50 to Normanton, he had to land at Donors Hill with a broken crankshaft. Wright had to fly an engine to him in pieces, then assemble it on the dusty ground using whatever lifting gear they could scrounge from the bush. Another time, Donaldson was flying a DH.50 between Longreach and Camooweal when a piston came through the side of the engine while he was over rough country at 1600 metres. He made for an open patch of Mitchell grass; it was uninhabited terrain, but fortunately he had on board twenty-eight blocks of ice to hold off thirst until help came.

Fysh stressed safety above all else and was incensed when his staff ignored his directions. He even frowned upon pilots having wives, as inevitably he would lose his best men when their better halves finally got sick of the harsh weather and primitive living conditions in Longreach. It always brought a smile to Fysh's face when he recalled Donaldson sending him a telegram that read, 'Can I get married please reply immediately.'

But it was Charles Scott who gave Fysh more headaches than the Longreach summer heat. After Scott flew Lord Stonehaven on a long tour of northern Australia in early 1928, he followed it later that year with an equally extensive tour of the north, this time flying Sir John Salmond, the British Marshal of the Royal Air Force. Salmond was visiting Australia as a guest of the Commonwealth government to advise on aerial defence.

The tour was about to start when Scott infuriated Fysh along with McMaster, who wrote 'it was common street talk' that before leaving Longreach for Hughenden in the high-powered DH.50J *Hermes* to pick up Salmond, Scott had been drinking heavily and engaging in conduct contrary to his 'own interests and the interests of commercial aviation'.[14]

The Salmond tour ended in Adelaide, and amid heavy fog in the early morning of 4 September 1928, Scott took off in the *Hermes* for home with his young mechanic George Nutson,[15] the son of a well-known and long-standing Longreach carrier.[16] Scott had placed fourteen tins of petrol in the cabin as, without telling Fysh, he planned to investigate a quicker route via Broken Hill and Thargomindah across sparse country where landing fields were largely unknown.[17] He never made it that far, though. The *Hermes* was airborne for only a few minutes when Scott, flying blind, ploughed into a fence in the Adelaide Hills. The aircraft crashed into rocks overlooking a gully and exploded.[18] In shock, with a broken jaw and with flames licking his handsome face, Scott bravely dragged Nutson out of the burning machine, but the youngster died on the way to Adelaide Hospital. His body was transported home to Longreach on the goods train, and Fysh was one of many grim-faced mourners at his funeral.[19] As well as the terrible human toll of the crash, the loss of the *Hermes* cost Qantas £3263 eight shillings and one pence, at a time when it was still too expensive to insure their aircraft.[20]

Dazed and confused, Scott appeared heartbroken in hospital when told of Nutson's death. After the badly injured pilot finally left hospital, Fysh sacked him for ignoring the Qantas rule about flying blind into clouds, despite repeated warnings.

Scott had been planning to quit Qantas in any case, but the heavyweight champ railed against Fysh's jabs at him and appealed to the board. Fysh met with Scott in Longreach on 10 October 1928, then reported to McMaster that from their discussion there was nothing that should make them take a more lenient view of his 'general behaviour and the Adelaide crash'. But Scott was back flying for the airline by the end of January, under even stricter conditions, and Fysh issued a set of 'Rules for the Observance of Pilots'.[21]

THE AFTERMATH OF THE disaster came as Fysh was embroiled again in a tense battle with Jimmy Larkin's company, Lasco. Qantas was still fighting to extend its route on the long hoped for Brisbane–Darwin connection, but the forcing of tenders – utterly unexpected by the airline – gave competitors the opportunity to grab a foothold at both the northern and southern extremities of Qantas's Charleville–Camooweal route. 'Our very existence,' wrote Fysh, 'was in jeopardy'.

On 9 May 1928, Qantas had started the first daily air service in Australia. It was an unsubsidised route between Brisbane and Toowoomba, 160 kilometres away. The journey took fifty minutes on a DH.50A and cost £2 15 shillings against the first-class rail fare of £1 1 penny and a journey of four hours. The daily service lasted only a few months, with most of the passengers curious about air travel rather than commuters, but the Brisbane–Charleville extension fuelled Fysh's continued ambition to link Australia with the rest of the British Empire.

Fysh won the tender for the Brisbane–Charleville route via Toowoomba and Roma, in September 1928, but only after he ordered two of the new seven-passenger DH.61 Giant Moth aircraft from de Havilland's Stag Lane Aerodrome in Edgware. The aeroplanes, powered by the nine-cylinder, 500-horsepower (370-kilowatt) Bristol Jupiter XI engine, were far superior to Lasco's older DH.50 and ANEC III machines. A total subsidy of £21,356 was to be paid over three years.[22]

Despite the success that Qantas was having with its mail route, Jimmy Larkin – who was desperate to remain in business – undercut Fysh on price to win the service for Camooweal–Daly Waters, even though it was only a short route thousands of kilometres from Larkin's base in Melbourne. He knew the price he tendered wasn't sustainable, given the conditions and the expense of fuel in the outback. The subsidy for the service amounted to £25,000, over three years.[23]

Fysh was crestfallen, as it seemed his hopes for an eventual Brisbane–Darwin air service route for Qantas would never materialise. 'This is unpleasant news,' he told McMaster. 'I feel we should make a systematic and strong protest without overstepping the mark.' Later, during a trip to Melbourne, he told his chairman: 'I have not been able to get to the bottom of the Camooweal-Daly Waters service. The whole thing is regarded amongst aviation people down here as a great blunder, and as a huge joke with the joke on Larkin.'[24]

The Lasco contract became even more problematic because while Daly Waters was chosen as the proposed railhead for the trainline from Darwin, the line extended only to Birdum Creek, eighty kilometres away. Larkin, sticking to the contract, refused to fly his aircraft there to collect the mail. This meant a mailman had to carry the post to Daly Waters by car, but during the wet season the roads were impassable. For three years, the Qantas pilot Joe Wilson flew a small de Havilland Moth over the eighty-kilometre gap at further expense to the government.[25]

Overseas airlines, meanwhile, expanded their operations significantly. Britain's Imperial Airways took over the Cairo–Baghdad route from the RAF as the British government looked to expand its international routes and sphere of global influence. The Dutch carrier KLM had taken off on its first intercontinental flight to Batavia (now Jakarta) in the Dutch East Indies on 1 October 1924, and under the direction of Albert Plesman[26] was about to begin regular scheduled flights from Amsterdam to Batavia in 1929 – this would be the world's longest regular airmail and passenger service. Plesman had plans to stretch the route to Australia, while Qantas remained hamstrung by bureaucracy and fierce domestic competition.

At least Fysh had Nell and his infant son John to help him cope with the stresses of running a business that often seemed besieged. Nell was lonely in Longreach. She felt dreadfully out of place and the social scene in the ladies lounges at the various pubs were not her cup of tea. She made friends as she could, flitting about in the first Qantas company car, but she dreaded the regular 11-day business trips Fysh made to Melbourne to talk to Brinsmead and the Civil Aviation Branch.[27]

AFTER CHARLES KINGSFORD SMITH had landed his big blue Fokker tri-motor *Southern Cross* at Brisbane's Eagle Farm on

9 June 1928, following his team's 12,000-kilometre flight from Oakland, California, he worked the media like a magician. When the *Southern Cross* landed at Mascot the next day, police estimated the welcoming party to be 200,000 people.

In August, after making the first nonstop flight across Australia from Point Cook to Perth, Kingsford Smith and Ulm targeted a first-ever crossing of the Tasman Sea. Their aim was to receive a government subsidy for their new airline Australian National Airways (ANA) to establish an airmail route between Australia and New Zealand. The last two airmen to attempt a crossing of the Tasman, Lieutenant John Moncrieff and Captain George Hood, had disappeared without trace.

Kingsford Smith and Ulm left Richmond airfield, north-west of Sydney, on the evening of 10 September 1928. With them was the navigator Harold Litchfield and the radio operator Thomas McWilliams. After a stormy flight through freezing conditions, they landed the next morning before a crowd of thirty thousand at Christchurch.

Fysh would find Kingsford Smith and Ulm a tough act to follow as the pair emerged as new formidable competitors in the Australian airline market. He would call the battle with them and others that followed 'a bitter life and death struggle',[28] one that could have ended his career in aviation after a decade of dedication and sacrifice.

Chapter 18

Qantas has pioneered commercial aviation in Queensland, having, to a large extent, built up public confidence which has resulted in the good support which is being received to-day ... That we are on the verge of further developments in commercial aviation is certain.

Hudson Fysh after shifting the Qantas headquarters from Longreach to Brisbane in 1930[1]

FYSH HAD WON THE Brisbane–Charleville tender by the skin of his teeth,[2] and Qantas made the most of it after ordering from England two DH.61s, christened *Apollo* and *Diana*, the company's first aircraft to feature an in-flight lavatory. On 16 April 1929, Lady Goodwin, the wife of the Queensland governor, christened *Apollo* in front of a large crowd at Eagle Farm, and the next day 'Skipper' Moody set off from the aerodrome for Charleville. Among his passengers was the 91-year-old Alexander Kennedy, wearing a three-piece suit and felt hat, and making his thirteenth Qantas flight.[3]

Despite the fierce competition for aerial routes, Fysh had told the Qantas board that prospects were bright for 1929, and that while the airline's work had been 'almost entirely of a pioneering nature', now the constant tendency for aircraft to become safer and more economical 'must indicate with certainty that commercial aviation, not only in Queensland but the world over, has a future at least equal to any other mode of transport'.[4] Qantas had just written to Defence Minister Bill Glasgow,[5] a former World War I general, 'pointing out the advantage and saving in time in continuing the proposed England to Australia air service down the Qantas route instead of down the West Australian coast'.[6]

The airline was finally able to operate with insurance for a fleet that had been valued at £29,000,[7] after Lester Brain made a trip to England and negotiated the cover with the Liverpool-based underwriters C. T. Bowring and Co. Qantas wasn't required to make a claim for years, but the aviation industry still struggled to convince the greater public that flying was a safe mode of travel. Accidents were always big news. Just two days after Moody took off from Eagle Farm in *Apollo* for the first scheduled service to Charleville, Lester Brain left for the west in a Qantas DH.50 called *Atalanta* to aid in the search for aviators lost in Australia's vast heart. With Brain was the mechanic Phil Compston[8] as well as the wireless operator Fred Stevens,[9] deputy director of the Brisbane radio station 4QG.

Kingsford Smith, Ulm, Litchfield and McWilliams had departed Richmond in the *Southern Cross* on 30 March 1929, aiming for the town of Wyndham, Western Australia. This was the first leg of a planned London flight and a quest to smash Hinkler's record of fifteen and a bit days for the journey in reverse. The crew wanted to negotiate aircraft deals in England for their company ANA.

A day into the flight, they made a forced landing on the flats of the Glenelg River estuary about 240 kilometres short of Wyndham, on an area the aviators dubbed 'Coffee Royal', because all they had to sustain them were flasks of coffee and brandy. They then lost radio contact.

By 3 April a major aircraft search was underway, but no trace could be found. Then on 7 April, Keith Anderson,[10] Kingsford Smith's estranged friend, and Bobby Hitchcock[11] took off from Richmond to join the search in a woefully inadequate Westland Widgeon III monoplane, the *Kookaburra*. After three days they had reached Alice Springs, then they headed for Wyndham with some sandwiches and two bottles of water. They disappeared too. Fear gripped the nation, and there was speculation that both crews would perish.

Flying the DH.61 he was using for charter work out of Mascot, the former fighter ace Les Holden[12] found the *Southern Cross* and her four thirsty, famished aviators on 12 April.[13] Nine days later Lester Brain, at the controls of the *Atalanta*, came upon the *Kookaburra* in the Tanami Desert, at a godforsaken spot about 120 kilometres from Wave Hill in central Australia. Circling low, Brain saw Hitchcock lying dead under the wing, black in the face from decomposition. The *Kookaburra* appeared undamaged and was standing in a large red patch of ground where the grass had recently burnt and was still smoking.[14] Brain correctly guessed that after making a forced landing, the aviators had started a fire to clear a runway but it had got out of control, burning the spinifex for thirty kilometres.[15]

Brain couldn't land the *Atalanta* because of the dangerous shifting sands, so he dropped a can of water attached to a small parachute in case Anderson was still alive. The Qantas pilot returned to Wave Hill before setting out again the next morning to look for Anderson – but the can of water had not been touched. After refuelling at Newcastle Waters, Brain was joined in the search by three RAAF machines. One of them, a DH.9, crashed near Tennant Creek and was destroyed, though there were no fatalities.

Anderson's body was located by a ground search party led by Aboriginal trackers on 29 April. He and Hitchcock had been so thirsty that they had even drunk the alcohol out of their compass before succumbing to the heat.[16]

Kingsford Smith and Ulm finally flew the repaired *Southern Cross* back to Richmond on 27 April, right into a nationwide storm amid rumours that their disappearance had been a publicity stunt. They were exonerated by an Air Inquiry Committee that sat in Sydney in May and June 1929. The two aviators proved that the forced landing was no stunt by producing an agreement between them and Sun Newspapers Ltd that guaranteed them £500 if they broke Hinkler's record.[17]

While Kingsford Smith and Ulm were being grilled, Brain and the *Atalanta* were in the air again, going to the rescue of Jim Moir and Harold Quinn, who had been making their second attempt at an England–Australia flight. On 18 May they had set off from

Bima for Darwin on the last leg of their trek in a Vickers Vellore. They came down short at the Cape Don lighthouse on the tip of the Cobourg Peninsula, not far from Melville Island.[18] Brain was awarded the Air Force Cross for rescuing them, and Fysh told the Qantas board that 'the helpful publicity which we have received through the success of the two search trips could probably not have been bought for money'.[19]

Despite this favourable publicity, Norman Brearley and his company now rebranded as West Australian Airways, were making bigger strides than Qantas as they both chased the grail of an airmail route from Australia to England. The opening of Brearley's Perth–Adelaide service, on 2 June 1929, was seen by many in the Australian government as an important link in the transport chain to Britain. Still, Fysh kept pushing for further extensions to the Qantas services because he knew that eventually tenders would be called for Australia–England flights. He continued to advocate for the idea that a route linking the eastern states to Darwin through central Queensland was a much safer and more reliable way to England than Brearley's proposed trek through Perth and Wyndham. But while Qantas had its mail subsidy cut to a year-by-year basis again, Brearley was awarded the government's first five-year airmail contract and the first to be based on payment per pound of mail carried instead of mileage.

The amount of subsidised mail on the route – much of it in bulky land sales catalogues – far exceeded expectations, costing the government a lot more than it had intended to spend. Subsidised mail runs began to fall out of favour in the new capital, Canberra. It was a nailbiting time for Fysh, who not only had to worry about keeping his airline in the air, but also how he would make a living if Qantas went under, especially now that Nell was expecting their second child.

Albert Green,[20] who had just succeeded Bill Glasgow as Minister for Defence, proposed closer cooperation between commercial airlines and the RAAF in case of war, with civil airlines to employ RAAF pilots, and the RAAF to carry the mail.[21] In the end, a compromise was reached: all Qantas pilots would have to be of British Empire nationality and members of the RAAF reserve.

A GREAT DEPRESSION WAS choking the world after the stock market crash in New York, but Kingsford Smith and Ulm were in buoyant moods as the 1930s dawned. Having brushed off the Coffee Royal disaster, they had gone on to eclipse Hinkler's record in the reverse direction. The aviators had landed the *Southern Cross* at London's Croydon Aerodrome on 10 July 1929, just twelve days, twenty-one hours and eighteen minutes after leaving Derby, Western Australia.

On 1 January 1930, Kingsford Smith and Ulm launched their company Australian National Airways with five Avro 618 'Tens'. The aircraft were basically Fokker tri-motors like the *Southern Cross* but built under licence by A. V. Roe. They were nicknamed the Tens because they could carry two crew and eight passengers. Kingsford Smith and his ANA copilot Scotty Allan,[22] recruited especially from London on a salary of £540 a year,[23] flew one of their machines, the *Southern Cloud*, from Mascot in Sydney to a rain-soaked Eagle Farm in Brisbane to begin a daily airmail and passenger service.[24] They had plans to open the Sydney–Melbourne route six months later. The Scottish-born Allan had never been to Brisbane before that first flight, but when Kingsford Smith told him to take the controls he simply followed the Australian coastline until he came upon the Queensland capital.[25]

Ulm, who was waiting at Lennons Hotel in Brisbane to make the return flight, told the press that the new service between Australia's three biggest cities was an epoch in the development of aviation, as the first important unsubsidised air route in the Commonwealth.[26] ANA had a staff of thirty with plans to expand soon. 'We shall fly day in and day out, hail, rain, or snow, 365 days in the year with clock-work regularity,' Ulm declared triumphantly. 'Given war time facilities, our aeroplanes could be converted into bombers within 24 hours should the emergency arise and they would be able to fly 1000 miles from their base and return without re-fuelling. We are determined to put commercial aviation in Australia on the world map and by that we don't want people to imagine that we are decrying the great pioneering work of Q.A.N.T.A.S. or the West Australian Airways – they have done yeoman service, the credit for which can never be taken away from them.'[27]

Ulm regarded Fysh and Qantas as useful addendums to his grand plans. 'So far as Queensland is concerned,' he said, 'we have great faith in her future, and we have entered into friendly agreements with Q.A.N.T.A.S. for our mutual benefit. I have had conferences with Mr. Hudson Fysh and we have decided to co-operate so far as Queensland is concerned.'[28] Brisbane's main air operations were soon to be transferred from Eagle Farm to Archerfield, and Ulm announced that Qantas and ANA would share a hangar. In the event of any ANA aircraft being out of commission, he said he could call on Qantas for one of its aircraft to maintain the service.

Fysh wanted Qantas to be much more than just a footnote in the ANA prospectus, but his plans to expand the airline faced threats on other fronts, too. Jack Treacy,[29] who had been a fighter pilot over France, started Queensland Air Navigation Ltd in 1930, buying a pair of Avro Tens for a Brisbane–Townsville run via Rockhampton.[30] Then 21-year-old Keith Virtue,[31] along with his future father-in-law, George A. Robinson,[32] established New England Airways with a regular service between Lismore and Brisbane.

While these intrepid aviators were giving birth to new services, Fysh's four-year-old son John entered the aviation industry too: the Qantas mechanics in Longreach passed him a broom and put him to 'work' sweeping the hangar floor. Before long John sat beside his father in Fysh's car, a Willys-Overland Tourer, as they motored from their little house in Emu Street to the Longreach Hospital. There John first cast eyes on his newborn sister Wendy, her wrinkled red face framed by a blue dressing-gown, as Nell carried her down the hospital stairs to the car.

Fysh continued to worry about his children's future – and that of his other baby, Qantas – as Ulm, the business brain of ANA, and Kingsford Smith, its charismatic public face, turned their airline into a roaring success within a few months. Fysh suspected much of their custom came from thrill-seekers rather than commuters with ongoing ticket purchases, and he sometimes rode on their machines when travelling interstate. In perfect conditions the Avro Tens would take five hours to fly between the southern capitals; if there was a strong headwind the flights could take seven hours, and even seasoned flyers such as Fysh would arrive partially deaf and exhausted. Sometimes they arrived even worse for wear. Once Fysh was sitting at the rear of the cabin, the place he knew to be the safest position if the machine hit a mountain, when a passenger in the front seat was violently ill into a receptacle. When he tried to empty the contents out of a window, the wind blew it back all over Fysh 'full blast'.[33]

Ulm's plans for ANA also made Fysh queasy. One of the brightest businessmen in aviation, Ulm wanted to extend the Brisbane–Melbourne route to Adelaide, where it could link with Brearley's service to Perth, and he was talking up a Brisbane–Townsville service linking with Rabaul. He told the press: 'We are hoping … that *we* may link Australia with Singapore in co-operation with the Imperial Airways when its service, now running to Delhi, is extended. If we are able to inaugurate this service we shall probably run via Broken Hill and Darwin.'[34] Imperial was suggesting an extension of its airmail service from India to further link the British Empire to Australia at Darwin. Fysh wanted Qantas to be the airline carrying the Imperial mail to the southern states, but Ulm was now in the box seat, operating among the capitals.

At the same time, Fysh was having all sorts of problems keeping his new Brisbane–Charleville service in the air. There was the global financial meltdown to contend with, as well as the frequent overheating of the high-pressure Bristol Jupiter Mark XI engines in the DH.61s. The engines became so unreliable, forcing so many unscheduled landings, that Lieutenant-Colonel Brinsmead threatened to ban the aircraft.

Fysh smoked more than ever, dragging one Craven A cigarette after another from their little red tins, and he tried to find solace as often as he could at the Longreach golf course. Eighteen holes on the dusty fairways was always a relaxing form of therapy while he worked on his handicap of fourteen.[35] But he knew that to help Qantas reach its full potential, he had to move.

AFTER TEN YEARS OF STRUGGLE in Longreach trying to establish and then grow his bush airline, Fysh was thirty-five and well aware that the future of both his company and his family lay in a big city. Qantas had made its name in pioneering subsidised routes between places of small populations, but Fysh could see the end of government subsidies looming. He knew that the survival of the airline centred on 'commercial aviation shifting rapidly from the interior to the cities and centres of population on the coast'.[36]

The Brisbane branch of the airline was now doing much more business than the Longreach head office as Qantas clocked up a million miles in the air. Fysh began planning the establishment of the company's head office in Brisbane. On 27 May 1930 he headhunted the dapper English war veteran[37] H. H. 'George' Harman, who had been the travel manager at the shipping firm Burns Philp, which had a close connection with the airline. Harman had enjoyed great success selling Qantas flights to Queensland's Gulf Country, which he promoted as a sportsman's paradise where holidaymakers could fish for huge barramundi, groper and swordfish, or shoot birds and crocodiles. The Qantas link to Normanton had also given commercial fishermen the ability to sell their morning catch in Longreach that afternoon.[38]

Fysh offered Harman £525 a year to be the company secretary and flew him in a two-seat Gipsy Moth from Brisbane to Longreach to show him the route. The weather was atrocious, and they had to land on a road at Mitchell as everything else around was covered in water. Fysh and Harman folded the plane's wings back and stored it in a garage before taking off the next morning down Mitchell's main street. The Gypsy Moth was an extremely versatile and reliable little machine; the British pilot Amy Johnson had just achieved worldwide fame after landing hers at Darwin, on 24 May, nineteen days after having left Croydon Airport. Harman arrived in Longreach wearing a bowler hat and carrying gloves and a walking stick – get-up that drew guffaws from the sun-bronzed locals, who had never seen anything like it before.[39]

It was almost as funny as when the Longreach fire brigade had rushed out to reports of a blaze at the Fyshs' home on Emu Street, only to find it was the backyard dunny going up in smoke.[40] A smouldering Craven A was the chief suspect.

But Fysh regarded Harman as a hard worker and 'a man of great integrity'.[41] On 16 June 1930, Harman was all business when he and the stenographer Ida Isaacs opened the Qantas headquarters in a suite of offices on the third floor of Brisbane's Wool Exchange Building on Eagle Street.[42] Ida would be Fysh's treasured assistant for another four decades. Six days later at Longreach, after extensive checks and rechecks, Fysh loaded one of the troublesome DH.61s, *Diana*, with Nell, John, baby Wendy and a nursemaid for the flight south. Nell and the children rode in the enclosed cabin, while Fysh took his place in the open cockpit behind the wings. Nell was glad to be leaving the heat and dust after seven years, but she and Fysh had heavy hearts as the curtain came down on what had been a big part of their lives. There had been a formal send-off at the Imperial Hotel with Fergus McMaster presiding over a tribute dinner,[43] and now out at the aerodrome, as they waved goodbye to their many friends, Fysh taxied the machine down the Longreach runway. He flew his family to Charleville, where 'Skipper' Moody took over the controls for the last leg into Queensland's River City.

Fysh arranged the sale of his house and all its goods, including kettles, cannisters, blinds, rugs, three wicker chairs, an electric fan and a pie dish with a firm of auctioneers who shared the Graziers Building with Qantas in Longreach.[44] He had paid £1000 for the house in Emu Street but despite improvements, Fysh agreed to sell it for less than £750. Money was hard to find in 1931 and any offer was better than none. Fysh made £79 from the sale of the household goods, though the lino in the kitchen and bathroom didn't sell, nor the verandah blinds.[45]

The family spent their first days in Brisbane at the grand Belle Vue Hotel[46] before moving into a smart house on Miles Street in the suburb of Wooloowin. It was a roomy 'Queenslander'-style home with concrete steps at the front, a wide verandah at the back, a leafy garden and a tennis court.[47]

Moody left Qantas to run his Charleville sheep station for a while, and Fysh appointed Lester Brain, in Brisbane, as chief pilot, head of sales, demonstrations and taxi trips. Russell Tapp was given the Brisbane–Longreach route to fly, and Eric Donaldson Longreach–Camooweal. Arthur Affleck stayed with the Flying Doctor Service, and Charles Scott, given a second chance after having mended his ways, became the instructor at the flying school. Fysh and Arthur Baird combined their other duties with work as reserve pilots.

Fysh continued doing aerial taxi work, often in a small de Havilland DH.80 Puss Moth with an enclosed cabin – the same type of machine that Bert Hinkler would soon fly across the South Atlantic from Brazil to Gambia, in the first solo crossing of that ocean. Fysh thought it an excellent vehicle for light duties, though he eventually decried it 'as faulty and too lightly built'.[48] Faults in the construction were blamed for Hinkler's death in a crash in Italy in 1933, and the death of Les Holden, who was a passenger in a Puss Moth when it crashed at Byron Bay in 1932.

FYSH WAS FLYING a Puss Moth once when one of his passengers was taken ill and puked into the pilot's brand-new hat that he had put in the back of the cabin. Air sickness was a common problem for early passengers, and it wasn't until the first *Qantas Gazette* of 1930 that Fysh could announce: 'Air sickness is much less common than sea sickness but during summer afternoons or flights in the far west air sickness is sometimes experienced. It is now generally recognised that chewing gum is of value in minimising the effects of all travel sickness and it has been recommended by the Air Ministry to the leading British Air Transport companies. Supplies of chewing gum, neatly got up in small envelopes, have been provided for the use of passengers on Qantas planes by Messrs Wrigleys Ltd., and are now in general use with good results.'[49]

In the decade since the airline had been registered after that first meeting at the Gresham Hotel, Fysh could report that Qantas had accumulated assets of £57,000,[50] and had carried almost 250,000 letters, 50,000 kilograms of freight and 8745 passengers on its scheduled routes, as well 6296 joyriders.[51] He said there had also been 1054 aerial taxi passengers who had covered more than 270,000 miles as the Qantas machines ferried, among other things, doctors and patients, urgent documents, diphtheria antitoxin, stores to flood-bound homesteads, urgent machinery, car parts, stock buyers, property inspectors, and police hunting criminals.[52]

There had been thirty-five accidents, two of them fatal, and eighty forced landings,[53] many of them as a result of the Jupiter XI engines, but the machines were improving all the time. The two original Qantas aircraft Fysh and McGinness had flown up from Sydney as well as the later mail planes had been sold or demolished for spares over time but Qantas now had a fleet of eleven newer machines: two DH.61s that could each carry seven passengers, six DH.50s that could each carry four, a pair of two-passenger Puss Moths, and a two-seat Gipsy Moth. The company was paying a dividend of seven per cent after recording profits for the year of £5770.

At every opportunity Fysh criticised a proposed route from England via Perth and Wyndham as being too long and slow. 'Practically every international flyer that has visited Australia,' he claimed, 'has used the "Qantas" route in reaching the Eastern cities, because it is by far the shortest suitable route available.'[54]

In Brisbane, golf again helped him stay focused, and there was also time for self-analysis and reflections on the journey so far. Four months after his move, he was back in Longreach where Qantas held a formal farewell from the west at the town's Café Elite in October 1930. Shareholders and staff also celebrated the company's tenth birthday a few weeks early. One of the Qantas directors, Guy Taunton,[55] a stock and station agent who was on his way to selling a million sheep around Longreach,[56] told the audience that after its difficult birth the company had become 'a child of the West … as much part and parcel of the West as are our rolling downs and gidyea scrubs'. The company had grown beyond that now, Taunton said, but it was still essential to 'pay homage to the founders of Qantas' and to celebrate their spirit – 'the spirit that inspired Columbus in his stormy voyage across the Atlantic … the spirit that sent Drake around the world … lifting mankind one step higher in the toilsome ascent from its primeval darkness'.[57]

Eight years had passed since Paul McGinness had resigned from the company for a life of hard slog in a remote corner of Western Australia, but in Longreach he had not been forgotten. 'To McGinness's enthusiasm and devotion to the cause I pay special tribute,' Taunton said. 'It was McGinness who first conceived the idea of Qantas.'[58] Fysh hadn't forgotten his old friend, either; he sent McGinness a clipping from the *Longreach Leader* newspaper, which ran a transcript of Taunton's speech and the recollections of how McGinness, Fysh and Baird had once travelled throughout the country like inspired evangelists 'preaching the gospel of Aviation' while overcoming difficulties, dangers and public scepticism.[59]

The fighter ace was now a struggling father of two daughters, Dorothy and Pauline, and still living on a hard-scrabble property with them and his wife. Apparently he was showing the same cavalier attitude towards his crumbling marriage that he'd once taken to facing the enemy over Palestine. Drought had ravaged McGinness's dreams and he tried prospecting for gold, making an aerial survey of places to explore, including one that became a prosperous mine – though not for him. He started drinking heavily, staying at local country pubs until the early hours on a regular basis before lurching home in the car. Surrounded by the sorrow of a missed opportunity with Qantas, McGinness wrote back asking Fysh to look after him. Could he return to the company he had envisioned beside a campfire all those years ago? He wanted 'a rather large salary with a grandiose title'.[60]

Despite the depressed economic conditions around the world, Qantas still had healthy passenger numbers. But western graziers, mortgaged to the hilt, were fighting drought and a slump in the value of wool, and Qantas fares had to be slashed to keep the propellers turning, so salaries, wages, and directors' fees were trimmed. Fysh was still making £1100 a year – more than three times the average wage[61] – but even though Nell still bought expensive clothes for the family, he brought a business frugality to their home at Wooloowin, cancelling the afternoon newspaper and restricting lavish entertainment to a small cut-price cinema in Fortitude Valley.[62] He knew that in such times, he couldn't afford any unnecessary luxuries at Qantas either – certainly not a pilot whose mind wasn't totally focused on safety and the corporate image.

Fysh had played second fiddle to McGinness in the war, but he was now a hard-nosed realist and a hard-driving businessman who demanded unwavering discipline from his staff and absolute adherence to the rules for his pilots. Charles Scott had decided to fly home to England,[63] and Fysh was done with mavericks. The Qantas aircraft of 1931 were far more advanced than the machines McGinness had flown in 1920. From his plush office on Eagle Street, beside the Brisbane River, Fysh wrote back to McGinness on 14 May 1931 to say, 'There is no place here for you.'[64]

ULM AND KINGSFORD SMITH were continuing to press their case for an ANA route from Brisbane to Darwin and then on to India to link with Imperial Airways. At the same time, Brearley was pushing for KLM to transport mail from Europe to his aircraft in Wyndham. ANA, KLM and Imperial all had the multi-engine aircraft and the experienced pilots to fly overseas routes. Qantas had only single-engine aircraft, and if another airline was able to fly the mails from Europe through to Brisbane, Fysh knew that the subsidised route he and McGinness had surveyed would be overtaken and Qantas would be closing its doors.[65]

ANA, though, was facing immense financial pressures on their unsubsidised capital city routes – pressures made worse after one of their most experienced bad-weather pilots, Travis 'Shorty' Shortridge, took off in the Avro Ten *Southern Cloud* from Mascot at 8.10 a.m. on 21 March with a copilot and six passengers bound for Melbourne. A few months earlier Shortridge had flown Australia's young cricket icon Don Bradman from Adelaide to Melbourne, then to Goulburn,[66] on what Bradman had described as a 'bumpy journey'.[67] For this Sydney–Melbourne flight, Shortridge had to negotiate heavy cloud and winds that tore into the aircraft at a hundred kilometres an hour. The *Southern Cloud* never made it to Melbourne, and its fate remained an enigma for twenty-seven years. In 1958 a worker on the Snowy Mountains Scheme found the wreckage atop a peak called 'World's End' near the town of Tooma.[68]

While the initial search for the *Southern Cloud* involving twenty aircraft was underway, the British Post Office announced on 1 April 1931 that two experimental round-trip airmail services would be made between London and Sydney, to leave London on 4 April and 25 April. Imperial Airways had just extended its route from Karachi to Delhi and was testing the waters for the long haul to Darwin, where the mail would be transferred to Qantas for delivery to Brisbane, with ANA then flying the letters and parcels on to Sydney and Melbourne. The new venture was hailed in the press as 'probably the greatest development in the history of commercial aviation within the British Empire',[69] and Fysh wanted Qantas to be in the prime position.

Chapter 19

Jogging down to Melbourne in the express last week a tall, spare, keen-faced passenger visioned aeroplanes, like giant dragon-flies, winging to and fro over the Timor Sea and all the island chains and ocean stretches between Australia and India.

PRESS REPORT ON FYSH'S PLAN TO LINK QANTAS WITH THE IMPERIAL AIRWAYS SERVICE FROM ENGLAND[1]

QANTAS BEGAN BOOKING its own tickets to London when an Imperial Airways aircraft carrying fifteen thousand letters and several diplomatic documents for the Australian government set off from Croydon on 4 April 1931 bound for Darwin.[2] Fysh saw this first experimental mail run as the green light for Qantas to expand its network to the other side of the globe. He wrote to the Commonwealth Director-General of Posts and Telegraphs, Harry Percy Brown, offering a permanent airmail link at Darwin with Imperial Airways for just £3000, using the tiny Qantas Puss Moth, an idea Fysh later admitted was 'nothing less than stupid'.[3]

There was an immediate hitch. A relay system of Imperial aircraft delivered the mails to Karachi, where Imperial's three-engine DH.66 Hercules, *The City of Cairo*, took off with them for Darwin on the morning of 13 April. It ran out of fuel and crashed six days later at Kupang in Timor, eight hundred kilometres short of Fannie Bay. There were no injuries, and the mail was intact, but as Fysh had no multi-engine aircraft to fly across the sea to Timor, it was left to ANA – still searching for the *Southern Cloud* – to save the day.

Kingsford Smith and Scotty Allan flew to Timor on the hard-working *Southern Cross*. Fysh wired McMaster, lamenting, 'Application made by ANA to operate permanent route Brisbane–Darwin and to India. Receiving considerable support. Position fairly critical to our interest.'[4] With a touch of desperation, Fysh wrote again to Harry Brown to say that Qantas could extend its interest in the mail run past Darwin, using flying boats.

Despite the *Southern Cloud* tragedy, Kingsford Smith remained the toast of Australian aviation. He had recently obliterated Hinkler's England–Australia solo record, covering the distance in just ten and a half days.[5] He and Scotty Allan arrived in Darwin with the London mail on 25 April. That same day, Russell Tapp flew the Qantas DH.61 *Apollo* out of Archerfield for Darwin carrying 28,406 letters and parcels that the *Southern Cross* would then carry to Akyab (now Sittwe) in Myanmar.

Tapp had hardly arrived in Darwin when he returned to Brisbane with the London mail to a noisy cheering crowd, including Fysh and McMaster, at Archerfield. He had covered six thousand kilometres in four days, flying low to avoid headwinds on the return trek. He buzzed over two cars broken down in the Territory. 'One fellow waved to me to stop,' Tapp recalled, 'but I couldn't. I was in a hurry.'[6] The first London–Brisbane mail took twenty-four days, the Timor crash delay included, while the return flight took nineteen.

A second experimental airmail run left Croydon on 25 April. Its five thousand letters for Australia were transferred to the *Southern Cross* in Akyab, and in turn transferred to the Qantas *Apollo* in Darwin. This run took eighteen days. For the return journey Fysh gave Russell Tapp a holiday at Southport after his exhausting schedule,[7] and on 17 May, Fysh and Dud Wright left Archerfield in the *Hippomenes*, one of the DH.50Js Qantas had built in Longreach. After reaching Darwin the next day, Fysh handed over 6952 letters and parcels to Imperial's Captain Roger Mollard, who flew them on the London run in a Hercules bought from Norman Brearley to replace the wrecked *City of Cairo*.[8] The return Brisbane–London mail run took just sixteen days, half the time of the journey by sea.

While Fysh was flying to Darwin, the Dutch had landed their KLM aircraft *Abel Tasman* in Brisbane[9] for a stopover as the airline made its own experimental airmail flight from Batavia to Melbourne. With the unemployment rate in Australia approaching thirty-two per cent[10] during the greatest financial crisis in recorded history, the federal government was facing steep financial pressure subsidising its internal mail routes, and Fysh was well aware that the Dutch flying international mail to Australia without subsidy threatened Qantas's future. *Aircraft* magazine editorialised, 'It would be most unfair if companies such as Qantas – which has carried on for years under difficult circumstances; which has opened up vast areas of Queensland that would otherwise never have been developed to the extent they are today; which has ushered in a new era for Outback man, bringing him medical attention, fresh food, fast mails, and making his life easier and happier – were to be forced out of existence at a time when they most need some financial support from the Government.'[11]

Nothing could help ANA, though, as the Depression and a loss of public confidence following Australia's worst air disaster starved its revenue. On 26 June 1931, Ulm and Kingsford Smith – a daredevil who was devoting more time to breaking aviation records than running the business – suspended their interstate operations.[12] Meanwhile, Fysh and McMaster moved to establish formal links with Imperial Airways and to secure tri-motor machines for Qantas to make flights to India or Myanmar if necessary.

As Qantas fought for survival, Fysh was negotiating several partnership offers. Deciding to concentrate on his core business of mail and passenger transport, he declined a deal with de Havilland to manufacture their aircraft in Australia, and he knocked back another offer for Qantas to become the agency for Ford motor vehicles in Queensland.[13] After Jack Treacy's Brisbane–Townsville service went into liquidation, Fysh tried to make the route work using Lester Brain and his *Atalanta*, but the passenger numbers still weren't there to sustain it.

Fysh worked long hours at his Eagle Street office from Monday to Saturday lunchtime, but he knew the importance of recreation and had always been a keen sportsman. He played tennis on his own court at Wooloowin on Saturday afternoons, inviting the Qantas staff over for backyard tournaments. He spent most of his Sundays tinkering around the aircraft at Archerfield. His son John had been enrolled at the Eagle Junction State School but usually accompanied his father to Archerfield and as a small boy helped him – often unsuccessfully – in selling tickets for taxi flights on the Puss Moth. Sometimes Fysh would give a local farmer two shillings so John could ride a horse around the open paddocks that surrounded the aerodrome. Fysh would sometimes take his children on flights from there and the cap-and-goggles would entirely cover Wendy's tiny head.

Fysh still worked as a back-up pilot and did charter work. He often took mining experts, brokers and financiers from Brisbane five hundred kilometres north-west to the Cracow goldfields. His passengers included the celebrated writer Randolph Bedford and the bookmaking entrepreneur John Wren, who Fysh collected from Brisbane's Mater Hospital after being told the gambling tycoon was in the chapel saying his prayers before the flight.[14]

Fysh once flew a doctor out to the Headingly cattle station south of Camooweal, damaging the wing tip, after the Chinese cook there had swallowed poison in a desperate bid to end it all when his opium supply had been confiscated. Another time, Fysh took off in a high-powered DH.50J from a flooded Lismore, sending sheets of water in all directions, when he was transporting Reg Robinson, the injured son of George A. Robinson from New England Airways. Over Tenterfield, Fysh found that the lever for the secondary fuel tanks was stuck. He had to land the big machine in a paddock to make repairs, then effect a death-defying take-off down a steep slope over telegraph wires, getting clear by just a few centimetres.

And once, after flying Queensland senator Harry Foll from Brisbane 650 kilometres north to Rockhampton, Fysh returned home only to make a dash for Coffs Harbour, 380 kilometres south, to collect passengers from a broken-down Hawk Moth.

Facing a strong headwind, he finally arrived back in Brisbane after dark. With an empty fuel tank, he was unable to find the unlit aerodrome. Horrified, he went into a cold sweat – but providentially, it seemed, the flares at Archerfield finally came on, and he touched down in their glow.

AWAY FROM THE STRESSES of the aviation business, Fysh would also find time to relax by writing, usually after Nell had put John and Wendy to bed. He said his love of the written word had grown from his 'hopelessness as a public speaker'[15] and a need to get his ideas down, instead, on paper. He compiled newspaper articles and radio talks on aviation, and as a companion to the pioneering work his airline had done across Australia, he wrote *Taming the North*,[16] a book about Alexander Kennedy and other settlers who had carved out their cattle and sheep empires in the outback.

Most of Fysh's writing, though, centred on forming an alliance that could see Qantas loosen the boundaries of Australia. Charles Scott, Jim Mollison[17] and then Arthur Butler[18] had all taken turns to break Kingsford Smith's record for a flight between England and Australia, and the globe was becoming smaller and smaller. With the support of Ben Chifley[19] – the former locomotive driver who was now Minister for Defence in the Scullin Labor Government – Fysh wrote on 16 June 1931 what he called 'the first definite approach'[20] to George Woods Humphery, the managing director of Imperial Airways. Fysh suggested merging their resources, and correspondence between Brisbane and London continued for months as he mounted a strong case for Qantas during a time of what he called 'anxious negotiation, lobbying, propaganda, and chess-like moves … in a war of manoeuvre and of nerves'.[21]

Kingsford Smith and Ulm, despite having cashed in most of their chips, had one last throw of the dice left: the Australian postal services authorised ANA to carry a special Christmas mail from Australia to England. On 20 November 1931, Scotty Allan flew out of Melbourne carrying fifty-two thousand letters and parcels on an Avro Ten, the *Southern Sun*. His only passenger was

Charles Ulm (left) and Charles Kingsford Smith proved to be tough business opponents for Fysh. *State Library of NSW FL1133534*

the Controller of Civil Aviation, Lieutenant-Colonel Brinsmead, who was heading to London to further discuss a regular airmail service to Australia.

Allan's centre engine failed while he was trying to take off with the seven hundred kilograms of mail from a waterlogged airfield in Alor Setar, Malaysia. Kingsford Smith flew another Avro Ten, the *Southern Star*, to Malaysia to pick up Allan and the mail, but Brinsmead had decided to take a different KLM flight to London. That aircraft crashed on take-off from Bangkok, killing the Dutch pilot and four others, and leaving Brinsmead in a critical condition with fractured ribs and skull.[22] He remained an invalid until his death in Melbourne three years later.

The mails made it to London safely. On the return journey Allan was attempting to land in the darkness, trying to find a beacon for the Croydon Aerodrome, when he crashed the *Southern Star* into an orchard in Kent. The wheels were torn off, the undercarriage was smashed and the fuselage pierced by a small tree.[23] Allan walked away, but ANA's run as a Qantas rival was all but over.

Woods Humphery wrote to Edgar Johnston, Brinsmead's replacement as controller, to decry ANA for 'creating nuisances' to their operation.[24] But Norman Brearley, upon realising that subsidies for his struggling West Australian operation were unlikely to continue, suggested a merger to Ulm that would involve his company, ANA, Qantas, the government and 'possibly the Commonwealth Railway Commissioner'.[25] 'There may be a certain amount of difficulty with Q.A.N.T.A.S. if this matter is opened up too early with them,' he wrote, '[but] in due course they could be persuaded to fall into line.'[26]

Fysh feared any merger would swallow the business he had started, a view Ainslie Templeton shared when he wrote to Fysh saying, 'if we merge with Ulm and Brearley, their contribution will be their machines and practically nothing else of advantage to us. We will contribute our organisation which has proved the only successful one in Australia. We are on good terms with Civil Aviation, we are not broke … I cannot see any good reason for going to the rescue of others at this stage.'[27]

THE NEXT BREAK IN THE clouds for Qantas came in January 1932 when Fysh's fellow Tasmanian 'Honest Joe' Lyons[28] became prime minister, and Sir George Pearce[29] – the one-time carpenter whose long term as defence minister included all of the Great War – was returned to the portfolio. Pearce had held that role since before powered flight in Australia; now, alongside Edgar Johnston, he planned to reshape Australian aviation to meet the demands of international travel.

The world was still in the grip of a financial meltdown, but Johnston wrote to Fysh to say that the change of government was expected to return confidence to the Australian business community. Fysh had 'every reason to hope that there will be funds to keep the existing services going after termination of present contracts … So far as I personally am concerned, the excellent record of Qantas and its high reputation both with the public and with those more intimately concerned with your activities will ensure that its claims for inclusion in any further development schemes will not be overlooked. I fully appreciate that your claims to operate to the Northern Territory have been considerably established by "right of conquest".'[30]

At the same time, Ulm wrote directly to the new prime minister, pointing out that ANA had flown 583,700 miles on its regular services in 1930 and a total of 681,220 miles – 'more than the combined mileage of the subsidised companies', Qantas and West Australian Airways. Ulm pointed out that while Jimmy Larkin's operation, WAA and Qantas had all been 'heavily subsidised for the past seven or eight years', his company had 'received no help from the federal government nor any other government'. [31] In order to 'carry on valuable air services, retain for Australia its large modern aircraft and equipment and continue to employ its highly skilled and specially trained personnel', Ulm asked for government help through a subsidy or even the purchase of half the company with an eye to establishing a regular weekly airmail and passenger service between Wyndham or Darwin and Delhi, Rangoon or Singapore.[32]

As Ulm's communications with the government continued, Sir Walter Nicholson, Britain's nominee on the directorate of Imperial Airways, arrived in Australia by ship. He was on a voyage around the world making contacts and contracts for aerial development, and he had meetings planned with Australia's Federal Ministry as the time was 'ripe for an extension of the Indian service to Australia'. Nicholson told the press that the air link 'should be established at the earliest moment that financial conditions both in England and in Australia permit'. 'The advantages of a regular service,' he said, 'not only in regard to normal relationships between the people of the dominions and the mother country, but in the interests of trade, are obvious.'[33] On 2 February 1932 Nicholson flew into Brisbane on a New England Airways tri-motor, but the only airline he wanted to discuss was Qantas.

Fysh greeted Nicholson on arrival at Archerfield,[34] and the next day sat down with him and Qantas director Fred Loxton to nut out an understanding. A few days later in Sydney, Fysh presented Nicholson with his proposal for a new company to operate to Singapore or Calcutta as part of the England–Australia route. Imperial and Qantas would be the majority shareholders, but ANA, Brearley and even the pugnacious Larkin could have representation on a new board proportional to their shareholdings.[35]

The following week, Fysh went to Melbourne to start a long series of meetings with Ulm, Brearley and Edgar Johnston. Fysh later told Brearley that his rival's idea to form an alliance with KLM was misguided and that the new route should be all British 'red',[36] though Brearley couldn't be dissuaded and continued pursuing a deal with the Dutch for a mail route through Wyndham rather than Darwin. At one meeting Ulm suggested a deal in which each party could fly the London mail on their existing routes but ANA would fly to Rangoon in Myanmar, giving his airline three times the mileage of Qantas, Brearley's West Australian Airways and Lasco. Fysh said nothing, 'being an avid listener', but knew his other Qantas board members wouldn't let that one fly. Brearley threw his lot in with KLM but went nowhere, while Ulm, floundering and flailing away in a financial freefall, continued submitting proposals heavily in favour of ANA.

It was a stressful time for every participant in the air battle, but Fergus McMaster also had weighty personal matters on his mind. Fysh had found work in the west for his younger brother Graham, and by 1932 he was the overseer and sub-manager of McMaster's Moscow station. Graham and McMaster's 22-year-old daughter Jean had fallen in love and were talking about marrying. McMaster was worried largely because of the unhappy childhoods Hudson and Graham Fysh had experienced as a result of their parents' divorce.[37] McMaster felt that Jean was too young and inexperienced to marry, and that Graham, even at twenty-nine, had few real prospects. The strain told on the old digger. In a Christmas letter to McMaster, Fysh advised his chairman, 'I hope you don't mind me saying that you need a good holiday, and it is your duty to take this in the interests of your family and your Pastoral Company. A holiday will cause less of a dislocation than an inevitable breakdown.'[38] It was advice that Fysh could also have taken to heart.

Fysh warned his board that Qantas could never be 'happy bedfellows' with the other Australian companies.[39] So it was that on 23 February 1933, Imperial's Sydney agent Albert Rudder,[40] a director of the NRMA (National Roads and Motorists' Association), appeared before the Qantas board with a plan to form a separate company to tender for the new air route. The company would be set up on a fifty-fifty basis between Qantas and Imperial, with Qantas to fly the Darwin–Singapore leg of the London route.

Fysh knew that the cost of operating Australia's first overseas air service would weigh heavily on the government during the ongoing Great Depression. He also knew that to sweeten any government tender, Qantas needed newer, more economical four-engine aircraft that would provide greater reliability and performance over the existing tri-motors such as Kingsford Smith's *Southern Cross*. On 12 April 1933 Fysh wrote to Imperial's George Woods Humphery to say that he had been trying for several years to have de Havilland build such a machine with a speed of 135 miles per hour (217 kilometres per hour). 'A machine such as this,' Fysh told his British counterpart, 'will secure the contracts.'[41]

The following month Imperial sent out to Australia one of its four-engine Armstrong Whitworth AW.15 machines, the *Astraea*. The promotional exercise was also designed to further survey the planned mail route and to test the machines in weather conditions from Singapore to Darwin.[42] The *Astraea* left Croydon on 29 May 1933 under the control of the company's air superintendent, Major Herbert Brackley,[43] who had made several record-breaking flights in North America and who was one of the last men to farewell Bert Hinkler on his final flight.[44] Australia had never seen a machine like the *Astraea*, a high-wing monoplane that could give seventeen passengers an uninterrupted view from their luxurious cabin, and that had the capacity to carry a tonne of mail and freight at a cruising speed above 160 kilometres per hour. The range was limited though, and the promotion started badly when the South African-born pilot Robert Prendergast[45] had to make a forced landing an hour short of Darwin, on Bathurst Island, when strong headwinds caused the *Astraea* to run short of fuel. Brackley sent a radio message asking for more petrol supplies to be sent by motor launch.[46]

The *Astraea* flew on to Darwin, Newcastle Waters, Camooweal, Cloncurry, Longreach and Roma, before Fysh and McMaster greeted the crew at Archerfield on 24 June[47] in a ceremony broadcast on the radio station 4BC.[48] Two days later Fysh and Rudder travelled in the *Astraea* on its five-hour flight to Sydney, where a huge crowd cheered the machine at Mascot,[49] before it powered on to Canberra and Melbourne.[50] 'The public in the South displayed great interest in the *Astraea*,' Fysh told the Brisbane press, 'and the few persons who were privileged to make flights in the big monoplane expressed their surprise that air travel could be so comfortable, and, above all, so quiet.'[51]

NELL PREPARED TO TAKE John and Wendy to stay with her family in Auckland as Fysh readied himself for a monumental expedition that would help transform Qantas. When he climbed aboard the Qantas mail plane at Archerfield on 4 July to meet the *Astraea* in Longreach, Fysh made history as the first Australian passenger to take off for London over the regular air route.[52] He brought a movie camera with him and wrote of the first part of the journey from the *Astraea*'s 'comfortable lounge seat' as though it was a 'magic carpet' ride. [53]

His report on the flight, published as a booklet[54] and in newspapers around Australia, highlighted the exotic sights along the way – and captivated the nation with the best publicity possible for selling Qantas tickets. 'The long stages are wound off each day under the wings of the big monoplane,' Fysh wrote. 'Each evening a new land is reached, with its different people, manners, and customs ... It is a perfect kaleidoscope of change.'[55] He also wrote, 'After the long stages one stepped from the machine fresh and clean, and without the usual experience of deafness and headache which is associated with more noisy types of aircraft.'[56]

The flight from Darwin to Calcutta took Fysh over jungles, impenetrable forest, and islands from which mountains rose to almost four thousand metres. The *Astraea* powered along the shores of Sumbawa, in what was then the Dutch East Indies, with dense jungle stretching away to the foothills of mountains lost in cloud and mist. The shadow of the aircraft raced across the water below, and a lone whale broke the surface in a shower of spray before disappearing into the inky depths.[57] Then it was on past 'the wonder island' of Bali to Surabaya on Java, and then Singapore with its 'maze of shipping, steamers, native junks, and boats in the harbour'. While at the military airfield there, Fysh watched a big KLM Fokker take off with the Amsterdam mail, carrying the American air racer Major Jimmy Doolittle and his wife Josephine to Bangkok;[58] a few years later, Doolittle would lead the American bombing raids on Tokyo. While 'Siam and Burma' were a mass of paddy fields,[59] the 112-metre Shwedagon Pagoda – 'entirely covered with gold leaf' – glistened in the sun as the *Astraea* came in to land at Rangoon, in a 'never-to-be-forgotten' experience.[60] The aircraft touched down at Calcutta's Dum Dum aerodrome, and then Allahabad, Cawnpore, Delhi and Jodhpur.[61]

At Karachi, Fysh transferred to an even bigger machine, the *Hannibal*, one of the massive four-engine Handley Page HP 42 biplanes flying the five-day London–Karachi route. It was the biggest airliner in the world, with a wingspan of forty metres, and it could carry up to thirty-eight passengers. At Sharjah, Fysh stepped off the *Hannibal* and was immediately introduced to the captain of the local guard, resplendent in a deep red headdress, well-filled cartridge belt, silver curved dagger scabbard, and sandalled feet. Next Fysh met the local ruler, Sheikh Sultan Ibn Saghir Al Qasimi, 'a most colourful person, with direct, piercing eyes and a rich dress set off by a beautifully worked, gold-mounted sword and scabbard' of solid gold.[62] Landing at Basra, Fysh said, was 'like descending into a huge oven with a forced draught of burning hot air'. It was 118 degrees Fahrenheit (48 Celsius) in the shade – more than equal to a Longreach heatwave. Then, on the way to Baghdad, the big aircraft passed over the vast marshes between the Tigris and the Euphrates.[63] Near Baghdad the ruins of ancient Babylon were far out on the left, and the captain came down and flew low over Ctesiphon and its colossal 1600-year-old arch. Baghdad was equally astounding, with what Fysh believed to be the most modern airport in the world; it boasted a well-designed and equipped central administrative building, booking offices, customs clearance, post office, spacious lounges, bookstalls, a bar and hotel accommodation 'all in modern, efficient style'. The Airport Hotel was full, so Fysh stayed at the Tigris Palace, paying for a drink at the bar in the sixth currency since leaving Australia.[64]

Back in Alexandria for the first time since the war, Fysh alighted from the Handley Page to continue the journey on the *Satyrus*, a large flying boat powered by four Bristol Jupiter engines. One passenger had to be jettisoned to keep the weight down as a strong headwind prevailed, but sitting in his comfortable lounge chair, watching the Mediterranean sail by, Fysh imagined he was cruising on 'some sumptuous private yacht'. When the aircraft refuelled on Crete, some of its passengers donned their bathing suits for a dip. In Athens, Fysh's senses were swimming as he marvelled at the ancient ruins he'd only ever seen in pictures. He said the trip from Athens to Brindisi 'surpassed in beauty any portion of the trip from Australia'.[65] The indescribably blue open bays were bordered with the palest blue where the water shoaled off to meet the rocky, timber-clothed hills, which changed colours as the cloud shadows floated past. Little white-walled villages nestled on the hillsides.[66] Before long Fysh was back on one of the huge biplanes, having a delicious breakfast over the English Channel as he neared Croydon.[67]

FYSH WAS APPROACHING his thirty-ninth birthday, and the shy, awkward boy from Launceston was barely a memory as the Qantas boss met with Woods Humphery in London and took advice from the former prime minister Stanley Bruce, who was now Australia's resident minister in the United Kingdom. Fysh sent a draft agreement with Imperial for McMaster to present to the Qantas board, and he also sent him an uncharacteristically curt missive in response to public criticism from Norman Brearley over Imperial's accident record. Fysh told McMaster that Brearley had apparently 'forgotten the opening of his own north-west route', which was marred by a fatal crash, 'and the fact that surplus subsidy revenue has not gone into his service but to enrich himself and his shareholders'.[68]

Brearley wasn't the only one, though, opposed to Imperial's involvement with the mail run. James Dunn – a NSW Labor senator who had added the middle name 'Digger' for his ballot entry to cash in on postwar patriotism – argued in Parliament that the heroic Australians Kingsford Smith and Ulm were 'to be left on the beach, while this profitable contract goes to Imperial Airways, to be run on British capital, manned by British airmen, and paid for with good Australian cash'.[69] But from London's Grosvenor Hotel, Fysh wrote to McMaster on 17 September 1933 to say there wasn't 'the slightest doubt but that control would remain in Australia', and that he was 'only more settled in my opinion that Woods Humphery is a great leader and that his executives are excellent'. 'In the early days,' Fysh wrote, 'you had your dreams for Qantas. Now all lies at our feet if we get Singapore–Brisbane. It will be no monopoly, however, as we have not yet done with Brearley and the Dutch.'[70]

Government policy dictated that from Darwin the airmail service should follow the Qantas route down to Charleville but then track further south to the NSW country town of Cootamundra, the halfway point in the rail link between Sydney and Melbourne. Brisbane was to retain its connection with Charleville, but there were to be no direct air connections with Sydney or Melbourne. Through the press McMaster argued that any England–Australia service should operate through Brisbane to Sydney, a view backed by members of the press who editorialised that by establishing the terminus of such an important link with London in Cootamundra, the government was taking a narrow vision of the 'new epoch about to open in Australian overseas communications'.[71] There were reports that European airmen were laughing at the route ending in a bush town and that the Australian government was retarding the development of international aviation for the sake of local railways.[72] But the government couldn't be shamed into changing direction.

On 22 September, forms for the tender specifying Cootamundra as the Australian terminus were issued. The following day, Fysh placed an order with de Havilland to build an aircraft for the Australia–Singapore run: the DH.86, a biplane with a twenty-metre wingspan and a cabin that could carry ten passengers. It would use four of de Havilland's new 200-horsepower Gipsy Six engines.

Confident he had the mail run safely in Qantas's hands, Fysh travelled to New York by ship and then flew aboard the United Airlines night service across the country in a revolutionary new machine, the Boeing 247. It was the first successful all-metal monoplane to be used in air transport, combining an anodised aluminium construction, retractable landing gear, and an autopilot system that Fysh observed in action as the machine cruised at 250 kilometres per hour. 'All these high speed machines here make everything look out of date in England,' he noted, 'except the Handley 42 – which is supreme in its class.'[73]

Fysh left Chicago on a dark and stormy night, bound for San Francisco. He felt depressed, even afraid, as he bounced along in the blackness 'on such a midnight horror', doubting he would see 'sunny Australia again'. But despite poor visibility outside, he could see the future clearly. United Airlines provided a service for a country bigger even than his own, using airline staff working around the clock. Chicago alone had forty-two air services every twenty-four hours, and there were revolving beacons every sixteen kilometres or so to aid night-time navigation.

Throughout this time, Ulm and the now Sir Charles Kingsford Smith continued to push the envelope as they tried to win the airmail contracts. Kingsford Smith flew a single-engine Percival Gull monoplane, the *Miss Southern Cross*, from the village of Lympne in Kent to Wyndham, in Western Australia, in a record seven days and four hours, in order to show how quickly he could deliver a letter if given the chance. A crowd estimated at a hundred thousand celebrated with him at Essendon on 15 October.[74] Ulm went one better: flying with Scotty Allan and P. G. Taylor in an Avro Ten, *Faith in Australia*, he arrived at Derby in Western Australia six days, seventeen hours and fifty-six minutes after leaving Feltham airfield in London's west.[75]

Fysh followed the progress of ANA through the American newspapers, but he knew that sound business proposals would overcome their death-defying stunts. 'I see Smith and Ulm are in the limelight,' he wrote to Woods Humphery from the Ambassador Hotel in Los Angeles, 'but we will win all the same so long as we go at it with undiminished energy, and that we will do.'[76]

In California, Fysh inspected the new Douglas DC-2 that was about to go into production for Trans World Airlines. He was told that it could reach 340 kilometres per hour. Douglas was advanced with their plans for the DC-3, which would become one of the most popular and durable aircraft of all time. While aboard the SS *Monterey* heading home, Fysh wrote to tell Woods Humphery that the DC-2 was 'the most ingenious, carefully thought out high speed aircraft ever put out'.[77] Its metal construction, he said, was even stronger than that of the Boeing, 'the skin taking most of the stresses', and it was undoubtedly the construction method of the future, 'simple, strong and lending itself to perfect streamline'. While the Douglas machines were only twin-engine and not suitable for the Singapore run, Fysh told Woods Humphery that Imperial should implore de Havilland to adopt the metal skin construction because cruising speeds over three hundred kilometres per hour were now 'safe, practical and essential'.[78]

FYSH WAS REUNITED with Nell and the children in Auckland after four months of travelling. Around the time they returned to their home in Brisbane, his campaign to win the airmail tender was given a lift by an article in the British magazine *Aeroplane*, which commented on Qantas completing two million miles of commercial flying: 'This is a remarkable record of sound commercial aviation made with a degree of safety and regularity which has not been surpassed by any operating company in the world. Mr Hudson Fysh, the managing director of the service, deserves well of the

Commonwealth.'[79] At a Qantas board meeting in Brisbane on 14 November 1933, a week after Fysh's return, a motion was carried that Qantas Empire would be the tendering company.

Ten days later Fysh screened a film of his travels in the King's Hall at Parliament House in Canberra. The esteemed geologist Walter Woolnough worked the projector. Members from both sides of the House watched the film and listened to Fysh's lecture on aviation, with Fysh nodding towards his crusty old antagonist Billy Hughes, who took a seat as close as he could to the screen. Controversy erupted the next day, and not because the screening was temporarily paused by a projector meltdown – members of the Opposition decried the film as nothing but propaganda for Imperial Airways to steal business from Australia.[80] Fysh brushed off the criticism, and soon he was joined in Brisbane by Sidney Dismore,[81] Imperial's London-based secretary. Together with George Harman and McMaster, who had just finished the shearing at Devoncourt, they began preparing the final draft of the tenders for a new entity.

On 18 January 1934, Qantas Empire Airways Limited was registered in Brisbane. Four days later Fysh and McMaster, representing Qantas, and Dismore and Albert Rudder, representing Imperial, approved the board of directors for the new airline QEA. They were: McMaster, chairman; Rudder, vice-chairman; Fysh, managing director; the former Victorian premier and one-time acting prime minister William Watt,[82] director; Fred Loxton, director; a Sydney accountant, Frederick J. Smith, director; and George Harman, secretary. Fifteen years earlier, William Watt had announced the £10,000 prize for the England–Australia air race in Parliament. The Qantas–Imperial agreement provided for a capital of £200,000,[83] with each airline providing forty-nine per cent and the remainder under the control of Sir George Julius, a prominent inventor and businessman who, Fysh noted, acted as 'an umpire who never had to blow his whistle'.[84]

At a Qantas board meeting on 24 January, Fysh told his fellow directors, 'The closing of Qantas operations marks the termination of a singular pioneering effort in commercial aviation which will not be repeated.'[85] That same day, Fysh and Dismore boarded a train for Melbourne with a large black tin box containing their twenty-two tenders for the various legs of the airmail run, which were to be submitted to a committee including Edgar Johnston by 5.06 p.m. on the last day of the month. Between Sydney and Melbourne, Ulm and Brearley of all people – heading to Melbourne to make their own bids – sat on a bunk for a yarn with Fysh and Dismore. They rested their feet on the tin box containing the Qantas secrets.

At 4.15 p.m. on 31 January, Fysh submitted the tenders that would involve five of the new DH.86 biplanes. The tender price submitted for the Singapore–Brisbane main trunk service and the Charleville–Cootamundra terminator was based on two shillings and 11 pence a mile, reducing in the fifth year by sixpence a mile, for a total subsidy of £339,486. The tendering process was fierce, with intense lobbying and jockeying for positions, as well as accusations that QEA was colluding with de Havilland at the expense of the other companies in the race.

Ulm was in such desperate financial straits that he considered restarting his career with barnstorming tours. But Fysh noted that he and his wife Jo still found time to go out of their way to give a great deal of comfort to Lieutenant-Colonel Brinsmead, who died at the Austin Hospital in Melbourne on 11 March 1934.[86]

Nerves were fraught in the lead-up to the announcement of the winning tenders in Canberra by Prime Minister Lyons on 19 April. Qantas Empire Airways was awarded the Australia–England air route between Singapore, Darwin, and Brisbane. Arthur Butler won the Charleville–Cootamundra section, while Ivan Holyman's[87] family-owned Tasmanian Aerial Services picked up the Melbourne–Launceston–Hobart contract. Norman Brearley failed in his bid for the Katherine–Perth route, which went to Horrie Miller's MacRobertson Miller Aviation. Miller had formed the company with backing from Sir Macpherson Robertson,[88] the chocolate manufacturer who produced such Australian favourites as the Cherry Ripe and Freddo Frog.

The prime minister told the press that the DH.86 was the fastest four-engine airliner in the world and that Qantas would commence the Singapore service within eight months. The flight from Brisbane to London would take no more than fourteen days, Lyons said, a length of time that would eventually be reduced with experience.[89]

The vanquished responded differently. Ulm was gracious in defeat, but Kingsford Smith complained bitterly of injustice, railing against the involvement of a British company in Australian affairs.[90] Brearley had not only failed with his overseas bid, but he had also lost the Perth–Katherine service. Soon he would sell his company and the rights to the Perth–Adelaide route, leaving Qantas as the only survivor from the pioneering days.

McMaster had two reasons to celebrate. He had also helped Graham Fysh secure the management of Toolebuc, a 140,000-hectare station about 120 kilometres south of McKinlay on Boulia Road. Now that the young man had a solid financial position, McMaster's concerns about his daughter marrying him were assuaged. Jean McMaster would marry Graham in July in Rockhampton, meaning that Fergus McMaster and Hudson Fysh were now family as well as business partners.[91]

As always Fysh, a cool head in a crisis, had kept Qantas flying through rough conditions. His airline was more than ready to take on the world. But as Lester Brain left for England on 23 May 1934 to oversee delivery of the new de Havilland four-engine aircraft, disaster was waiting.

Chapter 20

Another great factor in success in the risky business of aviation in the 1920s and '30s was pure survival, and thoughts are instinctively turned to those numerous leaders who gave their lives, or lost their lives, in civil aviation. The toll was indeed high.

HUDSON FYSH IN 1966, RECALLING LOST FRIENDS[1]

HAVING GROWN UP in a dysfunctional, broken family, Fysh set about, with Nell, giving his own children, John and Wendy, a much more solid foundation in life. To their children the couple became known not so much as Mum and Dad but as 'Hud' and 'Darl', the names they and their own children used more frequently for them.

The house in Miles Street, Wooloowin had a large wooden verandah that was ideal for Wendy to race her tricycle. Fysh would often take Wendy and John out to the backyard tennis court to give them lessons or to set up a cricket pitch. Their collie dog would field the ball. Almost ninety years later Wendy could recall all the times standing in front of the stumps with the huge cricket bat in her hands and her father telling her with a wry smile 'Good on you, Wendy, you got a duck!'. She was so proud and couldn't wait to score another.[2]

The children always knew they were loved and cared for, even if Nell sometimes took her love of fashion too far, dressing Wendy like the young princesses Elizabeth and Margaret in tweed coats and gloves during sticky Brisbane summers. Around the dinner table, Fysh would entertain his wife and children with funny stories about the characters he met at work, and about whom he invented nicknames, 'Ole Blow His Bags Out' Murray, a windbag who never stopped talking himself up, and 'Depreciation Duff', the accountant always going on about how to profit from a loss.[3]

Whenever Fysh could get a break from charting a course for Qantas he steered his family to Queensland coastal holidays at Noosa, and at the newly named but largely barren retreat of Surfers Paradise. For a few years Fysh owned a parcel of land there but offloaded it before the development boom.

Wendy's fondest memory of her father was what she described as 'the beauty and strength of his hands' with square palms and long, elegant fingers. She remembers always how as a child she always felt safe and secure holding one of those hands.

Safety was always Fysh's first priority at the helm of Qantas, too, even though aviation in the 1930s remained a hazardous pursuit.

The death of Lieutenant-Colonel Brinsmead still cast a pall over the industry when the first full board meeting of Qantas Empire Airways Ltd was held at Challis House in Martin Place, Sydney, on 1 May 1934, two weeks after the successful airmail tender. Fysh remained determined to make aviation a popular method of transport for all Australians.

On 14 June 1934 Fysh took off from Archerfield, flying a Gipsy Moth, to inspect the landing grounds and facilities along the mail run to Singapore. Arthur Baird and another pilot, flying a pair of Puss Moths, accompanied him as far as Darwin.[4]

While Fysh was in Darwin awaiting the steamer *Mangola* to Surabaya and then a KLM flight for Singapore,[5] Albert Plesman, the chief of KLM, announced in London that his airline intended to compete with Imperial for the service to Australia. For the Amsterdam–Batavia run, the company was introducing a new Fokker and one of the Douglas DC-2s.[6] After visiting Singapore, Fysh returned to Brisbane ready to fend off the Dutch as he awaited Lester Brain's arrival from England with the first of the new DH.86 airmail machines.

Australians would soon get their first look at KLM's DC-2, along with other high-speed aircraft, in an aviation extravaganza. The Centenary Air Race was a huge event featuring twenty of the fastest machines in the world. Organised as part of the celebrations for Victoria's centenary of European settlement, the race would take intrepid contestants from the Mildenhall RAF base, north of London, to the Flemington racecourse in Melbourne. There would be no limit to the size or power of the aircraft, and no limit on the number of crew. The chocolate baron Macpherson Robertson was providing £15,000 in prize money to entice many of the world's leading pilots and aircraft manufacturers. Kingsford Smith and Ulm were still dominating aviation records, but both would miss the event as they concentrated on separate crossings of the Pacific.

The race was due to start at dawn on 20 October 1934.

With the increased mileage for his merged company, Fysh hired new pilots, including 34-year-old Norm Chapman,[7] who had honed his skills flying the Melbourne–Hobart run. Fysh had chosen him among others to eventually fly the DH.86 aircraft on the Singapore route, via Darwin. Chapman had suffered a traumatic start to his aviation career in 1926 when he crashed a Curtiss biplane in Essendon, an accident that killed his older brother Geoff.[8] The grieving pilot named his first-born son in Geoff's honour.

As the aircraft were being readied in England for the Centenary Air Race, Norm Chapman took off from Longreach at 5.40 a.m. on 3 October flying Qantas's trusty *Atalanta*, a reliable machine for five years under the control of Lester Brain. Chapman was ferrying the mail to Darwin and had two passengers: Henry Henrickson, a sales manager for Shell Oil, and William McKnoe, the manager of the Sandalwood Cutting Company of Winton.[9]

Chapman circled the homestead of Fergus McMaster's property, Moscow, to drop a parcel of papers. Flying low, he continued towards Winton, where he was due at 7 a.m. He never made it. The following day, a search party found the charred remains of the aircraft and the three men about twenty-five kilometres away at Vindex Station.[10] McKnoe's watch, smashed in the accident, had stopped at 6.57.

Fysh was devastated. He issued an immediate statement to the press: Qantas 'extremely regretted the unfortunate accident to the *Atalanta* and the sorrow caused by the loss of life'. The only possible consolation was, he said, that since the inception of the company in 1920, month in and month out, Qantas aeroplanes had been 'winging their way on errands of mercy, conveying the sick and doctors and so being the means of saving many lives'.[11] Fysh pointed out that Qantas had flown more than two million miles (3.2 million kilometres) and until now had gone six years without a serious accident. 'Accidents, unfortunately,' he told the press, 'are part of every transport endeavour, but aviation is inclined to suffer more from the consequences of accidents as the spotlight of public imagination is still centred on its every doing.'[12] Privately he wrote to Chapman's widow Ella, left with two small sons, to say, 'There is no authentic evidence whatever which throws any blame on Norman. We mourn the loss of a good officer.'[13]

Ten days later, Lester Brain landed the first DH.86 in Brisbane, having spent nineteen days in unrushed transit. The new QEA mail service was due to start in two months, and while Fysh already had Brain and Russell Tapp for the over-water Darwin–Singapore route, he also hired Scotty Allan. The former ANA pilot was well experienced in flying across the Timor Sea, which at 824 kilometres was still the longest regular over-water route in the world, and Fysh offered him £650 a year. To replace Norm Chapman on the Longreach–Darwin leg, Fysh hired Bill Crowther, a former jackaroo, who, while on horseback, had watched the Smith brothers fly over the railway line outside Charleville fifteen years before. Brain had taught Crowther to fly, and Crowther had gone on to become an instructor for the Queensland Aero Club.

Fysh himself, closing in on his fortieth birthday, was nearing the end of his commercial aviation career, having clocked 1699 hours and fifty-five minutes at the controls, with just three more hours to come for Qantas in 1935.

The 1930s was a decade when many other pioneering aviators also flew for the last time. The Holymans had ordered two DH.86s for their service across Bass Strait, and the first, *Miss*

Hobart, made its debut on 1 October, linking Melbourne, Hobart and Launceston. Just eighteen days later *Miss Hobart* disappeared over Bass Strait after leaving Launceston's Western Junction Aerodrome at 9 a.m. On board were pilot Gilbert Jenkins, pilot and co-owner Victor Holyman, and nine passengers. The next day patches of oil were discovered sixteen kilometres off Victoria's Wilsons Promontory and a seat was washed up further along the coastline at Waratah Bay.

CHARLES SCOTT AND HIS copilot Tom Campbell Black were rocketing their way across the globe towards Melbourne as the search for *Miss Hobart* survivors continued. Scott had flown the London–Australia route twice, but in nothing approaching the speed of his streamlined twin-engine DH.88 Comet monoplane, sponsored by London's Grosvenor House Hotel. It was one of three Comets in the Centenary Air Race, and when it flashed in a scarlet blur between the finishing pylons at Flemington Racecourse on the afternoon of 23 October, Scott and Campbell Black were £10,000 richer. They had taken just under seventy-one hours for the flight from England. A hundred thousand people on the racecourse below cheered wildly as the former Qantas pilot made headlines around the world.[14]

Only twelve of the twenty race starters made it to Melbourne. The British pilots James Baines and Harold Gilman were killed in a crash at Palazzo San Gervasio, Italy, while Amy Johnson and her husband Jim Mollison, flying a Comet called *Black Magic*, suffered engine failure over India and had to withdraw. Second place in the race went to the KLM DC-2 *Uiver* (meaning 'stork' in Dutch), but only after the crew averted a near-death experience on the final leg from Charleville to Melbourne.

The DC-2 was pitched into an electrical storm. Lightning wreaked havoc on the navigation equipment and radio communications, and the pilot became lost in the darkness. Three local residents went to the Albury electrical substation and began flashing the town's streetlights to signal A-L-B-U-R-Y in Morse code. There was no local airport, but just before 1 a.m. Arthur Newnham, the 2CO radio announcer, sent out an urgent message for cars to assemble at the Albury Racecourse. About eighty cars arrived, and the big KLM airliner touched down safely at 1.17 a.m. thanks to the headlights shining on a makeshift landing strip.[15] The next morning three hundred locals gathered to haul the DC-2 out of the mud as the Dutch continued to Melbourne, winning the handicap section in just over ninety hours.

An American crew under the skipper Roscoe Turner, in a Boeing 247, took third spot. True to form, Ray Parer – fifteen years after making his first England to Australia effort at tortoise pace – eventually made it to Melbourne as the last finisher, four months after Charles Scott's victory.

Sadly, the Dutch DC-2 crashed in Iraq only a few weeks later on a Christmas flight from Amsterdam to Batavia, and all seven people on board died in the desert. Two years later, Tom Campbell Black also died in an aircraft crash.

Despite such disasters and the downpour of morbid headlines about the dangers of flying, Fysh's company continued to grow. He moved the Qantas offices from the Wool Exchange Building to bigger premises at 43 Creek Street, Brisbane, and approved 27-year-old Dubbo-born Cedric Turner[16] as the company's chief accountant. Turner had worked in the 1930s in London, Paris, and Rotterdam, and was in Berlin when he was recruited by Imperial Airways to be chief accountant for the new airline formed with Qantas. He left Berlin during the Nazi killing spree known as the 'Night of the Long Knives' in 1934. Turner could be abrupt and abrasive, but he was destined for big things at Qantas.

THE COMMONWEALTH Air Accidents Investigation Committee was in session in Sydney on 15 November 1934, still looking into the Holymans' crash, when a few kilometres from Longreach on the Ilfracombe road, two kangaroo shooters eating a campfire breakfast saw the early morning sun glisten on the aluminium-painted sides of the newest Qantas DH.86 as it flew low.

The DH.86 had just arrived from England and was on its way from Darwin to Fysh at Archerfield. At the controls was Robert

Prendergast, Imperial's chief pilot and the man who had brought the *Astraea* to Brisbane the previous year.

As the aircraft flew towards the kangaroo shooters, they shaded their eyes and stood slack-jawed as the machine, just a few hundred metres away, began a slow flat turn to the right, then started to spin and fall wildly. Prendergast and his two crewmen were killed instantly. Their passenger Emory 'Bunny' Broadfoot, a rotund and jolly Shell Company representative,[17] died soon afterwards.

The loss in quick succession of not one but two of the new de Havilland aircraft, 'in reasonably good weather for no apparent reason',[18] sent the Qantas plans for overseas expansion haywire. Three days after Prendergast's death, P. G. Taylor told the press in England that while he was about to travel to Singapore to join the Qantas international service, he was having second thoughts.[19] 'The sorrow is increased,' he said, 'by the fact that I had personal friends whose skill was undoubted on the *Miss Hobart*. I also knew Captain Prendergast by repute as an excellent pilot.'[20] On 21 November, with the new Qantas Singapore service due to start in less than three weeks, Taylor wired Fysh that the second DH.86 accident proved the unsoundness of the aircraft and that he wouldn't fly them 'under any circumstances'.[21] The DH.86 was withdrawn from flying in Australia pending further investigations and rigorous testing by Lester Brain.

Meanwhile, the federal government insisted that the new airmail service should still start on 10 December. Not only was this the anniversary of the arrival in Darwin of the Smith brothers in 1919, but there had been extensive planning of a grand opening of the service at Archerfield by the King's son Prince Henry, the Duke of Gloucester, who was in Australia on a royal tour. In a telephone call between Fysh in Brisbane and Imperial's Herbert Brackley in London, it was decided the service could still proceed but that until the DH.86 problem was solved, Imperial would fly the Singapore–Darwin route with their four-engine Armstrong Whitworths, while Qantas would fly their single-engine DH.61s and DH.50Js from Brisbane to Darwin.

They weren't the only ones making do. Charles Ulm's loss to Fysh in the tendering process hadn't dissuaded him from trying a different route to corporate success. Ulm was a fighter, and he had outfitted a new aircraft, *Stella Australis*, a British-built twin-engine Airspeed Envoy light transport machine. In September he had established a new company, Great Pacific Airways Ltd, and he planned to operate a San Francisco–Sydney airmail service.

On 3 December, as Australia's aviation industry was still reeling from the DH.86 crashes, Ulm was flying from Oakland, California, to Honolulu, Hawaii, to test his new route. He sent a series of radio messages in Morse code to Hawaii advising that he, his copilot George Littlejohn and their navigator Leon Skilling were lost and low on fuel. Ulm was not heard of again despite an extensive week-long search[22] by aircraft and twenty-three naval ships. Two years later Ivan Holyman registered a new Australian National Airways to control the air routes between Perth, Adelaide, Melbourne and Sydney.

THE STRESSES OF THE QANTAS accidents and doubts over the company's controversial new aircraft were taking a fierce toll on Fysh's health. He began to complain of heart palpitations, anxiety and insomnia, but he was all smiles when he and his family were at Archerfield Aerodrome on the morning of 10 December 1934 to see the launch of a new era in Australia's relationship with the rest of the world. Sporting a glowing suntan, the Duke of Gloucester had arrived in Brisbane early that morning from the NSW North Coast, riding a special train with a squadron of aeroplanes overhead as an escort.[23]

Prime Minister Lyons told the crowd that the opening of the service to Singapore marked the Commonwealth's entry into the sphere of international aviation.[24] Exactly fifteen years after he shook hands with Ross Smith after the flight of the Vimy, a beaming Fysh watched as Lyons handed the King's son a pair of golden scissors in the shape of propeller blades. The duke then cut a length of red-and-blue ribbon joining the Qantas Empire Airways' DH.61 *Diana* to a post in front of the official stand.

The *Diana*, which had its engines running, taxied away at 10 a.m., piloted by Lester Brain, and was followed closely by *Hippomenes*, piloted by Russell Tapp. They took off as part of a spectacular aerial display by nineteen aircraft. Together they were carrying half a tonne of mail for England, and by the time they reached Darwin they had collected even more mail brought up from the railhead at Cootamundra to Charleville by Arthur Butler's airline. Fysh's two aircraft arrived in Darwin with a total of 55,967 letters and parcels that were loaded onto the Imperial Airways aircraft *Arethusa.* The mail reached London's Croydon airport just after midday on 24 December 1934, fourteen days after leaving Brisbane. The saving over land and sea transport was about twenty days, something Fysh recalled as being 'quite spectacular'.[25]

After this success, and after the rigorous testing of the DH.86 and modifications to its rudder tab, Fysh had a late fortieth birthday present when the Civil Aviation Branch approved the aircraft for service on 25 January 1935. Then the fins on two new machines from England were found to be faulty, needing repairs and modifications, which delayed Qantas's start as a truly international carrier.

Fysh gave his new machines the prefix RMA, for Royal Mail Aeroplane. The first of them to enter scheduled QEA service, RMA *Canberra*, finally left Brisbane for Darwin on 29 January 1935, with Lester Brain in command. It was the start of multi-engine operations for Qantas and of a new two-day service, with the DH.86s cutting a day off the old schedule. But it continued to be a difficult time for the airline, with Imperial discussing plans to scrap the DH.86s just as they started operating, in favour of flying boats.

There were tensions on the QEA board over Imperial's policy of secrecy: the company gave out information on their plans only to select executives. McMaster was furious that Imperial's manager Woods Humphery was sending Fysh correspondence marked for his eyes only, and an agitated Fysh complained to Woods Humphery, 'It is no exaggeration to say that, as things are, board meetings are impossible at present.'[26] Such a state of affairs, he feared, could be 'crippling to the proper conduct of the whole business' and on top of all the problems with the DH.86 could be the 'last straw'. 'At present I enjoy the full confidence of the Board,' Fysh told his English counterpart, 'at least in regard to the Q.A.N.T.A.S. side, but I must say, while the present position prevails, I am not anxious to receive any letters from anyone marked "Private" or "Confidential", and I am not anxious to write any letters under these headings.'[27] Fysh had been imploring McMaster to move to Brisbane and relieve Fysh's workload; he confided in Woods Humphery that McMaster was away on his western Queensland properties most of the time and only came to the capital for meetings once a month. Fysh said he was often left to effect important decisions in Brisbane without McMaster's assistance.[28]

The strain continued to damage Fysh physically. Sleep became a stranger, and he would rise pale and bleary-eyed in the morning for work. He had chest pains, too. Nell feared her husband would run himself into the ground. Soon his heart started to beat irregularly, and he smoked more and more. The Civil Aviation Controller Edgar Johnston urged Fysh to take a long break.[29]

Fysh's brother Frith and his best man Jack Butler had died young, and Fysh planned to live a lot longer than his forty years. Under doctor's orders he gave himself a two-week holiday to enjoy the sun, sand and fishing just off Brisbane's coast. An aircraft took him and Jack Mehan, a friend who worked for Shell, onto Stradbroke Island, and they made their home at the Moreton Bay lighthouse as guests of its keeper and his wife. She made memorable curries from goats shot on the island and cooked superb seafood dinners from the bounty that Fysh hauled out of the sea. Fysh and Mehan stayed hydrated with longneck bottles of beer kept cold in a deep well. Fysh would spend his mornings watching seabirds circle over the cliffs and sea turtles frolic among the waves, and he and Mehan would walk down a winding cliff path to swim in warm rockpools.

After a fortnight in the bracing sea air and foaming waters, Fysh was back at his desk.[30] He promised Nell that he would never let his health suffer from overwork again, though he continued to drive himself hard.

ON 25 FEBRUARY, the day after Fysh returned from his vacation, Lester Brain flew the first international flight for Qantas Empire Airways, taking off in RMA *Melbourne* from Singapore for

Darwin with the London mail. The following day, just after dawn broke over Fannie Bay, Scotty Allan taxied the RMA *Canberra* to the far corner of the dew-covered grass runway in Darwin, accelerated over the slab marking the place where the Vimy had first touched down in 1919, and lifted the big biplane majestically into the air above the Timor Sea, its four engines creating a blur in the first overseas departure from Australia of a Qantas flight. Fysh made another inspection tour of the mail route to Singapore, then declared all was in order for Qantas to carry its first international passenger out of Australia.

Major Harold Pedro Joseph Phillips,[31] 'Bunnie' to his friends and lovers, was a good-looking, raw-boned young giant who had family connections to the British nobility and to the President of Peru. In 1935 he was engaged in a torrid affair with Lady Edwina Mountbatten,[32] the wife of the celebrated Lord Louis Mountbatten,[33] a close relative of Britain's royal family. Edwina's daughter remembered Bunnie, an officer with the Coldstream Guards, as 'thrillingly handsome, with perfect posture … and being half South American, [he] rode like a dream'.[34]

At 8.05 a.m. on 17 April 1935, the 195-centimetre British officer became Qantas's first overseas passenger when one of Fysh's new pilots Herbert Hussey[35] took off with him from Archerfield in the RMA *Melbourne*.[36] Publicly, Phillips said he had been 'escorting' her ladyship on a pleasure cruise from Tahiti to Sydney, and was now accompanying her to Malta where Lord Louis was commanding his first ship, the destroyer HMS *Daring*. The Qantas flight collected Edwina in Charleville,[37] after she had been holidaying in Sydney, and she and Bunnie flew together to Singapore before changing onto an Imperial aircraft. Being the first passengers on a Qantas international flight had excited the couple's love for adventure.

Edwina and Bunnie lived together on and off for years. Lord Louis considered divorcing her, but it was said he had too much admiration for Bunnie to publicly scandalise him – and he was perhaps too busy having his own affair with Yola Letellier, a French socialite and the wife of a newspaper baron. Years later, Lord Louis's nephew Philip married the future Queen Elizabeth II.

The DH.86 RMA Melbourne was the first Qantas aircraft to make a scheduled international passenger flight. *John Oxley Library, State Library of Queensland, Libraries Australia ID 66921357*

THE WEEKLY BRISBANE–LONDON airmail route left Archerfield for Croydon every Wednesday just after 7 a.m. It was by far the longest air service in the world: 20,525 kilometres spread over twelve and a half days. It involved not only five different types of aircraft, but also two types of railways and as many as thirty-one stops. A first-class ticket cost £244 at the tail end of the Great Depression, when an inner-city Brisbane house could still be snapped up for less than £1000.[38] Qantas passengers would fly with the mail on the DH.86 to the British air force base at Seletar Aerodrome in Singapore, 7018 kilometres away. That journey took three days and, depending on the amount of mail, could be extremely cramped. There was no cabin staff, but the first officer would usually hand out sandwiches and pour tea from a flask.

The first day's bumpy flying through the unpredictable air currents over the hot, arid outback included six stops, with a long stay at Charleville to collect the Sydney and Melbourne mail from Arthur Butler's Cootamundra service. The day's flying usually ended with an overnight stay at Cloncurry's Post Office Hotel. Between Longreach and Cloncurry there were six revolving beacons a hundred kilometres apart to aid with night-time navigation if the service was running late. The next day's flying began as kangaroos were chased off the Cloncurry airfield, and it involved six more stops before Darwin, with a long break at Daly Waters, where the West Australian mail was collected from Horrie Miller's aircraft.

Qantas was still servicing the small towns that had spawned it while playing a major part in the world's most advanced international air route. For a time, Fysh lamented, as the international business began to dominate, 'these old friends and our Qantas shareholder founders got less and less consideration; they were sacrificed to the juggernaut which they had created'.[39]

In Darwin passengers stayed at a clifftop cottage that Fysh rented from Lord Vesty's meatworks. It looked out over the Timor Sea, which they would tackle the next day. Kupang, on Timor, was always seen as the gateway to the world for Qantas, as the locals – in their bright clothing, and leading their sturdy ponies – were vastly different in appearance and culture from what most European Australians were used to. The sight of them was a captivating entrée to all the exotic sights the Qantas passengers would encounter on their trip across the globe.

After the three and a bit hour flight from Darwin to Kupang, there was always a sumptuous breakfast of fried eggs and bacon with fried bananas. As the passengers continued to Singapore, they found the view of the Javanese islands spectacular, and the pilots sometimes flew low enough for them to see whales leaping out of the water or large fish and sharks just under the surface. On the island of Lombok, the passengers bounced along a bumpy road, dodging monkeys on the way to their overnight stay at a rustic Dutch hotel in Selong that was surrounded by a jungle and its 'babel of bird cries'[40]. The hotel had only three beds, so passengers would sometimes have to share, three to a bed. Above all this was the smoking 4000-metre volcano Mount Rinjani, its sharp peak enveloped in cloud.

From Singapore's Seletar Aerodrome, the passengers flew on a nine-seat Armstrong Whitworth, operated by Indian Trans-Continental Airways, to Rangoon, with its golden pagoda, and then on to Karachi. From there they flew to Cairo on the 24-seat Handley Page 42Es that were notorious for running out of fuel in headwinds. At Sharjah the aircraft landed on the desert sand, and the passengers spent the night surrounded by barbed-wire entanglements and armed guards on the lookout for warring tribesmen. Passengers then flew up the Persian Gulf to Basra and Baghdad, before soaring over the Pyramids of Cairo.

Then they caught an Imperial Airways flight on a Short S17 Kent class flying boat from Alexandria across the Mediterranean to Brindisi on the heel of Italy, often with a refuelling stop at Mirabella on Crete. Once when they were stopped there, Fysh and Herbert Brackley dived from the aircraft into the vivid azure waters, before the flight continued to Brindisi. Because diplomatic problems often prevented British aircraft from flying across Italy and France, Qantas passengers would ride on Italian and French railways to Paris, where they would board the Handley Page 42W Heracles model machines for London. The Heracles aircraft could carry thirty-eight passengers at a sluggish 160 kilometres per hour, giving them time to savour the last word in luxurious travel on the way to Croydon, with armchairs, stewards in white coats, and silver service for meals.

BY OCTOBER 1935, the Qantas DH.86s had flown almost half a million kilometres on the Brisbane–Singapore route without mishap. Then the future of the machines was again thrown into doubt. On the morning of 2 October, the pilot Norman Evans, who had once flown for Qantas, left Essendon in another of the Holyman DH.86s, *Loina*, heading for Flinders Island en route to Launceston's Western Junction airport. The *Loina* crashed into Bass Strait, and all five people on board were killed. Australia's new defence minister, Archie Parkhill[41] – known as 'The Archduke' and 'Sir Spats' for his habit of overdressing – ordered a committee of inquiry into the machines, which had now been involved in three fatal accidents within a year.

A month later, Sir Charles Kingsford Smith, his copilot Tommy Pethybridge and their Lockheed Altair, the *Lady Southern Cross*, disappeared over the Bay of Bengal in the early hours of 8 November. The pair had been trying to top Charles Scott's England–Australia record when they ploughed into the dark ocean.

Their bodies were never recovered, although eighteen months later, Burmese fishermen found an undercarriage leg and wheel, with its tyre still inflated. The tragedy continued to fan negative sentiment for Qantas's international business, and a cigarette was always in Fysh's mouth.

While rigorous testing of the Qantas machines continued, on 13 December 1935 yet another Holyman DH.86, *Lepina*, made a forced landing on Hunter Island off northern Tasmania while carrying eight passengers across Bass Strait. Edgar Johnston suspended all DH.86 flights. Fysh had promised Nell that he would never let work matters get on top of him again, but he considered this the most stressful period of his time at the Qantas helm.

The DH.86 finally passed its structural and directional tests and went back into service. By 16 May 1936, demand was such that Fysh made the Brisbane–Singapore service twice weekly. At that stage he was already planning for the replacement of the troublesome machines in favour of a faster service using the flying boats that Qantas and Imperial had been discussing for the past few years.

Fysh and Brackley had started surveying a new route from Singapore in a great lumbering RAF seaplane, the Singapore III model. It was built by the Short Brothers, who had been in the aviation business in England since the balloon days of 1897 and had become specialists in flying boats. Fysh wasn't so confident, though, in this offering from them, nor in their new autopilot system; thinking of Nell, John and Wendy back in Brisbane, he took out a £5000 life insurance policy before he and nine others in the survey party took off.

The seaplane reached Darwin and then headed towards Townsville, with Fysh advising the pilot to land on the wide Roper River. As they neared the river's mouth on the Gulf of Carpentaria, Fysh was startled by a loud crack as a cabin window sheared off and was blown apart by one of the propellors. But they landed safely and anchored in ten metres of water as kangaroos nearby bounded away and white cockatoos screeched overhead. While the men were inflating their dinghy to go ashore, they saw several sets of eyes and nostrils poking out of the water nearby –

and realised the Roper was hosting a crocodile convention. Several of the crew immediately decided they were too busy to go ashore, but Fysh and a few others braved it with one man standing guard in the stern with a gun.

The survey continued to Townsville, Bowen and Brisbane, where a length of the Brisbane River at Pinkenba became a Qantas flying boat base. The survey ended in Sydney, where Brackley decided that Rose Bay would be the perfect terminus for QEA's overseas operations.

The survey represented a monumental shift in direction for Qantas and the clear change in eras was emphasised in April that year by the death of their pioneering passenger Alexander Kennedy at the age of 99. When Kennedy's 89-year-old widow Marion died in September, their ashes were carried on a Qantas aircraft to Cloncurry from where Eric Donaldson flew them to McMaster's Devoncourt Station so they could be interred with a copy of Fysh's book *Taming of the North* inside a three-metre high monument.[42]

IN WESTERN AUSTRALIA, Paul McGinness was desperate to get back into aviation as everything else in his life crashed and burned. During his aerial surveys for potential mine sites, McGinness staked a claim and registered the name Randwick Gold Mine in 1935.[43] He hired an office in St Georges Terrace, Perth, and tried to float shares in the venture. He even posed beside his wife Dorothy and a new autogiro – an early helicopter – at Perth's Maylands Airport, hoping the publicity would help to sell shares in his mining business. Instead, he soon found he couldn't pay the rent or phone bill at his office. He only avoided bankruptcy with the help of his wife's father, Charles Baxter, who bought out the company's shares. McGinness retreated back to the farm to drink and to brood.

In January 1937, Fysh was determined to spend more quality time with his family and less with Qantas headaches. John was now at the Toowoomba Grammar boarding school and Wendy at Brisbane's Clayfield College and he and Nell took the children for a wonderful beachside holiday at Nambucca Heads in northern

New South Wales. But while McGinness's former observer from Palestine now had a loving family, a plush office, and the ears of governments in Australia and England, the old fighter ace faced the wrath of two women on his isolated property north of Perth.

A decade of tension erupted between McGinness and his wife. Dorothy telephoned her mother Jessie in Perth, asking her to visit and provide reinforcement – but when Jessie arrived, McGinness ordered his mother-in-law out of the house, and told her to take Dorothy and their two girls with her. Mother and daughter refused to leave, so McGinness, as impetuous as ever, stormed out instead, and Dorothy didn't hear of him again for years. She later told a divorce court that arguments had arisen over McGinness 'carrying on with other women',[44] and that after he cleared out, she had decided to stay on the farm, only to learn her husband had sold it over her head.[45]

McGinness went home to the old family farm in Victoria and managed to rustle up a job across Bass Strait as temporary caretaker of Launceston's Western Junction Aero Club. He started work on 2 March 1936 at £246 a year plus an extra £26 per flying course as a teacher.[46] But he soon ran afoul of authority: he refused to hand back the fee for a flying course to a dissatisfied student, then drew the wrath of bureaucracy for hiring a typist and asking the Department of Civil Aviation to pay her wages.[47] On 18 November 1936, the department told McGinness his services were no longer required.

BY SEPTEMBER 1937 WAR clouds were darkening over Europe, but Fysh flew to London to sort out the final details for the start of the Sydney–London flying boat service. The British were pushing civil aircraft into the background, working feverishly on fighting planes, and Fysh inspected the 'wonderful Spitfire'.[48] He had his own battle of sorts in Holland, when he finally met the KLM director Albert Plesman in his palatial office only to be harangued over government subsidies and free trade in the air.

Fysh flew on a silver KLM DC-3 back to London, where by March 1938 seven Qantas captains, as well as Arthur Baird and two ground engineers, had been trained in the complexities of the Empire flying boats. The machines were built by the Short Brothers at their Rochester Seaplane Works on the River Medway in Kent. The company built thirty-one of the S.23 C class machines, and Fysh initially bought six: *Coolangatta*, *Cooee*, *Carpentaria*, *Corio*, *Coogee* and *Coorong*. The enormous aircraft were likened to flying ships, with an expansive wingspan of thirty-five metres and four huge Bristol Pegasus XC engines that each produced 920 horsepower. The aircraft could carry fifteen passengers in great comfort with two air crew and a cabin staff of three. Cruising speed was 265 kilometres per hour with a range of 1225 kilometres.

On 18 March 1938, Scotty Allan and first officer Bill Purton[49] took off from Southampton in Qantas's first Empire flying boat, *Coolangatta*. Before long the *Cooee* joined it in Brisbane.

With Sydney now the terminus for the airmail service to London, Fysh uprooted his family after eight years in Brisbane and drove them to their new home in Sydney through a rain storm that caused severe flooding in many places they passed. At first the family lived in a boarding house at Double Bay before renting a flat overlooking Sydney Harbour. The kitchen window afforded magnificent views of the naval ships docked there and sometimes Wendy would watch what seemed like romantic fairytales as parties were hosted on board, women in glamorous dresses and sailors in crisp white uniforms.

Qantas moved into smart new offices at Shell House in Margaret Street, near Wynyard Station. Twenty years earlier Fysh had thought of making a few quid from rabbit trapping; he now earned £2000 a year[50] – the price of an average Sydney home – and headed a company that employed 228 people. Way ahead of his time, he began distributing a company code of conduct and notes on ethics.

On 5 July 1938, the nine and a half day Sydney–Southampton flying boat service began when Welshman Patrick Lynch-Blosse[51] and Bill Crowther took off in the *Cooee* from Rose Bay. Eighteen days later, Prime Minister Lyons signed a fifteen-year agreement for Qantas Empire Airways to fly Australian mail to London three times a week.

The Empire flying boats offered passengers a whole new world of leg room and luxury with a promenade deck, a smoking cabin, and a games area to play quoits or miniature golf. Breakfasts were sumptuous and dinners divine, with roasts, pressed meats and ox tongue, peach melba or cherry flan, and cheese and fruit before coffee, all served by stewards who were 'aerial pioneers of personal care and service'.[52]

Paul McGinness, meanwhile, was looking for a very different flying experience. With the Japanese invasion of China dominating the news and Adolf Hitler looming large over Europe, McGinness's thoughts returned to the days of Palestine when he and Fysh had fought the Germans in the desert skies. Now in his early forties, and worn down by stress and drinking, McGinness wrote to a senior member of the Chinese air force, outlining his distinguished wartime record and offering his services as a mercenary. The Chinese consul-general sent McGinness a letter on 3 August 1938 with his 'sincere appreciation' for an Australian's willingness to help China, but he politely declined the offer.[53]

FYSH'S JOY AT THE SUCCESS of the Empire flying boats was tempered by another tragedy. On 25 October 1938, the Holymans' DC-2 *Kyeema* crashed into Mount Dandenong when approaching Melbourne in poor visibility. Eighteen people died, including Fysh's old schoolmate Charlie Hawker, who had been the Federal Minister for Commerce but had quit on principle over a bill to raise politicians' salaries. There was press and public outcry following the crash due to the lack of radio safety aids provided by the Civil Aviation Department.

Another of Fysh's old friends from Geelong Grammar, the federal politician Jimmy Fairbairn – a former fighter pilot and prisoner of war who would soon be made the Minister for Civil Aviation – wrote to Fysh to express his sorrow over the tragedy and to declare that it had 'certainly put the press and public into a frame of mind to demand a clean-up of civil aviation, which has been due for a long time. Things may happen, I think, quickly.'[54] They did, and Fairbairn offered Fysh the job heading civil aviation in Australia, replacing Edgar Johnston.

Fysh declined the post. He had enough on his plate, and once again a cigarette was rarely out of his mouth as the anxieties mounted. He was constantly battling the Dutch for improved flight facilities in their territories on the route from Darwin to Singapore. Kupang was the worst of the Qantas stopovers because the aircraft often landed amid floating logs and other debris, and the Shell refuelling barge and other boats were unwieldy and unsafe.

At the end of 1938, McMaster was due to join the accountant Cedric Turner in London for discussions over subsidies with Imperial Airways. But on the eve of his departure the now 59-year-old chairman suffered a heart attack and his health never really recovered.[55] 'He will be out of action for some months,' Fysh told Edgar Johnston on 12 November. 'I must now leave for England in his place.'[56]

Once again Fysh said goodbye to Nell and the children and set off on the *Coolangatta* to wage war in Europe. Not long into the journey the *Coolangatta* flew through a hailstorm, and Fysh felt that the aircraft was being assailed by a 'machine-gun fusillade, or a thousand urchins hurling stones against a tin fence'. But the storm abated, and a lovely rainbow shone below the plane, above the blue sea.[57] It reminded Fysh that there was always light at the end of a tunnel, though he spent most of 1938 and 1939 worn down by overwork and business battles.

British Prime Minister Neville Chamberlain had returned to London from Munich, after meeting Hitler in September 1938, to declare 'peace in our time' – but few believed him. Fysh worried himself sick over what another war would do to humankind and how Qantas could survive it. He visited the Bristol company with Cedric Turner, trying to organise deliveries for large orders of spare parts for the big Pegasus engines in anticipation of a national emergency. Because of his Calvinist approach to work expenses, Fysh stayed at the crowded and rundown Cumberland Hotel.

He met with the imposing new Imperial boss Sir John Reith, the 198-centimetre tall former head of the BBC. They got on famously, and Reith offered Fysh a job as head of operations at Imperial. Fysh talked it over with Nell, who was undecided, and in the end he told Reith that his heart lay in Australia. He even showed Reith and other Imperial directors his lantern slide show about the birth of Qantas in western Queensland.

As Fysh displayed the pictures of the company's early biplanes, he did so with an immense sense of gratification over the way he had steered his airline to become a world leader. The London to Sydney airmail service had become so popular that Christmas 1938 created the kind of rush that neither Fysh nor anyone else involved in air transport had ever experienced. It was estimated that twenty-one million letters and parcels weighing 240 tonnes were sent on aircraft from Southampton over the Christmas–New Year period while Fysh was in England, and a total of sixty-five aeroplanes flying 1.8 million kilometres were needed to transport it all.

Fysh spent a white Christmas in London with Herbert Brackley and his wife Frida at their Knightsbridge flat – but not all of Fysh's European tour was as pleasant. In Germany he met the heads of the civil airline Lufthansa, and he inspected the extraordinary work on the grand Tempelhof airport at Berlin and the autobahns that led the world in roads. As officials clicked heels and raised their arms with a 'Heil Hitler' salute everywhere he went, and as portraits of the Fuhrer glared at him from almost every wall, Fysh knew that this German militarism was dangerous to the 'sleeping democracies' of the world. 'Never since the days of Napoleon has our future been so challenged,' Fysh wrote in his diary.[58]

He returned home in February 1939 only to face another crisis. At 7 a.m. on 9 March the *Capella* glided out of Rose Bay and headed for Singapore,[59] marking the hundredth flying boat service to depart Sydney. Three days later when Captain Hussey was taxiing the aircraft in to dock at Batavia, its hull was ripped apart by the remains of a wrecked frigate just under the surface. The *Coorong* had been swamped at Darwin, but it was repaired and put back into service. The *Capella*, though, was a write-off.

Fysh had long been fighting the Dutch over the dangerous nature of the East Indies seaplane ports. Soon, though, as he had long feared, there would be a much greater danger to Qantas aircraft than the debris of wrecked ships.

Chapter 21

Fellow Australians, it is my melancholy duty to inform you officially that in consequence of a persistence by Germany in her invasion of Poland, Great Britain has declared war upon her and that, as a result, Australia is also at war.

Australian Prime Minister Robert Menzies in his address to the nation, 3 September 1939[1]

THE QANTAS CAPTAIN Bill Crowther was winging his way from Sydney to Townsville in the Empire flying boat *Champion* on 1 September 1939, as the pilots of German Stuka dive bombers began their lightning war, the blitzkrieg, on the ill-equipped defences of Poland. Like screaming birds of prey, the Stukas sounded the start of Nazi terror, their howling, near-vertical bombing raids made all that more terrifying by specially designed propeller-sirens called 'The Trumpets of Jericho'.[2]

As the Nazis unleashed hell on Poland, Crowther, unaware of what was happening a world away, was admiring the jewels of Queensland's tropical coast and the seemingly endless chain of emerald islands surrounded by a sapphire sea. He arrived in Townsville to find the world on a knife edge, and Fysh notified him on 2 September that the Civil Aviation Department had ordered the *Champion* to return to Sydney.

The next evening, as Australians gathered in anxious silence around wireless boxes everywhere from the outback to the ocean, Prime Minister Bob Menzies was seated in a Melbourne radio studio, reading and re-reading the most important speech of his career. Menzies had been the nation's leader for only four months, since the death in office of Joe Lyons had been followed by the nineteen-day caretaker prime ministership of Earle Page, leader of the Country Party.

Australia was hushed, a nation waiting nervously, counting down the minutes to Menzies' scheduled 9.15 p.m. broadcast. Most of Australia already knew what was coming. When the moment arrived, Menzies spoke in a low, slow, sombre voice, declaring that Australia was again at war with Germany. He knew that thousands of Australians would die, and the nation's very survival as a jewel of the Empire would be threatened. 'No harder task can fall to the lot of a democratic leader than to make such an announcement,' he told Australia. 'We are therefore, as a great family of nations, involved in a struggle which we must at all costs win and which we believe in our hearts we will win.'[3] Many of the listeners had already lived through the ordeal of one global war, and this threatened to be a second.

Japan and Italy were slow to show their hands, though. On 6 September, the federal government decided that it was still safe for the Empire flying boats to continue with the London–Sydney route. A twice-weekly service resumed with a surcharge to reduce the volume of mail and make more room for essential supplies for the war effort.

Fysh was in the reserves for both the Light Horse and the RAAF, but the government felt he was much more useful running an airline that would be vital to the war effort than fighting on the frontline again, especially after McMaster had been incapacitated by his heart attack. Paul McGinness was keen to fight, though, even at the age of forty-three. He had found work at Burroway Station at Narromine in western New South Wales when war broke out, and he tried to re-enlist – without success, due to his age.

McMaster was in Sydney convalescing, and he, Fysh and Baird met up with McGinness on the rooftop above Fysh's office in Shell House for a twenty-year reunion of their trek to Darwin.[4] While Fysh was glad to shake the hand of the man who had been his mentor in the Great War, and who had been in charge during the early days of Qantas, he still found his old friend unpredictable, unreliable and unsuitable for anything requiring responsibility. He knew about the drinking and the marital disaster.

In November 1939, McGinness wired McMaster for help, hoping the Qantas chairman could come to his aid in the same way that McGinness had helped drag McMaster's car out of the Cloncurry River all those years ago – a gesture of generosity that had led to the formation of an international airline. He told McMaster that he had also wired the grazier's old political friend Sir Donald Cameron, who was now chairman of the NSW recruiting committee for the RAAF, but that the response was 'No vacancies in air force over the age limit'.[5] He asked McMaster to intervene on his behalf. McMaster told Fysh he would write to Cameron backing McGinness for an air force role. 'McGinness has a lot of faults,' he told Fysh, 'but, in the RAAF drive, I cannot see that he can do any harm, and it is possible that he, with his splendid war record, would make a very much more effective appeal than others ... I suggest that, if you can do so conscientiously, you should support his application also.'[6]

'I am sorry, Fergus,' Fysh wrote back, '[but] you will be well advised not to recommend him for any job, as he is quite unreliable and uncontrollable – and will only land up in a mess. Mac has had chance after chance but always ends up the same way. I have seen him lately and Arthur Baird and I both realise it is useless trying to do anything for him which entails employment in a job where average common-sense is required. In a fighting aeroplane he would be good even yet, if his health would stand it.'[7]

IN THE FIVE YEARS SINCE it was launched, Qantas Empire Airways had flown ten million kilometres by the end of 1939. The ailing McMaster saw Fysh as the logical successor as chairman.

In April 1940, representatives of the British, Australian, and New Zealand governments met in Wellington to reach a deal for a trans-Tasman mail service. At 6 a.m. on 30 April, Fysh and Edgar Johnston were among nine passengers on the Short S.30 flying boat *Aotearoa* that left Auckland with forty thousand letters for the nine-hour, fifteen-minute flight to Rose Bay. The Tasman Empire Airways' regular weekly service between Sydney and Auckland had been launched.

Ten days later, as the Nazis stormed across Europe, Winston Churchill became England's prime minister. By 26 May he had ordered the evacuation of his besieged British troops from Dunkirk. Britain faced bombing raids and imminent invasion, and with Italy having joined the war as a German ally, Qantas's link with London across the Mediterranean was severed.

A new weekly service, using sixteen flying boats, commenced on 19 June 1940. It traversed a horseshoe route from Sydney to Cairo and then Durban, where the mail was loaded onto ships for England. It became a twice-weekly service on 17 August. That was just four days after Fysh's old schoolmate Jimmy Fairbairn, the Minister for Air and Civil Aviation, died in the crash of an RAAF Lockheed Hudson bomber near Canberra. Nine others died when the aircraft stalled on approach to the aerodrome, including General Sir Cyril White, Chief of the General Staff, who had planned the brilliant evacuation of Gallipoli in 1915, and two other cabinet ministers: Brigadier Geoffrey Street, Minister for the Army and Repatriation, and Sir Henry Gullett, the Minister in Charge of Scientific and Industrial Research. The loss of three senior ministers eroded the Menzies power base.

Amid the turmoil of war and calamity, Fysh tried to retain a quiet and ordered private life. His son John was now at Fysh's old school Geelong Grammar, while livewire daughter Wendy was trying to fit into the sedate confines of the posh non-denominational Ascham School in Edgecliff. Fysh had moved the family into a new rental property nearby, making a new home on the second floor of the three-storey harbourside mansion *Ardenbraught* on the tip of Point Piper. The mansion overlooked Rose Bay and the base for the flying boats. Before school Wendy would often accompany Fysh on his visits to the base to make sure all was well. She would swim lengths in the local baths there with Fysh timing her, and pencilling the times on graph paper which he would stick behind the kitchen door. Wendy would always ask her father for some chewing gum that Qantas kept in small packets for passengers to deal with air sickness, but he always refused her saying the gum was for passengers only and it would be wrong to take it from the company. Arthur Baird was easier to sway, though, and in between working on the flying boat engines he could always find her some spare gum. At home Fysh was always concerned about the amount of housework Nell had to perform. He always cleaned the bathroom for her each morning before work and he and Wendy were always in charge of doing the dishes.

The world outside *Ardenbraught* was not so placid though, and Fysh had to call on all his calm resolve as he became well used to tragedies claiming friends and colleagues.

IN THE 1941 NEW YEAR'S honours list, Fergus McMaster was created a Knight Bachelor in recognition of his long services to Australian air transport and the pastoral industry. But it was Fysh who was foremost in the government's sights; he was asked to start a QEA service to Dili, in Portuguese Timor, where the Japanese had established a disquieting presence. They had not yet declared war on the Allies, but their rape of China gave the world a taste of their huge military power and ruthlessness.

Fysh flew to Dili and undertook scouting missions there, reporting to Menzies. He told the prime minister that the local governor was apparently pro-British but would trade with whoever would buy the products of his people – and the Japanese were spending big. 'The Japanese penetration has been carefully planned,' Fysh wrote, 'and had been most effective [and] the Japanese at Dili are annoyed at the concessions granted the British flying boat service … Summarised, the Japanese can be expected to increase their hold on the country unless Australia is prepared … to put money [in].'[8]

Across the Pacific in San Diego, Lester Brain and his crew were preparing to bring home the first of nineteen Catalina flying boats ordered by the Australian government for the RAAF. At the Consolidated Aircraft plant, Brain watched twenty thousand employees building the Catalinas, as well as the four-engine B-24 Liberator bombers. The factory turned out sixty big aircraft a month. 'At this moment there are literally acres of aircraft under construction out of doors,' Brain wrote. 'As far as the eye can see there are lines and lines of big, modern warplanes. The sight is one I shall never forget.'[9]

Menzies was in Britain at the time, in the middle of a four-month visit discussing war plans with Churchill and other Empire leaders, while his power in Parliament waned at home. As the lack of support worsened, Menzies resigned as prime minister on 27 August 1941. After forty days with the Country Party leader Arthur Fadden in the role, Labor's John Curtin took office just as the war was approaching Australia's doorstep.

Brain, P. G. Taylor and a new Qantas pilot, Tasmanian-born Aubrey 'Aub' Koch,[10] had made the final delivery flight of the Catalinas, landing on Rose Bay on 24 October 1941, having stopped off in Suva, Fiji. They commemorated their mission by flying the first airmail delivery between Fiji and Australia – a delivery that caused conflict with the American airline Pan Am, which had the contract to convey the mail from Fiji but wasn't due to start operations until two weeks later.

Fysh had much greater conflicts on his horizon. He left Sydney by flying boat on 4 December 1941 for talks in Singapore with Leslie Runciman,[11] the director-general of the British Overseas Airways Corporation (BOAC), which had been formed by a merger of Imperial and British Airways. With Fysh at the helm, QEA was now flying more than two million kilometres a year, with most if its operations being the flying boat service between Sydney and Singapore. In 1941 it had carried 7600 passengers, 260 tons of freight and 296 tons of mail.

On the day that Fysh checked into Singapore's Raffles Hotel, the Japanese attacked the American naval base at Pearl Harbor. America was neutral at that stage of the war, but 353 Japanese aircraft rained fire on the pride of its navy, killing 2400 people and destroying much of its Pacific fleet.

The next morning, as the alarm clock in Fysh's hotel room showed it was a quarter past four, the unmistakable bark of anti-aircraft guns, and the roar and shudder of bombs falling on Singapore shook him from his slumber as the Japanese began their charge on that island city. Two days later in the South China Sea, Japanese air raids destroyed the two mightiest battleships of the Royal Navy, the *Prince of Wales* and the *Repulse*. Fysh flew out of Singapore on 14 December for Batavia, where hurriedly, given the speed of the Japanese war tsunami rolling south, he arranged with KLM and its local subsidiary to share mail duties in these critical times.

AIRMAIL FLIGHTS WERE still being made to Singapore and Rangoon in January 1942, but increasingly they carried arms and ammunition, while the return flights, especially from Singapore, mostly carried refugees. On 30 January, Aub Koch set out from Darwin on a special flight to bring back refugee women and children from Surabaya. He was in the cockpit of the Empire flying boat *Corio* with eighteen people on board. Koch was an experienced pilot who had honed his talents in New Guinea, and he had taken some of the last photos of the American aviator Amelia Earhart at Lae aerodrome before she disappeared in a flight across the Pacific.

The *Corio* was just reaching the shores of Timor's south coast when there was a peculiar rattling in the fuselage. Koch was stunned to see the air filled with tracer bullets. Seven Japanese fighter planes had swooped on the defenceless aircraft. With his heart racing faster than the aircraft, Koch opened the throttle and dived to water level, steering for the nearest beach more than twenty kilometres away. The noise of the bullets crashing through the machine became terrific, and Koch tried vainly to outrun the Mitsubishi Zero fighters just above the water. Incendiary bullets exploded through the fuselage and killed some of the passengers. Two engines caught fire, and the cabin filled with smoke. Passengers, alive and dead, were hurled about the cabin. Then the aircraft's nose plunged beneath the water, and Koch was thrown over the instrument panel and out into the sea.

When he came to the surface, Koch had a broken right leg and bullet wounds to the left leg and left arm.[12] He saw the *Corio* partly submerged with its torn wings on the water. The Zero fighters circled, about five hundred metres above, keeping

The Qantas Empire S.23 flying boat *Corio* was shot down by the Japanese off Timor's south coast. *State Library of NSW FL755278*

watch for a few minutes before they flew off to destroy an RAAF Hudson in nearby jungle.[13]

Six other men had survived the crash. The copilot Victor Lyne had a bullet graze to the neck and was clinging to a floating basket. Lyne saw the radio operator Archie Patterson trying to stay afloat after having taken a cannon shell through the knee; Lyne gave Patterson the basket, but the man died soon after. The *Corio* began burning to the waterline, and the survivors swam as best they could for the shore. Only five made it.

Asked for a statement in Sydney, a crestfallen Fysh told the press: 'Our crews have carried on, maintaining Empire communications under trying conditions during more than two months of the war in Malaya. Qantas deeply regrets the loss of life amongst the passengers entrusted to its care and extends sympathy to the bereaved relatives.'[14] Prime Minister Curtin suspended the Batavia–Darwin section of the Empire route the next day, but Qantas pilots and passengers still risked their lives flying refugees between Tjilatjap (now Cilacap), on the south coast of Java, and Broome in Western Australia.

Two weeks later, Singapore fell, and on 19 February the Japanese landed at Dili. That same day at 9.58 a.m. they began bombing Darwin in the largest ever attack on Australian soil. More than 114 tonnes of bombs fell on the port city and about 250 people were killed, but of more importance to the Japanese was the sinking of three warships and six merchant vessels, and the damage inflicted on another thirteen. The 242 Japanese aircraft also destroyed fuel reserves and aircraft at the RAAF base and at the civil aerodrome that Fysh had once surveyed. Two Catalina flying boats were sunk at their moorings and many major buildings were devastated.

That morning, Aub Koch was recovering from his wounds in the Darwin hospital when he heard the sirens and the roar of the Japanese planes almost simultaneously. Three bombs exploded nearby, then the walls shook and pieces of the ceiling crumbled as another hit a hospital wing. Koch limped his way to the beach for safety, and watched Japanese machines diving low and machine-gunning buildings.

The Qantas captains Hussey and Crowther were also in Darwin. They bravely decided to rescue the QEA flying boat *Camilla*, which was shrouded in thick smoke at her moorings as ships burned around her. The civil aviation launch, which had been ferrying survivors to safety from the bombed ships, took the two pilots out to the flying boat. Crowther, who had been shaving when the first bombs hit and still had soap around his ears, started the engines, and they took off flying south. A burning ship at the wharf beneath them blew up, and the QEA passenger launch and all small craft nearby were sunk. Crowther and Hussey headed for the flying boat base at Groote Eylandt, 650 kilometres away, and – despite the heavy odds against them – landed at midday. It was a rare bit of good news for Qantas during a disastrous time for the company, and for the world.

Just one day after the bombing of Darwin, the forty-year-old Qantas captain Cecil Swaffield,[15] a second cousin to Kingsford Smith,[16] left Archerfield in a DH.86 carrying the inland mail for Darwin. With him were his copilot Lindsay Marshall and seven passengers. Soon after take-off, Swaffield lost control in heavy cloud. The machine crashed into Mount Petrie, twenty kilometres away. All on board were killed in the first fatal accident on QEA's seven-year-old mail service. Fysh believed the cause was a structural failure, as the plane's tail was found a kilometre and a half from the rest of the wreckage.

A week later, on 27 February, Flying Officer Robert Love was piloting the Empire flying boat *Coogee* that the RAAF had requisitioned from Qantas when he ploughed into the waters of Cleveland Bay near Townsville. Love and seven other RAAF personnel died.[17] The very next day, as a huge Japanese convoy approached Java from the east, Captain Bill Purton took off with three crewmen in the flying boat *Circe* from Tjilatjap, Java, on a mission to evacuate sixteen passengers to Broome. Purton was due to be married in Sydney in a few weeks.[18] Two hours into the flight, he encountered a twin-engine Mitsubishi 'Betty' bomber. Neither the *Circe*, nor the twenty souls on board, were ever heard of again.

Less than a week later, on the morning of 3 March 1942, nine Japanese Zero fighters attacked Broome, where more than a thousand Dutch refugees sheltered after their evacuation from Java. As the fighters flew back and forth with machine guns blazing, they destroyed at least twenty-two Allied aircraft – including the QEA Empire flying boat *Corinna*, which was being refuelled as twenty-five passengers and crew waited on the wharf with their baggage for the flight to the eastern states. The *Centaurus*, which QEA had put under charter to the RAAF, also was destroyed, and a US B-24A Liberator full of wounded personnel was shot down. At least eighty-eight people were killed in the hour-long carnage, the youngest just one year old.

As the Japanese raced south over the Owen Stanley Range in New Guinea towards Port Moresby, all available air transport was needed to supply the troops holding them back. But at 1 a.m. on 22 March, after a night flight from Groote Eylandt for the RAAF, the *Corinthian* – one of the three surviving QEA flying boats – crashed on landing at Darwin and was wrecked. It was under the command of Captain Lewis Ambrose, with Russell Tapp acting as first officer. There were twelve passengers on board, all US personnel, and a heavy military load of gun barrels and ammunition. A rescue launch picked up survivors, but two were missing, believed trapped in the rapidly sinking hull. Despite his injuries, Tapp dived under the water and swam in the darkness into the interior of the aircraft's wrecked and sinking hull, but the two bodies were never found. Although it was possible that Ambrose had been blinded by a searchlight, Fysh said the most likely cause of the accident was that the flying boat struck some semi-submerged object. Ambrose spent the next five months in hospital but later became a Qantas operations manager.

QEA had only two flying boats left, with three others under charter to the RAAF, and after the Japanese attack on Darwin, that city had faded in popularity with paying passengers. Qantas was almost back to its humble beginnings, while Ivan Holyman's remade ANA flew routes throughout Australia. Fysh lamented that Qantas was now 'in an obscure and weak position'. Charter flights were made to Noumea, two of which carried on to Vila, and the RAAF used Qantas to carry troops and equipment to Port Moresby. But regular services to Darwin and Noumea were frequently interrupted when the flying boats were needed for more urgent trips.

Then the possibility of reopening an air link to Britain across the Indian Ocean emerged. On 27 April, Fysh raised the matter with the new director-general of Civil Aviation, Arthur Corbett,[19] but found no support. He was told that the route was unsafe and that the very idea of flying there was 'little short of murder'.[20] This despite the fact that Corbett had asked Qantas to send unarmed flying boats into parts of the East Indies in order to evacuate civilians and troops, resulting in the *Corio* and *Circe* being shot down with the loss of thirty-three lives. Fysh desperately needed to keep his airline's employees working. On 7 May he wrote to Air Vice Marshal William Bostock[21] to say that Qantas 'now find ourselves struggling to keep operating'.[22]

At the time, Fysh was busy organising the rescue work for the fearless Father John Glover,[23] a 39-year-old Catholic missionary with a pilot's licence, who for some months had been evacuating people from New Guinea but couldn't take all ninety hopefuls from the Mount Hagen mission station. Fysh organised for two DH.86s, quickly fitted with long-range tanks, to base themselves at Horn Island just off the tip of Cape York. From 13 May, the Qantas crews – led by Captain Orme Denny – navigated the soaring peaks and deep valleys of that wild terrain. Denny asked for help from the local population to harden the surface of the take-off strip, and soon two thousand people were singing all night and most of the next day as they stamped their feet to flatten the grass and compact the soil. Over the next five days, the Qantas flights evacuated seventy-eight people, including forty-five troops.

ON THE NIGHT OF 31 MAY 1942 three Japanese midget submarines, each with a two-man crew, entered Sydney Harbour and attempted to sink Allied warships. All three crews perished in the attack but not before sinking the converted ferry HMAS *Kuttabul*, and killing 21 sailors.

A week later, on the morning of 8 June, much larger Japanese submarines bombed Newcastle and Sydney's eastern suburbs, with shells falling in Fysh's neighbourhood, at Rose Bay, Woollahra, Bellevue Hill and Bondi. There were no deaths this time, but a general feeling of panic gripped Sydney and with Fysh's blessing, Wendy's school packed her off for a year with 20 other pupils in the safer confines of a country home near Cooma, 400 kilometres south of Sydney. Wendy wrote regularly from there to her parents, though her mail was always monitored. When her schoolmistress saw one of her letters home addressed to 'Dear Hud', Wendy was admonished and ordered to amend it to 'Dear Father'.

While Wendy was safe and secure far from the Japanese targets, Fysh feared that unless overseas routes were opened again, nothing could save Qantas – especially as America's growing aerial dominance in the Pacific threatened to continue in peacetime through Pan Am. Fysh outlined his plans to the press for an Indian Ocean link to Britain,[24] and he wrote to McMaster on 24 July to say that the future of British aviation, and with it the life of Qantas, hung in the balance. 'Our leaders are either hopelessly deaf and blind or hopelessly tied up,' he told his chairman. 'The papers are full of the great strides of American air transport, while the British remain silent and inactive. Unless we can obtain new aircraft and new routes, QEA will inevitably come to the end of its tether as an overseas operator.'[25] Fysh then lamented that air transport in Australia was in the hands of minister Arthur Corbett: 'an average post office official, retired as having completed his usefulness, old, crochety, dogmatic, unwilling to listen to the advice of his subordinates or the old operators'. Federal Treasurer Ben Chifley, Fysh said, had 'no qualifications for anything that I have ever heard of, except that he may have been a good engine driver'.[26]

Fysh was also angry about the methods of the young go-getter Reg Ansett,[27] whose small Victorian airline had boomed during the war as he concentrated on engineering work and charter flights for the US Army Air Corps. Fysh wrote to Holyman with the opinion that Ansett 'has been out after our pilots and is offering them jobs at advanced remuneration and position, not only for the war but after it, when his scouts say he is backed by unlimited American capital for big expansions in this country. Unless he can be trimmed he will be a big menace later on …'[28] Holyman, who had turned ANA into the largest carrier of air freight in the Commonwealth and a busy shuttle service between major centres, replied: 'Ansett is not a bit concerned with any of the other operators; to the contrary, he is not even concerned with his own company — he is only interested in Reginald Miles Ansett and will sacrifice anybody and anything and go to any lengths to achieve notoriety and financial gain for himself.'[29]

QANTAS CREWS CONTINUED to operate on dangerous missions to Milne Bay and New Guinea, carrying commandos, troops, guns, ammunition and bombs, and flying home the sick and wounded. But on 8 August 1942, another Qantas flying boat that had been requisitioned by the RAAF, *Calypso*, was lost – it sank after hitting submerged debris in the Gulf of Papua. The *Calypso* had been attempting to pick up survivors from the MV *Mamutu*, which had been destroyed by a Japanese submarine with the loss of 114 lives. An airman on the *Calypso* was killed, while one *Mamutu* survivor joined the crew aboard their two life rafts. It took them six days to reach civilisation.

Fysh continued pressing the Australian government to open the Indian Ocean route to London, but he was getting nowhere. Air Minister Arthur Drakeford,[30] a former locomotive driver like that other Labor Party heavyweight Ben Chifley, told Fysh he was 'hammering at the door of an empty house' and 'an Indian Ocean crossing by unarmed aircraft is not at present a practicable proposition'.[31] In desperation, Fysh breached protocol: he wrote personally to Governor-General Lord Gowrie,[32] saying that a route between Australia and Ceylon (now Sri Lanka) could be covered by Catalinas, and asking if the vice-regal could help Qantas. He couldn't, and when Civil Aviation Controller Corbett heard about the approach he thundered that seeking assistance in such a way 'appears to me improper'.[33] McMaster's meeting with Prime Minister Curtin – in which he outlined a Pacific Ferry Plan by which Qantas could fly five hundred bombers a year from the United States to Australia – was also a waste of time.

But then QEA was offered three Lockheed Lodestars that the United States had bought from the Dutch after the evacuation of Java. Qantas would use two of them in supply missions to New Guinea as the Allied forces drove the Japanese back. In poor weather, with landings hazardous, supplies were often dropped into the jungle from treetop level. On 13 December 1942, two of the Qantas DH.86s were damaged in a Japanese bombing attack on Port Moresby but were quickly repaired.

Fysh visited New Guinea to see the Qantas operations. He then sent a memorandum to senior executives, urging awareness of coming competition and the need for continued effort. 'The year 1943 is going to be a hectic year for Qantas,' he told them, 'and if we are to survive we have got to show we are alive and virile.'[34]

On New Year's Day 1943, Captain Crowther flew *Camilla* north from Milne Bay to the Trobriand islands to pick up the crew of a US Boeing B-17 Flying Fortress that had been forced down. Four days later he flew to the same area to rescue the crew of a Liberator bomber that had crashed. Then on 6 and 7 January, Captain Hussey, escorted by three fighter planes, flew *Coriolanus* to Wamea Island on a search for the crew of another B-17. They weren't found, but Hussey came across another group of stranded airmen grateful for the rest of their lives for his efforts.

FYSH HAD BEEN LOBBYING BOAC in an effort to start the Indian Ocean route in the Catalinas. On 16 March 1943 he left for England via the United States, flying on a US Liberator transport from the RAAF base at Amberley, outside Brisbane. His trip would take five months, and with McMaster still unwell, George Harman was appointed acting general manager of Qantas.

Fysh flew via Noumea, Canton Island and Hawaii, then over the Golden Gate Bridge in San Francisco. Some of the Americans on board started to cry at the sight of the great span and the relief of finally coming home to loved ones. At one of the Lockheed plants in Burbank, near Los Angeles, Fysh heard about the company's revolutionary new four-engine Constellation, born from an idea by the major shareholder of TWA, the reclusive millionaire and aviator Howard Hughes. With its pressurised cabin, the Constellation would allow passengers to fly above most instances of bad weather for the first time.

From Montreal, Fysh flew to Prestwick, Scotland, aboard a Liberator that climbed to more than five thousand metres, leaving him and the other nine passengers half frozen and huddled on mattresses on its bare floor. This was a fitting introduction to the austerities of wartime Britain, with its bombed-out buildings and crowded air-raid shelters. He lodged at the RAF Club in Piccadilly and had a drink with aviator friends most nights at the Royal Aero Club, thirty metres away; the club was still going strong despite a German bomb having sliced through the roof and several storeys before exploding in the basement. But Fysh found that civil aviation in England was floundering. BOAC was in turmoil, with its chairman and chief executive resigning. Fysh wrote that he had been freely canvassed for the job as BOAC chief executive but had declined because it might have caused trouble with other dominions.[35]

On 22 April 1943, England's Director-General of Civil Aviation, William Hildred, gave Fysh the news he had been eagerly awaiting: the British government had approved an Indian Ocean service using four of its Catalinas, with Qantas as the operating agent. A fifth Catalina would be added later.

Fysh's joy, though, was curtailed. On that same day, Captain Aub Koch – recovered from his crash in the *Corio* – took off from Townsville for Port Moresby just after midday in the *Camilla* with three crew, twenty-two RAAF personnel and five Americans. Seven and a half hours later, Koch was still circling Port Moresby at low altitude in cyclonic winds, hoping for a break in the weather. Finally, running low on fuel, he tried to land. The *Camilla* stalled at ten metres and crashed into the water. Eleven passengers and two crewmen drowned. Koch and his first officer Bill Peak were picked up the next day after swimming for eighteen hours.

In England, Fysh visited all the big aircraft manufacturers. He watched Avro turning out Lancaster bombers in Manchester and was a passenger as Geoffrey de Havilland Jr showed off the world's fastest warplane, the Mosquito. The pilot did vertical turns at almost four hundred kilometres an hour, making Fysh so dizzy that he almost blacked out. Fysh was still in England when his mother, Mary died at her home in Kew, Melbourne on 28 May. Fysh missed the funeral but for the rest of his life he made sure that her grave in Fawkner Cemetery was well maintained.

Bill Crowther was put in charge of new top secret Indian Ocean flights that would take the Catalinas 3513 miles (5654 kilometres) from the QEA base at Matilda Bay, on the Swan River in Perth, to Lake Koggala in Ceylon. It would be the longest nonstop regular passenger air service ever attempted, and with the Japanese still patrolling the skies, crews would have to operate in radio silence and rely on rudimentary navigation by compass, sextant and stars.

The first flight from Perth to Ceylon with a Qantas crew, under Captain Russell Tapp, left at 4.30 a.m. on 29 June 1943. It carried the returning RAF crew who had delivered the machine. After twenty-eight hours and ten minutes of nervous tension, they moored at the RAF terminal at Lake Koggala. On 22 July, Tapp commanded the first scheduled service westbound from Perth to

Ceylon, beginning a weekly run in each direction. Departures were timed so the flights would cross Japanese occupied territory during darkness, and crews would observe the sunrise twice, which led to the service being known as 'The Double Sunrise'. Although these trips were officially airline flights and the aircraft carried no armaments, their cargoes of weapons and military dispatches made the crews liable to face beheading by the Japanese if they survived being shot down. But still they flew on and on, week after week.

In August 1943, Fysh left England and flew east across the Mediterranean, the Middle East and India. He joined Bill Crowther, who was flying the eighth Catalina service from Lake Koggala for Perth on 30 August. Fysh called the Indian Ocean crossings by Catalina 'the most fascinating and romantic undertaking ever performed by Qantas',[36] and he formed 'The Secret Order of the Double Sunrise' with an elaborate certificate for passengers who completed the flight, though for a long time they had to keep it a secret.

In fact, such was the secrecy around the missions in Perth that crews weren't allowed to wear their Qantas uniforms away from the base, and the young fit men were sometimes accused of being cowards for not doing their bit. Captain Rex Senior once tried to buy cigarettes in a Perth shop but was told to 'go and join up' and that there were no cigarettes in the store 'for bludgers'. On another occasion he received a white feather, the traditional symbol of cowardice, in the mail.[37] All of the Qantas pilots and crews risked their lives every time they took off on the Double Sunrise flights, which from November 1943 were extended another 2500 kilometres to Karachi. In 1944, Fysh added a kangaroo emblem to the Qantas logo for its flights across the Indian Ocean with the design based on the kangaroo engraved on the Australian penny coin.

The service eventually made 271 crossings of the ocean, covering more than two million kilometres. It delivered more than four tonnes of mail and carried about eight hundred passengers, including the World War I correspondent turned newspaper executive Keith Murdoch and the British MP Edith

Summerskill. The radio silence meant that most of the flights were completed unmolested. On 14 February 1944, Tapp's Catalina was being refuelled on the Cocos Islands where he had stopped to collect a naval officer, when a Japanese aircraft dropped a bomb that missed. To everyone's relief, the enemy pilot flew away, most likely out of ammunition.

WHILE THE WAR HAD curtailed much of Qantas's air traffic, the airline's workshops were kept busy repairing and maintaining aircraft for the war effort. Qantas employed nearly a thousand people in Brisbane by August 1944, operating in Archerfield and at an engine overhaul workshop in nearby Moorooka. Down in Sydney, four hundred Qantas workers were employed at the flying boat base in Rose Bay and another hundred at the instrument shop in Double Bay, and hundreds more at workshops in Randwick and Mascot. By now Fysh had moved his family again, this time to a house in Wallaroy Road, Woollahra. Wendy, now 13, was back from Cooma and she would often accompany Fysh on walks with their Welsh terrier dog around nearby Cooper Park. Fysh would always make Wendy laugh with his nicknames for all the other dogs they would encounter, usually naming them after German generals. Australia's mood was now confident. The Japanese were in retreat in the Pacific and the Germans under siege.

In October 1944, not long after the Empire flying boat *Coolangatta* crashed on Rose Bay with the loss of two lives, Fysh headed back to London for conferences concerning what civil aviation would look like after the war. There were plans in Australia to nationalise air service in the same way governments controlled trains.

At a blacked-out Victoria Station, Fysh was met by Sir Keith Smith, who twenty-five years after being the copilot on the Vimy was now Vickers' Australian representative. Sir Keith had just had his hat blown off by a German V-2 rocket, and Fysh could never relax in London knowing that at any moment he could be 'blown to smithereens … utterly without previous notice'.[38] He also knew that while the Rolls-Royce Merlin engines were powering the ever-present British warplanes to victory over Germany, he still faced a huge battle for the future of Qantas after the war.

The idea of global government ownership of international airlines was put forward at the November 1944 International Civil Aviation Conference in Chicago, with the Allied victory in sight. The conference backed an international air transport agreement as a way to promote peace through the collective minds and labours of different nations working together to share knowledge for the brotherhood of man. Ever the pragmatist, Fysh saw it as 'a utopian but impractical policy'.[39]

Fysh was still in London discussing the future of civil aviation with British and Australian authorities. They included the industrialist Essington Lewis, the head of BHP; Dicky Williams, now an air marshal in charge of the RAAF; and the Australian Minister for Air, Arthur Drakeford.[40]

Fysh spent Christmas 1944 at the Sunningdale mansion of Air Commodore Alfred Critchley, one of the BOAC chiefs, where he met Lord Brabazon,[41] the first Englishman to make a powered flight. Fysh played golf on a snow-covered course with Critchley and his wife Diana, a former English champion, and spent most of his game wishing he'd played with a red ball rather than white. On New Year's Eve 1944 he joined the revellers in Piccadilly Circus, watching a parade of euphoria as cars loaded with raucous hangers-on drove by honking their horns. As Big Ben tolled the start of a new year, and with peace on the horizon, Fysh watched a scene that could have been in *My Fair Lady*. A bedraggled elderly flower girl came out of her dreary booth and, in a spirit of friendship with a new age dawning, stuck her hand out towards an elegant toff in top hat, with a similarly elegant woman on his arm, dressed in dazzling white furs. 'Shake 'ands, Guv'nor? It's New Year. Shake 'ands,' she implored. The 'Guv'nor' was having none of that, though; he turned up his nose and turned his back. The flower seller replied, 'Well – take that, you old bastard', and gave the toff a boot in the backside worthy of a football star.

A week later Fysh celebrated his fiftieth birthday, playing billiards with Sir Keith at the RAF Club in Piccadilly. The next day he left for a stopover in Montreal, and on 16 January

1945 he took the TWA overnight flight from Chicago to Los Angeles, checking in at the Beverly Hills hotel at 5 a.m. on the seventeenth. Later that day at the Lockheed plant in Burbank, he flew in an early version of the Constellation, which was then going through its initial trials. With its triple tail fin and dolphin nose, the Constellation has often been described as the most beautiful aircraft ever made. All on board wore parachutes just in case, but the trial went well – Fysh was moved to write in his diary, 'extraordinarily good aeroplane'.

BACK HOME, FYSH OVERSAW the start of a Qantas service with DC-3 aircraft to Rabaul on 2 April 1945. In the hive of activity, McMaster's health went into serious decline, and once again Fysh had to carry the extra responsibility. On 26 April 1945, he wrote to Lord Knollys, the chairman of BOAC, to say, 'Sir Fergus has now been in bed here in Sydney for some five weeks. I expect it will be some months before he is back to any real active participation.'[42]

Qantas Captain Bill Crowther (left), Albert Rudder and Fysh listen as Captain Bert Hussey explains the airline's war-time international routes in June 1945. *State Library of NSW FL7552808*

Around that time, Qantas needed to expand its operations from Shell House in Sydney. The company moved into the real estate realm, buying a three-storey building on Bridge Street from Burns Philp for £25,253 – a canny deal, considering that twelve years later the building sold for £115,000.

On 2 June, Qantas reopened an Australia–England service from Sydney in conjunction with BOAC, this time flying Avro Lancastrian aircraft that had been developed from the Lancaster heavy bombers. Fysh said goodbye to Nell and the children yet again and left for England on 23 June, making his first trip on a Lancastrian to attend a meeting of a committee of Commonwealth air transport operators. He would also help Dan McVey,[43] Australia's new Director-General of Civil Aviation, at a meeting of the Commonwealth Air Transport Council. Fysh found the catering from Sydney on the aircraft 'very good', but he thought the metal trays were useless – curry ran off his plate and accelerated across the tray onto his lap, and most of the passengers to England suffered from stomach trouble 'brought on by eastern foods, eastern heat and lack of exercise aboard'. Passenger seats and leg room were good, he noted, and the bottom bunk was adequate, with passengers going to bed and reading before turning off their lights. But only six passengers could travel in comfort.[44]

Back home, Fysh's son John had joined the RAAF as a trainee pilot. John achieved 120 hours on Tiger Moths, but with the war winding down he was discharged before his training was completed. He applied to Qantas to become a flyer, then made a trip as a supernumerary pilot in a DC-3 to New Guinea as part of his training. But he failed his medical because of a slight astigmatism, and he instead joined the company as a junior booking officer at Shell House, sometimes riding his motorbike out to Rose Bay at 4 a.m. to check in flying boat passengers and weigh them and their baggage. If there was a headwind he often had to soothe upset passengers because they were off-loaded to allow for extra fuel.

Despite Fysh's concerns, Paul McGinness had been accepted into the RAAF in July 1940 at the age of forty-four with the rank of flying officer. On his application he wrote that he was one of the founders of Qantas and had kept in flying practice on Moth aircraft.[45] He was recognised by his old commanding officer from World War I, Air Marshal Richard Williams, and given the job of helping to train pilots at Evans Head, New South Wales, and Point Cook, Victoria, before being posted to Parkes, Townsville and New Guinea. After being demoted for a while to barracks officer as he sorted out his drinking problem, he eventually became a temporary squadron leader in a non-combat role. In 1943 he married a forty-year-old Melbourne woman, Irene 'Rene' Searls, who had inherited a profitable engineering business from her father. McGinness was starting to suffer heart problems, though, and he was eventually discharged 'being surplus to the RAAF's requirements'.[46] When Rene travelled to America on a business trip as the war was coming to a close, she told her husband that he was surplus to her requirements, too.[47]

The Double Sunrise Flights ended in July 1945 as the war came to a halt. For their extraordinary feats across the Indian Ocean using primitive navigational methods, Captains Crowther, Tapp and Ambrose were awarded the prized Johnston Memorial Trophy in England, which previously had been awarded to such aviators as Bert Hinkler, Jim Mollison and Jean Batten. Qantas handed back the Catalinas to the RAAF in Perth, and under a 'Lend Lease' agreement with America, the machines were towed out to sea and sunk. 'It was a dismal fate,' Fysh said, 'for those splendid boats which for two long years saw us through on our most hazardous operation ever.'[48]

While Fysh was in England, Australia's Prime Minister John Curtin, worn down by wartime stresses, died in office aged sixty on 6 July 1945. He was at first succeeded by his deputy Frank Forde, but seven days later a Labor Party ballot installed Ben Chifley as its leader and prime minister. Chifley planned on acquiring and nationalising key industries, and the passing of the Australian National Airlines Bill in August 1945[49] gave the government a monopoly on domestic flights, in the way the government also controlled railways, public hospitals and the post office. Despite the support that Ivan Holyman's aircraft had given Australia during the war, Labor wanted to rein in ANA, which was now a conglomerate owned by British and Australian shipping interests. Holyman wasn't prepared for his company to go under without a fight – he began a legal challenge backed by big business and the Liberal opposition under the rejuvenated Bob Menzies.

Fysh realised that nationalisation of the aviation industry would cost Qantas its historical Brisbane–Darwin route. But while that had sentimental value, his company's focus was now global.

Chapter 22

Hudson Fysh's contribution to Qantas and therefore Australia was much more than just the beginning. Many, many companies are 'founded' but it was the on-going stewardship which to me was more important. He guided Qantas through many obstacles for 46 years and made it both one of the world's great airlines and an Australian icon.

ALASTAIR FYSH, HUDSON'S GRANDSON, SYDNEY, MAY 2022.[1]

ON 26 SEPTEMBER 1945, just three weeks after Japan's formal surrender aboard the USS *Missouri* ended World War II, Fysh managed to get a pass for Nell so she could at last accompany him on an overseas flight. They flew from Mascot to San Diego aboard an RAAF Transport Command Liberator, on their way to Montreal for the first meeting of the International Air Transport Association (IATA).

Hudson Fysh was fifty years old, but with aviation technology evolving faster than ever there was no chance of him slowing down. During the war the Germans had launched their V-2 rockets on London, and Messerschmitt had unleashed jet-propelled fighter aircraft on the Allies, capable of reaching nine hundred kilometres per hour. At the same time Geoffrey de Havilland, with an Allied victory in sight, had started work on his DH.106 Comet passenger jet after the British government had asked him for a pressurised transatlantic aircraft that could carry a thousand kilograms of mail across the Atlantic at a cruising speed of 640 kilometres per hour.[2] Over the twenty-five years since Fysh's first biplane flights for Qantas, the developments in aviation technology had been staggering. Although it would take time before jet propulsion would drive his passenger aircraft, he wanted Qantas at the forefront of the field.

From November 1944 through to the end of 1945, Fysh was rarely at home in Woollahra. In the space of twelve months he had travelled to England three times and twice to North America. His tastes remained simple, though. He and his family still lived in a rented home and preferred a quiet life to the social whirl. He worked long hours but when in Sydney was always seated at the table for Nell's dinner at six. He drank moderately but puffed endlessly on his Craven As.

Civil aviation in Australia and Great Britain had been used mostly to support the Allied war effort, and Qantas's business had often been sacrificed for military needs. The Americans, though, with the fighting a long way from their backyard, had continued building their domestic air services while developing long-range transport aircraft. Australia and Britain both had Labor governments at the end of the war, and both were backing the government ownership of international airlines – the Americans, however, were supporting unrestricted free enterprise in air traffic. The hamstrung British demanded a degree of international control since they had neither the money nor the machines to compete with the United States.

In the months after the end of the war, Fysh was torn between his loyalty to the Crown and all things British that had caused him to enlist three decades earlier, and his firm belief that the Americans were now making far superior aircraft. While the British government was imploring Australia and its other dominions to re-equip their fleets with the ponderous sixty-seat Avro Tudor II machines, Fysh knew the Lockheed Constellations ultimately would be the best aircraft for Qantas and for Australia. Amid clouds of cigarette smoke, Fysh agonised for months over the choice, as Avro and the British government of new Prime Minister Clement Attlee continued to lobby the Empire to buy British. Finally, two days after Christmas 1945, Fysh's recommendations to the QEA board favouring the Americans were accepted, although the Australian government still had to support the move. Fysh described his call in favour of the Constellations as 'one of the most momentous decisions in the history of QEA'.[3]

Ivan Holyman, meanwhile, had taken his fight over nationalisation to the High Court and eventually the Privy Council in London. In December 1945 the High Court ruled that parts of the new *Australian National Airlines Act* of 1945 governing nationalisation were invalid. The Commonwealth government, it said, did not have the power to prevent the issue of airline licences to private companies. The government could set up an airline, it ruled, but it could not legislate a monopoly, which sections of the press portrayed as a form of socialisation by stealth.[4] But the government pressed on with its plans, installing the retail king Arthur Coles – a former Mayor of Melbourne and a politician who had helped to topple the Menzies Government in 1941 – as chairman of the Australian National Airlines Commission in January 1946.

Also in January, an international conference in Bermuda resolved that while international carriers would not engage in cut-throat competition, prices would be 'kept as low as possible in the interests of the travelling public'.[5] Fysh was well aware that no country was likely to benefit more from these agreements than Australia because of her geographic isolation.

As a result of aviation developments during the war, faster and more streamlined international air travel was born. It lacked the romance of a flying boat landing beside a tropical island, but eventually Fysh would see to it that overseas travel became available and affordable for millions of Australians who might otherwise never have had the opportunity to visit distant parts of the globe. 'Out of the ashes of the past,' Fysh wrote, 'had sprung some new and wondrous instrument for the use of mankind.'[6]

In deciding to align himself and Qantas with American prosperity and know-how, Fysh procured the best possible long-distance passenger planes for Australia, as well as the technical skills of American designers to teach Arthur Baird and his engineering staff, a decision that would serve Qantas well for decades. The sleek

Constellations were the world leaders in passenger aircraft, with four Wright Cyclone engines each pumping out 2200 horsepower. The aircraft had a maximum cruising speed of 318 miles per hour (511 kilometres per hour) and a service ceiling of 26,000 feet (8700 metres), and they could each carry thirty-eight passengers. Fysh soon sent ten pilots and six first officers under Scotty Allan to Burbank for a five-week technical course at the Lockheed training school, while others did a course in the new technique of instrument landing at the Pan Am World Airways flying school at LaGuardia Airport in New York.

Fysh was amazed at the rivers of money flowing through America after the war. It seemed everyone owned new and shiny cars, and many people were forsaking cooking to buy frozen dinners, a fact he found extraordinary.[7] Before long, he introduced a Qantas flight kitchen – first at Rose Bay and then at St Marys, in Sydney's west – to prepare and freeze fine-dining meals to be served with Australian wines by white-coated stewards on Qantas flights. They would need all the meals they could make as passenger numbers soared.

AT THE END OF THE WAR, Qantas Empire Airways was in charge of the 'Kangaroo Route' from Mascot to Karachi, a section of the flight path that involved the Avro Lancastrians travelling between Sydney and the RAF base at Hurn, near Bournemouth, about 160 kilometres south-west of London. Qantas's kangaroo logo was first used on aircraft flying to Karachi, where the BOAC crews took over for the rest of the journey to the United Kingdom. The total flying time between Sydney and London was reduced to sixty-seven hours, and the Australian leg involved a stop at Learmonth, on the north-west coast of Western Australia, before the aircraft set off over the Indian Ocean. QEA also used Consolidated's Liberator aircraft twice a week between Perth and Colombo, and flew DC-3s on a Brisbane–Manila route and from Mascot to Lae, New Guinea. In addition, there were flying boat charters to Singapore, Fiji and Noumea. The DC-3s were also flying the old Qantas route from Brisbane to Darwin and Cloncurry to Normanton, while the Flying Doctors were using the DH.84 Dragon biplanes.

As negotiations continued over the reinstatement of the old Australia-to-London airmail route via Singapore, Qantas suffered another tragedy. All ten people on a BOAC Avro Lancastrian – piloted by Captain Frank Thomas, one of QEA's most able commanders – disappeared in bad weather between Colombo and the Cocos Islands on 23 March 1946. The converted wartime aeroplanes often smelt strongly of fuel, and investigators suspected the big Lancastrian had exploded when struck by lightning. No trace of it was ever found despite a search by thirteen aircraft that covered more than seventy-two thousand miles of ocean.[8] One of the five passengers on board was John Dobson – the son of Sir Roy, the Avro chief.

There was no stopping progress, though. Almost simultaneously with news that the search for the Lancastrian had been abandoned, Fysh announced that the Australia-to-London route via Singapore would be reopened with Liberator aircraft after three years of diversion across the Indian Ocean, the world's 'longest air hop'.[9]

In May 1946, QEA introduced the Short S.25 Sunderland flying boats and a variant known as the 'Hythes' after a village in Kent. They were bigger and better than the prewar Empire flying boats, and passengers to London could choose either a 67-hour express ride on the noisy Lancastrians or a leisurely flying boat journey that took five and a half days via Bowen, Darwin, Surabaya, Singapore, Rangoon, Calcutta, Karachi, Cairo, Basra and Marseilles.

That same month, after twenty-two years of a close working relationship that had started in remote and humble beginnings, Fysh lost Lester Brain to a new airline the government had started in opposition to Holyman's. It was called Trans-Australia Airlines, or TAA.[10] Reg Ansett had offered to run TAA and sell his airline to the government as a going concern, but Arthur Coles decided Ansett's asking price was too steep. Coles offered Holyman the huge salary of £10,000 a year; when Holyman said no, Coles offered to buy ANA, and that offer was knocked back too. Instead, Brain took the job running TAA for £3000 a year, well above what he was earning as operations manager at Qantas. In mid-June 1946 he acquired TAA's first two aircraft, both Douglas DC-3s, with more of them to come over the next few months, all ex-RAAF machines. Brain appointed Qantas captain Aub Koch as his senior pilot.[11]

In July 1946, Fysh and Nell flew to war-ravaged Lae. Fysh made plans for a Qantas service with DH Dragons to start that year, opening up the jungles, forests and mountain routes around New Guinea in the same way he and McGinness had done in western Queensland all those years before.

Fysh was still waiting for the federal government to endorse his decision in favour of the Lockheed Constellations over the British machines, as Clement Attlee continued to press Chifley over loyalty to the Crown and British manufacturing. On 31 August 1946, Fysh told readers of the *Sydney Morning Herald* that suitable British airliners for long-range operations on the Australia–England route weren't available and that Qantas couldn't wait for their products when superior American aircraft were on offer. 'Customers who would like to buy British may have to go elsewhere for heavy long-range airliners, until the much publicised British jet airliners come into operation, in six years or more,' he wrote.[12]

A few days later Fysh was granted an audience with Chifley at the Victoria Barracks in Melbourne. The prime minister had his ever-present pipe in his mouth and an urgent telegram on his desk from Attlee, imploring the Australian government not to support Fysh's order for the Constellations. John Curtin had formed an alliance with the Americans in favour of Britain during the war to defend Australia from the Japanese, and Avro and Attlee were asking Chifley to remember that the mother country needed as much postwar manufacturing income as possible in order to rebuild her devastated economy. It needed companies like Qantas to buy its products to make Britain great again.

Fysh knew the future of Qantas hung in the balance, and Chifley kept him nervously waiting. The prime minister sent plumes of smoke skyward over both the Qantas boss and Attlee's telegram. Finally, Chifley smiled. He told Fysh that he had decided to ignore Attlee's plea and give Fysh and his American planes 'a go'.[13] On Fysh's recommendation, he felt the

Constellations would be best for Australia. Fysh was heady with excitement as he caught the St Kilda tram back to the city and sent McMaster the joyful news in a telegram from the GPO.[14]

Sir Keith Smith, who would soon become a director of Qantas, sent Fysh a wire declaring, 'Now you have really started something. Congratulations on your personal triumph.'[15] Chifley then wired the Dominions Office in London on 10 September, advising the British government, 'All my experts and Qantas Empire Airways are satisfied that, notwithstanding the favourable figures anticipated for Tudor II operation, the Constellation remains a superior and more economic aircraft from the airline operator's point of view.'[16] The Tudors were to be built at Fishermans Bend in Melbourne by the Department of Aircraft Production, and the Qantas resolution cost the department dearly. But Fysh's assessment of the Tudors was spot-on – before long they were withdrawn from service after two crashes.

Fysh's acquisition of the Constellations would change the direction of Qantas forever. The aircraft became the key to the airline's growth after the war, and within a decade they would help Qantas establish more routes to Africa and America as it became the first airline to operate a scheduled service covering the entire globe.

Fysh estimated that to put four Constellations in the air and keep them serviced and flying would cost more than £2 million, and that any airline spending that money needed government backing at a time when airlines, including BOAC, were being nationalised around the globe. He understood that the end of Qantas, as he knew it, was just around the corner.

On 6 December 1946, a bill called the *Qantas Empire Airways Agreement* was passed in Canberra to buy out BOAC's half share of Qantas Empire Airways, which had existed since 1934.[17] This looked like a sound investment for the government, as under Fysh's management Qantas had become one of Australia's best performing businesses; in fact, the airline had just returned an annual profit of more than £54,000 with a dividend of seven per cent. It was quickly expanding, too.

Fysh celebrated the success by sending 30 guineas to the Finke River Mission at Hermannsburg in the Northern Territory for a painting of ghost gums by the increasingly popular Arrernte artist Albert Namatjira. On 21 March 1947,[18] the day the painting was posted to Fysh by the pastor in charge of the mission, Pan Am began its service to Australia when the DC-4 *Kit Carson* arrived in Sydney four days after leaving San Francisco. Fysh wanted Qantas to fly to America as well, but he saw the need to first establish regular services with Manila, Hong Kong, Japan and China, where he saw great opportunities. Since November 1945, the RAAF had been running a courier service with Dakota aircraft between Australia and Japan.

McMASTER'S FAILING HEALTH finally forced him to officially resign as chairman on 8 April 1947. A month later, on 26 May 1947, the Qantas board accepted that the government would buy out the Qantas shareholding and provide Fysh with a £1 million overdraft from the Commonwealth Bank to buy the Constellations. He told the press that Qantas was the oldest name in air transport but had to accept changes 'which time prescribes in all things'.[19] Three days later he left for London for meetings with BOAC and Britain's Air Ministry.

Fysh became so busy with long-range plans for Qantas that he was unable to attend the IATA annual general meeting held near Rio de Janeiro in October 1947 but he still had time to approve a new Qantas logo, this time with a winged kangaroo hopping the globe.

At the Qantas board meeting at Brisbane's Wool Exchange Building on 2 October 1947, Fysh endorsed the recommendation that the company be liquidated. At the time it comprised 261,500 £1 shares. Fysh held just 1235 of them, and Nell held 400. In the government buyout, there was no grand compensation for Fysh or McMaster for their pioneering work in building the airline from scratch and their steady control of the company for more than a quarter of a century. The sale to the government was, Fysh noted, both historic and harrowing, but he believed that Chifley was taking the correct course given the lack of resources for a private

Fysh sat for this portrait by the renowned photographer Max Dupain in 1947. *State Library of NSW FL8585672*

company to compete on the postwar world stage. Large sections of the media protested Chifley's nationalisation plans, though, especially as he turned his eyes to the banks.

The influential magazine *The Bulletin* declared that Qantas, 'the brainchild of Hudson Fysh and Fergus McMaster', had died from asphyxiation 'caused by socialism, a virulent disease transmitted by raucous demagogues'.[20] Fysh, the magazine reported, would continue as the chairman and general manager of this new government airline, while the ailing McMaster was retiring with a £2500 send-off after more than a quarter of a century of work as chairman that came with 'modest remuneration' despite often leaving him out of pocket. It credited Qantas as being primarily responsible for Australia's air routes internally and around the world, and heaped praise on the heroism of its pilots, who during the war had been the 'the providores of Milne Bay in overstrained flying-boats without even a toy pistol aboard to ward off Zeros'. It warned Chifley that if the government was to give service 'as good, as courteous and as characteristic as did Qantas in its prime' it would need to let 'Fysh and others of his kind have a free hand, which it won't'.[21]

It did to a point, however. In the first year under government ownership, Qantas turned a profit of £79,900. Then the air minister, Arthur Drakeford, engineered another £2 million of government funds into the airline for expansion. While Fysh was well paid at the helm of the business, heavy postwar income taxes ensured he would never become rich from the airline he had created.

One of the emergency measures Fysh instigated immediately after the government took over Qantas was a fortnightly service using Lancastrian aircraft to help the eight hundred inhabitants of Norfolk Island, 1600 kilometres north-east of Sydney. Their isolation had been made all the more manifest by inadequate shipping to the island at the end of the war. A service to the remote Lord Howe Island began soon after, with Catalina flying boats obtained from the RAAF.

FOUR QANTAS CONSTELLATIONS rolled out of the Burbank plant. They were named the *Lawrence Hargrave*, *Harry Hawker*, *Ross Smith* and *Charles Kingsford Smith*. Two others followed shortly, the *Bert Hinkler* and *Horace Brinsmead*.

At 10.06 p.m. on 1 December 1947, Captain Ken Jackson piloted the *Ross Smith* as it took off from Sydney for London with twenty-nine passengers. At Karachi, Captain Don MacMaster took over, landing in London on schedule at 12.35 p.m. on 5 December. The machine was carrying half a tonne of food parcels donated by Qantas staff for their colleagues at BOAC.[22] The one-class fare was £325 one-way or £585 return – air travel was still the preserve of the wealthy, and a return fare represented about eighteen months on the average wage in the days when three years' pay could buy a suburban home in Sydney. Male passengers generally wore suits and hats on board; women favoured hats, gloves and smart day wear.

Eleven days later, a Qantas Lancastrian left Mascot on the opening weekly flight from Sydney to Japan with stops in Darwin and Manila. It arrived in the depths of winter at Japan's snowbound Bofu airbase on 18 December with six passengers and half a tonne of Christmas mail for the Australian troops stationed there. The news of the Qantas aircraft arrival was captured in rolls of film that were placed in tiny metal capsules and taken by carrier pigeons to newspapers in Tokyo almost a thousand kilometres away. Before long the Qantas planes were using the renovated aerodrome at nearby Iwakuni, and the service was soon extended all the way to Tokyo. A regular Sydney–Hong Kong service started on 26 June 1949.

Fysh was very much a man of his time in that he believed men were breadwinners and women homemakers. He wasn't even that keen for his bubbly daughter Wendy, whom he adored, to pursue her physiotherapy studies at the University of Sydney.[23] But he decided to follow the trend of other airlines in employing air hostesses after years of using only male stewards. The advertisements for the role reflected the mores of another age. More than two thousand women applied for the first nine jobs offered, which paid just £7 a week. Applicants were to be twenty-two to twenty-seven years old, British subjects, 158 to 168 centimetres tall, no heavier than sixty kilograms, and with a nursing or St John's first aid certificate. The stylish Nell was on the selection committee, along with three men, and on 8 May 1948 the first air hostesses to serve on the Constellations left Mascot for London.

Three months later, Nell and Wendy flew on a Constellation to London. By then Fysh was plotting an Australia–South Africa service. A survey crew in a Lancastrian left Sydney just before midnight on 14 November 1948 for a six-day flight to Johannesburg, via Perth and the Cocos Islands. The long ocean crossing to Mauritius was made at night to take advantage of navigation by the stars with sextants, as the crews were on their own without radio support, just as the great explorers had been for centuries. The Lancastrian passed over the French island of Reunion and landed at Johannesburg on 20 November.

It was a heady time for Qantas. Even though Fysh still resisted his wife's pleas to buy a family home, preferring to rent given the loss he suffered on his Longreach home and his mother's real estate woes, he began an ambitious project of company real estate purchases to provide hotel accommodation for Qantas passengers at various destinations along the air routes. Eventually Qantas took over the Wentworth Hotel in Sydney, partly to counter the rooms being offered by rival services Pan Am and KLM.

On 2 April 1949, Qantas handed over to TAA its Queensland and Northern Territory routes, along with the Flying Doctor operations. Forty-one Qantas staff transferred to TAA, which paid £26,000 for four de Havilland Dragon aircraft and some old hangars. The plan was that from now on, Qantas would operate solely as an overseas airline.

WITH THE CONSTELLATIONS now flying three times a fortnight to London and Qantas's services expanding quickly, there were reports of petty cargo thefts on that run, pilfering at Mascot, and vandalism on the Catalina flying boats that flew the South Pacific. On 23 June, a QEA Catalina was destroyed under suspicious circumstances in the lagoon at Lord Howe Island.[24]

But that shock was nothing compared to the blast at 2.22 a.m. on Saturday, 27 August 1949, which shook Fysh from his bed at Wallaroy Road in Woollahra. One of the empty Catalinas, having recently flown in from Noumea, exploded at its mooring a kilometre away in Rose Bay. The blast could be heard miles away. When the shattered wreckage was hauled from the bottom of Sydney Harbour, detectives found the remains of a crude time bomb,[25] complete with alarm clock, lantern battery and a vibrator coil that may have come from an old Model T Ford. The device was designed to produce sparks, and the aircraft had been vandalised in such a way that fuel had leaked into the engineer's compartment. Bryan Monkton, the managing director of Trans Oceanic Airways – a rival airline also flying to Lord Howe – went on trial but was acquitted of destroying the £24,000 aircraft after he produced an alibi that was corroborated by P. G. Taylor, among others.[26]

Fysh had little time to deal with the investigation into a crime that was never solved, though security was beefed up. Only a few days after the explosion he left for England again, and at the Farnborough Air Display in Hampshire on 7 September he marvelled at the power and beauty of the stunning new de Havilland Comet, powered by four DH Ghost engines, the world's first gas turbine engine to enter airline service. Writing for Sydney's *Sun* newspaper, Fysh claimed that 'the difference between the conventional aeroplane driven by a piston engine and a propeller and a jet aeroplane is going to be as great as that between a horse and cart and a motor car', and that the high-pitched 'singing whine of the jet engines would be the song of the future'.[27] 'This marks the development we are entering into now,' he wrote, 'and is the next step before the probable futuristic use of rocket power, and, finally, the ultimate in atomic power, which glimpses the possibility of interplanetary communication.'[28]

His own family's future looked bright too. Wendy was just nineteen and at university when she began contemplating life as a working wife. While Fysh found that notion hard to comprehend,[29] he gave his blessing when she gushingly told him that a young accountant named John Miles, the son of Lieutenant General C. G. N. Miles,[30] had proposed to her. Nell thought Wendy was far too young to be talking about marriage, but the idea of organising a society wedding for the daughter who looked so like her quickly changed her mind.

FYSH WAS FIFTY-FIVE and busier than ever, though age was catching up with many of those around him. Arthur Baird, the company's chief engineer, retired on 30 June 1949 after almost thirty years of painstaking work overseeing the Qantas fleet. Baird had been Fysh's ever-reliable mechanic since World War I. 'I cannot think of anyone who has done more in laying the foundations of Australian air transport engineering,' Fysh wrote.[31]

In Queensland, Fergus McMaster had been rendered an invalid. He had moved permanently from his outback properties to a home on Rees Street in the Brisbane suburb of Clayfield, not far from Fysh's old home in Wooloowin. After managing Toolebuc Station for twelve years, McMaster's daughter Jean and her husband Graham Fysh had bought their own station, Answer Downs, in 1947.[32] The letters from the Fysh children in the outback gave McMaster great comfort at the end of a life that had been scarred by the early death of his first wife and the loss of a brother and son in war.

On 8 August 1950, McMaster died peacefully at home from congestive heart disease, aged seventy-one. Fysh and Baird were among the large Qantas contingent at his funeral at Brisbane's St Andrew's Presbyterian Church, and three aircraft from the Royal Queensland Aero Club dipped their wings in salute over the procession as it made its way to the Mt Thompson Memorial Gardens and Crematorium. Despite decades of struggle on the land and holding the chairmanship of Qantas for almost all of its life, McMaster did not die a rich man, leaving an estate of £32,000.[33] Fysh described the father figure who had guided him throughout the entire life of Qantas as 'a man of rugged character'. 'He possessed great tenacity of purpose,' Fysh wrote. 'This, plus a code of personal integrity and public duty which set a very high ethical standard, was the spirit which for so many years had inspired Qantas, then QEA.'[34]

There was little time to mourn. Later in August, Fysh had to hammer out further details with the government over the purchase of the Wentworth Hotel. The next month he was back at Farnborough with his son John, now almost twenty-five. Fysh then flew in the Comet at forty thousand feet (about thirteen thousand metres) over de Havilland's aerodrome at Hatfield. The world's first commercial jet airliner rocketed on, crossing European country after country in seemingly no time – though, ever the pragmatist, Fysh thought the aircraft's comparatively short range made it unsuitable for the Qantas long hauls. This was a prudent assessment, especially as BOAC and the Comets went on to be plagued by a series of crashes in the 1950s.

FOR THE THIRTIETH anniversary of Qantas, in November 1950, the staff organised a portrait of Fysh painted by the renowned Bill (later Sir William) Dobell. His stylised portrait style, accentuating features almost to the point of caricature, had caused controversy when he won the 1943 Archibald Prize.

Fysh sat for Dobell in his Kings Cross studio after a hard day's work. Dobell asked him to take off his glasses, and the resulting image is of a skeletal, haggard, exhausted man with a bad squint. Dobell entered the painting for the Archibald Prize, and Fysh was glad it didn't win.

The possibility of his unflattering likeness being shown around Australia as a prize-winning portrait was the least of his worries at the end of 1950, though. His overseas travels had convinced him that World War III was imminent, and Chinese involvement in the Korean War made him fear for the implications elsewhere. On 30 November 1950 he wrote to Air Marshal Williams about the Qantas route to London, telling his old commander, 'In regard to our route being broken, I am personally most frightened in regard to Singapore, eighty per cent of the population of which is Chinese, and who have no love for the British.'

At the time, Fysh and Cedric Turner, the new general manager of Qantas, were locked in fierce negotiations with de Havilland. It was over a new model of the Comet, set to cost more than £500,000, and Turner was firing all guns in his abrasive way.

Fysh's little Wendy married John Miles at St Mark's Church in Darling Point on 3 April 1951. Nell was smiling even more broadly than usual at the reception, held at Royal Sydney Golf Club where Fysh was known to hit a pretty good drive whenever he had the chance. The University of Sydney had warned Wendy that as a married woman she would probably have to forsake her physiotherapy course, but she graduated the following year. Wendy's marriage – which would last more than half a century, until John's passing – was a source of immense pride to Fysh, but much of 1951 was a time of turmoil for him, and the Craven As were being smoked constantly.

In May, Fysh's old friend Reverend John Flynn died from cancer in Sydney. A month later that great supporter of Qantas, Ben Chifley, now in Opposition, suffered a fatal heart attack at Canberra's Hotel Kurrajong. Then, in the space of five months, Qantas suffered three fatal crashes in New Guinea that resulted in eleven deaths. The first, on 16 July, occurred when a near-new Australian-built DHA-3 Drover tri-motor slammed into the

Huon Gulf near the mouth of the Markham River after the centre engine's propeller failed. The pilot and six passengers were killed.[35] Gold bars worth £36,000 were never recovered. The other two crashes involved de Havilland Dragons lost in the mountainous country of New Guinea's Central Highlands.[36]

PAUL McGINNESS, FYSH'S OLD partner in the skies over Palestine and the man who had conceived the idea of Qantas, had stumbled from one disaster to another after his discharge from the RAAF. An ambitious plan to produce frozen beef in the Northern Territory and sell it to Asia and England had been ruined by the end of his second marriage and continuing heart problems. An attempt to grow tobacco at Mareeba in northern Queensland was again thwarted by his poor health, and he finally returned to Western Australia in 1951 to grow tobacco on a soldier settlement farm at Northcliffe, near Albany.

For a time, McGinness was admitted to Hollywood Hospital with heart problems, then he was discharged. His daughter Pauline, who had just borne a baby girl, invited him home for Christmas Day, but he spent the holidays back in hospital – though he did not tell his daughter lest he spoil her young family's celebrations.

McGinness was walking slowly along a corridor at Perth's Hollywood Repatriation Hospital on 25 January 1952 when the man Fysh called 'the bravest of the brave' suffered a fatal heart attack. Four days later he was buried at Karrakatta Cemetery in Perth. His coffin was draped with an Australian flag, but there were only two mourners: his first wife Dorothy and their daughter Pauline. The only wreath came from his second wife Rene.[37]

McMaster once said that McGinness 'did more than any other single person in the establishment of Qantas', and Fysh wrote of him that while he 'was an adventurous spirit, not given to the humdrum and everyday aspects of life', he was a 'staunch wartime friend, with lots of guts'[38] and the original 'spark' of the airline. But McGinness died in obscurity, while Fysh now commanded a workforce of four thousand[39] and had transformed Qantas from a bush airline offering joy rides to a global colossus.

The world was changing rapidly in the postwar boom. Fysh and Nell had visited Japan together on the Qantas service in 1951, where they had seen a serene country resplendent with cherry blossoms and emerald fields quite at odds with the savagery of its military only a few years before. In 1952 they flew from Sydney to South Africa on the first flight of what became a regular Qantas route known as the 'Wallaby service' in a nod to the rugby union bond between the two countries.

Nell had seen Fysh rise from his life as a battling bush pilot in a mechanic's shed to become one of Australia's most important businessmen. But the best – and worst – was yet to come.

Chapter 23

So one comes to a sad part of one's life – the unwanted stage, the stage when influence is waning and one sees old ideals and urges go, in the march of time … All things pass. Times change. Ideals change – and my Qantas goes ahead under the new people to new victories.

FYSH ON BEING UNDERMINED AS THE DRIVING FORCE AT QANTAS[1]

THE BEAMING YOUNG Queen Elizabeth II sailed into Sydney Harbour aboard the Royal Yacht *Gothic* on 3 February 1954, becoming the first reigning British monarch to visit Australia. Her arrival in brilliant Sydney summer sunshine just before 11 a.m. at Farm Cove, where Captain Arthur Phillip had first raised the British flag 166 years earlier, attracted an estimated one million spectators, or more than half of Sydney's population at the time.[2]

Three days later, in front of a smaller but no less enthusiastic crowd of five hundred inside the ballroom of Sydney's Government House, Fysh knelt in front of the radiant young royal. He was now greying, and he wore thick glasses and a morning suit that Nell had ironed for him, while the Queen was dressed in a sleeveless daffodil yellow frock that she wore later to the Randwick races and a surf carnival at Bondi.[3] An aide unsheathed a ceremonial sword from a black-and-gold scabbard, and as the blade glistened in the ballroom's light, the Queen tapped Fysh on the shoulder, making him a Knight Commander of the Order of the British Empire in 'recognition of outstanding services in pioneering and developing Australian aviation'.[4] The Queen then hung the accolade ribbons around Fysh's neck and held him by the hand for a few minutes as he told her about the early days of Qantas and how he was the last man standing from that pioneering era in western Queensland.

Approaching his sixtieth birthday, Fysh was the chairman and managing director of an organisation that now carried 118,000 passengers a year, employed more than 5000 staff, and made an annual profit of more than £13 million. When his knighthood had been announced, he had told the press that he was still astonished by the magic of aviation and the benefits it could bestow to humankind. He said he could foresee a time when planes would leave Sydney at 8 a.m. local time and arrive in London at 7 a.m. the same day, because of the time difference. 'An experimental aircraft produced by the Douglas Aircraft Company has already flown 1500 miles an hour [2400 kilometres an hour],' he said. Looking still further ahead, he was confident that man would step on the moon 'in our time': 'People will eventually fly to the moon in rocket conveyances, which will possibly first be used for travelling between distant places on the earth. This may sound fantastic to some people, but it is no more fantastic than the suggestion that man would fly between England and Australia in 24 hours would have appeared to people 50 years ago.'[5]

While the Queen and Prince Philip were in Australia, they became Qantas frequent flyers, clocking up 6500 kilometres visiting various parts of the country on a specially fitted Constellation, the *Horace Brinsmead.* Fysh had overseen a new and enduring colour scheme for Qantas's aircraft: the top surface of the fuselage was painted white, the sign 'Qantas Empire Airways' was replaced by 'QANTAS' – to signify Australia standing on its own two feet – and the thin blue stripe below the cabin windows was changed to a dark red band. During the coronation year the Constellations, the flagships of the fleet, were also decorated with a commemorative crown badge.

On 1 April 1954, Qantas followed the experiment of airlines offering lower cost fares on flights across the North Atlantic: they introduced tourist-class airfares between London and Sydney on the Constellations. These seats were offered at a 20 per cent discount off the first-class fare, with less leg room and pared-down catering. The trial would prove immensely popular and redefine international travel. Soon tourist-class-only Constellations were flying to London and the South Pacific, making the flying boat services there less popular.

Six weeks later, on 15 May 1954, with Qantas having taken over the British Commonwealth Pacific Airlines operation in the South Pacific, Fysh and Nell were taxiing down the Mascot runway in the new $US2 million Qantas Super Constellation, a bigger version of their existing Lockheed aircraft. It was the start of the new Qantas service to San Francisco and Vancouver, with stopovers in Fiji and Honolulu. Development of Qantas's Wentworth Hotel was underway, and soon it would be the first hotel in Australia to have bathrooms in every room. Meanwhile, a new eleven-storey Qantas head office building was being constructed on the corner of Hunter and Elizabeth streets; Fysh had paid £87,000 for the block in 1949.

FYSH'S KNIGHTHOOD WAS the crowning honour of his public and professional life, recognising thirty-four years of diligence, tenacity, courage and loyalty to a little airline that he had grown into an Australian institution. But instead of being able to bask in the glow of his achievements, Fysh now became embroiled in twelve years of bitter infighting at the top of the organisation.

Dan McVey, a Qantas board member, was knighted on the same day as Fysh. While they congratulated each other publicly, privately they quarrelled. Fysh would write of the former Director-General of Civil Aviation that he had been jealous, subtle and cunning with him.[6]

At times, Fysh felt that the whole Qantas board was against him, even his old friend Sir Keith Smith. He complained of antagonism[7] and sometimes even bullying.[8] At the end of 1954, four of the six men on the Qantas board were knights, and some felt simply that the old bush pilot was yesterday's man, a relic of the pioneer days. The chairman they saw swapping horseracing tips with the lift driver or playing golf in old pants with a tie for a belt and a cigarette dangling from his bottom lip didn't fit their ideal for an international aviation chief in a world of high finance and political machinations.[9] Fysh was a simple man at heart who still didn't own a home and, according to his family, much preferred the company of 'the average Joe Blows of the world than the big business chiefs'.[10] His Victorian sense of morality was offended when married politicians asked Qantas for tickets – not for their wives but their mistresses.

Fysh saw the continuing challenge to his authority as having perhaps been 'engendered by my own personal deficiencies' – the lack of formal education and business cunning.[11] He suspected that board members Norman Watt[12] and vice-chairman Bill Taylor had tried to undermine him when Chifley's government had bought out the airline, and Watt, in turn, believed that while Fysh was 'an able man', he was not 'a modern man'.[13] Fysh bristled when his hard-drinking, pugnacious and ambitious general manager Cedric Turner, a taciturn accountant with curly white hair, kept secrets from him, referring 'as little as possible to his board'.[14] 'The fact also was, of course, in my long pioneering service, that I was receiving far too much personal publicity and public applause for the triumphant new happenings taking place,' Fysh wrote, 'whereas it was the board collectively – and the management – which, in the nature of things, had done most of the work. I tried hard to minimise this, but jealousies arose.'[15]

In July 1954, a new board member was appointed: the soon-to-be-knighted Dr Roland Wilson.[16] He was Secretary to the Department of Treasury and known at the time as one of the 'seven dwarfs', a group of powerful postwar Commonwealth public servants who were of short stature.[17] Wilson treated Fysh with 'increasing contempt and hostility'.[18] On a visit to the Qantas office in New York, Fysh was introduced to a glamorous new staff member, Joyce Chivers, who had been given the job by Wilson; Fysh was disturbed that the appointment had taken place without his knowledge, and he suspected that Wilson and Turner were undermining his authority. He brooded over the matter and made it worse.

Fysh became increasingly isolated on the Qantas board, and his great contribution to the airline and Australia became deliberately forgotten in the jockeying for positions of influence. Scotty Allan, Turner's assistant general manager, told the Qantas historian John Gunn in 1982 that while he greatly admired Turner's business nous during this time of massive expansion for Qantas, the general manager was 'remarkably selfish … incapable of dismissing anybody face-to-face, but from weakness, nor soft-heartedness … He was nasty when there was nobody there, but he would not be nasty to somebody to their face' and he had no close friends.[19] Allan said the man he called 'Huddy' was 'a nicer kind of human being' who despite the limits of his education built the airline on the back of his strength of character and determination, and engendered a spirit of loyalty among the staff because of his own loyalty to the brand.[20]

As part of the executive committee of the IATA, Fysh began making two overseas trips a year. He was also a member of the Tasman Empire Airways Limited (TEAL) board, chairman of the Wentworth Hotel board, and deputy chairman of the Australian National Travel Association. But he knew he was losing his grip on his own airline. He found he was 'getting jumpy under the pressure of it all'.[21]

His old commander Dicky Williams, who had succeeded McVey as Director-General of Civil Aviation, and Williams' assistant Edgar Johnston were both nearing retirement. On 7 December 1954 Fysh wrote to Bob Menzies' new Minister for Air, Athol Townley,[22] suggesting he should relinquish duties as chief executive on 30 June 1955, when he would be sixty, but continue as a full-time chairman of the board.[23] Townley approved, but he told Fysh that his salary would effectively reduce from £5000 to £3539 per year. Fysh could keep his company car and driver, George West, and would have an entertainment allowance of £500 per year. Fysh and Nell were now living on posh Fairfax Road, in Bellevue Hill, but Fysh thought the price of houses too steep, and he continued to rent. With the price of a bungalow in the area at almost £10,000,[24] how much higher could they go?

As chairman, Fysh would now be the channel of communication with the Minister for Air and the director-general on all policy matters, while Turner would be both general manager and chief executive. Fysh was insisting that the word 'Empire' was no longer appropriate in the airline's official name 'Qantas Empire Airways' – for one thing, American passengers just didn't get it. 'Meanwhile, we are stressing on our aircraft and in publicity material,' Fysh wrote, '"Australia's overseas airline".'[25]

With the changes at the top of Qantas, Turner released a full-page press release about his appointment through a media company headed by Charles Ulm's son, John. The release gave Fysh only passing mention in two brief sentences.[26] Fysh complained to Turner that it created the impression Fysh was retiring from the airline when in fact he would be an active and busy full-time chairman. He suspected Turner was trying to write him out of the Qantas history. 'I ask you to watch this sort of thing more closely in future,' Fysh wrote, 'especially in matters where it is right and proper for me to be consulted.'[27] Still, Fysh was made to feel less and less welcome in the business he had built.

QANTAS CAME TO OWN sixteen of the magnificent Super Constellations, but the jet age for passenger transport was fast approaching. Sydney was agog with the arrival of a BOAC Comet 3 on Sunday, 4 December 1955. It was the first jet airliner ever seen in Australia, and its initial approach to land had to be abandoned when several hundred people ran onto the runway at Sydney Airport for a closer look. But Allan and other Qantas technical experts believed there was an even better jet suitable for their requirements.

In September 1956, Turner placed an order for seven new aircraft from Boeing. The four-engine American jets were called 707s and cost $US5.15 million each. 'These aircraft,' Townley declared, 'will cruise at 550 miles per hour [885 kilometres per hour] and will operate over ranges up to 3500 miles [5632 kilometres]. They will be placed in operation on QEA's Kangaroo route to London and on the Pacific route to America late in 1959. The jets, 707-138 Stratoliners, will reduce flying time from Sydney to London from 48 hours to 27, and from Sydney to San Francisco from 28 hours to 16.' The new aircraft could carry seventy first-class passengers in full recliner slumber chairs through to 120 in tourist configuration.[28] Qantas would be the first business in the world outside America to operate American jet airliners.

In November 1956, Qantas carried the Olympic flame on a flight of 8600 miles (13,840 kilometres) from Athens to Darwin on its way to the start of the Melbourne Games. It was the longest flight ever made by the Olympic torch, and the first time it was carried south of the equator. Qantas was hard-pressed to keep up with demand for flights to the great sporting event, and on more than a dozen occasions, the entire Super Constellation fleet was in the air at the same time. Revenue for the year skyrocketed to almost £20 million, and passenger numbers grew to more than 161,000.

Early in 1957 Fysh wrote to Turner to congratulate him on the 'splendid results achieved by the Company during the past year by you and the Management Group under your control and guidance'.[29] Although he did not get on with Turner, Fysh pressed government ministers in Canberra to give the chief executive a pay rise from £6000 to £7500,[30] and he assured Turner of his 'continued sympathetic support'. 'I feel that this will be fully reciprocated by yourself,' Fysh told Turner. 'For the good of the organisation it is necessary that we should have the closest collaboration.'[31]

Collaboration, though, was not Turner's style. Four months later, on the eve of Anzac Day 1957, Fysh wrote in his personal diary that he felt increasingly distanced and out of touch with the senior executives at Qantas, and he was making notes to let off steam. He was disturbed by Turner's temper and his drinking. Turner was an intensely shy man who overcompensated with a gruff exterior, but he would sometimes fall asleep at public functions, and when he woke would continue conversations as though nothing had happened.[32] 'Mostly we are poles apart and antagonistic,' Fysh wrote. 'Never once in all the years do I remember Turner responding to a suggestion by me ... [He is] completely, absolutely money and power mad. Out of sympathy with me … out of touch with [those] lower down. Ruthlessly ambitious. On the other hand, his mind is brilliant. A great organiser and morally courageous. The great fault is that [he was] never in the services.'[33]

At the time, Qantas pilots were in the midst of a nine-day strike that cost the airline £500,000.[34] Fysh wrote that the staff under Turner were treated like pieces on a chequerboard instead of human beings, with older employees being dismissed with no regard to their long service or company loyalty. Others were shuffled around the business at short notice and played like piano keys. Turner, Fysh wrote, was 'adamant and unyielding in discussions … will not listen to the other fellow and cannot convey the impression that he has some understanding and sympathy with staff in their problems'. Yet Fysh respected Turner as 'a brilliant man [who has] done great things for the Company and Australia … If the pilots have had to put up with him, then so must I for the good of the Company'.[35]

On 28 October 1957, the pair posed awkwardly with Prime Minister Menzies at the opening of the £1.3 million eleven-storey Qantas House, just around the corner from where Fysh was married. Turner listened stone-faced as Fysh compared the magnificent curved building with the company's headquarters in Longreach thirty-seven years before: a weatherboard house with a hitching rail for horses at the front, and a barn at the local showground for use as a hangar. He pointed out that Qantas was now the seventh biggest airline in the world, covering an unmatched route of 65,597 miles (105,568 kilometres).

Addressing the five hundred guests, Menzies said it was remarkable that Hudson Fysh, one of the two men who flew on the original flight, still headed the airline. 'What pride he and the other people must share when they think over the past, from the days of Qantas' trifling background to the vast achievement which has been brought about. Thank God we still have pioneers in Australia, men who were prepared to accept risks in order to achieve success for their country. The day will come when, as well as having statues in honour of the explorers, such as Burke and Wills, people like Sir Hudson Fysh will also be similarly honoured.'[36]

A far cry from the early offices in Winton and Longreach, Fysh stands in front of the 11-story Qantas House on Hunter Street in Sydney. The curved structure was judged the best new building in the British Commonwealth by the Royal Institute of British Architects in 1959. *State Library of NSW FL8644129*

WHILE THE QANTAS 707s were being built at Boeing's plant in Seattle, Reg Ansett was also rapidly expanding his domestic operations by buying out ANA following the death of Sir Ivan Holyman in 1957, and forming Ansett-ANA. Ansett was exploring the possibility of entering overseas markets, but instead stayed locked in competition with TAA on domestic routes as the Menzies government enforced its Two Airlines Policy, a legal barrier to prevent new entrants coming into the Australian domestic markets.

The relationship between Fysh and Turner continued to decline, and at times the chief executive of Australia's national airline wouldn't even speak to his chairman. Shortly before handing over the head management role to Turner, Fysh had issued his senior staff with his booklet *Ethics and Other Things*, in which he outlined the principles he said had guided Qantas since its inception and which he believed should continue to light its way forward. He touched on twenty-one areas for his staff to follow, chief among them giving good service to the public, and highlighted the importance of such qualities as integrity, loyalty, diplomacy, discipline, self-respect, and the satisfaction that comes through work, achievement and well-earned rest and recreation.[37]

Fysh had scorned McGinness's fondness for drink thirty-five years before, and now Turner's drinking distressed him. Next to his ninth-floor office, the chief executive had a room with a bar set up where he would socialise and talk business. Fysh labelled it a 'drinking den' and said that at nightly gatherings, 'the spirits often flowed freely with the top staff and visitors, and far too much was drunk'.[38] Turner felt ill at ease at public functions, though, and Fysh – the Gallipoli veteran, war hero and pioneer aviator – continued to be the public face of the company.[39]

Edgar Johnston, who had known Fysh since the days at Longreach, saw Turner as 'a strong man, a bit difficult, a bit outspoken ... storming up and down the room', but also as the undisputed brains behind the new-look airline. Some on the board now shared Turner's opinion that Fysh was an unnecessary cog in a huge machine.[40]

Despite the simmering tensions, Fysh backed Qantas's foray into investment in overseas airlines. The company bought stakes in the Singapore-based Malayan Airways and Fiji Airways. The latter had been founded by the Tasmanian Harold Gatty,[41] who in 1931 had been the navigator on a record-breaking round-the-world flight with the American Wiley Post.

Five days before Christmas 1957, the Super Constellation *Southern Aurora* left Sydney with thirty-two press, radio and television representatives for a remarkable goodwill flight of twenty-seven thousand miles (forty-three thousand kilometres) around the world. John Ulm had arranged interviews for the media contingent with the prime ministers of New Zealand, India and Thailand, the presidents of Pakistan and India, Singapore's chief minister, Pope Pius XII, the Archbishop of Canterbury, a British cabinet minister, and Lockheed's chairman and chief executive.

This was a prelude to Qantas inaugurating its round-the-world service from Essendon Airport in Melbourne three weeks later, with two Super Constellations heading to New York. Fysh and Nell hosted a function for three hundred people as the wives of Victorian Premier Henry Bolte and new Air Minister Senator Shane Paltridge[42] cut ribbons to launch the aircraft.[43] Both Super Constellations flew to Sydney before separating, the *Southern Aurora* flying east via Fiji, Honolulu and San Francisco, and the *Southern Zephyr* flying west through Djakarta, Singapore, Bangkok, Calcutta, Karachi, Bahrain, Athens and Rome. Both arrived back in Sydney six days later. It was the start of two weekly complete round-the-world services for Qantas, one from Melbourne and one from Sydney. With two other Qantas flights connecting with BOAC in San Francisco, there were now four Qantas round-the-world services each week.

To cap off the extraordinary publicity the flights generated, the Queen Mother flew on the Qantas Super Constellation *Southern Sea* during her tour of Australia in February 1958. She flew back on it to London, and Fysh was always impressed with the way Her Majesty decided to stay with the flight despite engine troubles that caused delays in Mauritius and Entebbe before she finally boarded a BOAC flight in Malta.

Fysh launched Qantas's around-the-world-service on 14 January 1958 with Super Constellations flying above Sydney Harbour. *news.com.au*

THE TURBULENCE INSIDE Qantas House continued, though, and Fysh found respite in the great outdoors. He returned to the fly-fishing that had so enthralled him as a young boy beside the waterwheel near his childhood home. He began travelling to the Snowy Mountains whenever possible, enveloping himself in the solitude of the high country, breathing in the icy wind 'as the mountains' gift'.[44] Fysh usually stayed at The Creel, a rambling old accommodation house on the rushing Thredbo River near its junction with the Snowy, alongside fishermen who rode into the area with packhorses.

He faced a far different challenge in the corridors of power at Qantas. His concerns escalated about the relationship between the married Roland Wilson and the glamorous Joyce Chivers in New York, and he suggested to Turner that it might be best for all concerned if she left the company. Turner took no action. In January 1959, Wilson confronted Fysh; he said he had witnesses who were prepared to state that Fysh had defamed him in suggesting he had been indiscreet. Wilson claimed that Joyce Chivers was nothing more than an old family friend from Tasmania – and he threatened Fysh with legal action. Then he raised the matter at a board meeting on 19 February 1959, demanding an apology. Fysh offered one, and he also wrote a letter of apology to Joyce Chivers.

Years later Fysh said that he still should have been asked about the posting, that it was a bad look, and that while he believed his stance was correct, he had suffered for it. Wilson eventually married Joyce Chivers[45] after the death of his first wife. Turner continued to rile him. Fysh said the Qantas general manager gradually built up a reputation for 'ruthless aloofness, later to develop into a form of egomania and an attitude of absolutism which I found it impossible to deal with effectively'.[46] Fysh claimed that Turner was 'constitutionally unable to drink without showing the effects of it'. He said that once during a business trip to Wellington, New Zealand, Turner had to be led away from Senator Paltridge's table in a stupor, having slumped forward. 'The good name of the Company suffered, to my critical annoyance,' Fysh wrote, 'and at times the Chief Executive so worried the

Board that, on a number of occasions, members were detailed to see him, and curb his actions.'[47]

Turner insisted on having two seats for himself when he travelled. A New Zealand passenger complained that on a BOAC jet at Tokyo, Turner objected to being seated beside him; a row developed on the plane, which, as a result, departed late. Fysh wrote that on another occasion, when Turner was in Perth on his way to South Africa, a Qantas first officer had to be off-loaded because the chief executive wanted the seat next to him left vacant.[48] There were tense scenes in Qantas House, too; Fysh had once used a machine gun to shoot down Germans, but he tried diplomacy against the enraged, shouting Turner, seemingly without effect.[49]

Then Turner's outlook brightened dramatically, at least for a while. He arrived for work one morning to be met by a cordon of reporters. They told him that he had the winning ticket in the £100,000 Opera House Lottery, and that he would pocket it, tax free.[50] 'This has changed him,' Fysh wrote, 'more for his good. He will come good now and this whole windfall [is] like a golden leaf fluttering down at his feet.'[51]

IN MAY 1959, FYSH AND NELL headed to an IATA executive committee meeting in Lac Ouimet near Montreal. They then travelled to Boeing's headquarters in Seattle for the ride of their lives on the *City of Canberra*, Qantas's first 707. Fysh felt he was walking into a plush restaurant when the Qantas captain Bert Ritchie showed him and Nell around the aircraft. Its reclining seats were covered with persimmon-and-turquoise fabric made from Australian wool, and its walls were decorated with Australian wildflowers, wattle and bottlebrush. Fysh's senses were overwhelmed by the finest aircraft in the world – a 'magic carpet,' he called it, 'to waft its passengers at the speed of sound'.[52] With its four huge Pratt & Whitney engines, the 707-138 model could reach 595 miles per hour (958 kilometres per hour), almost double the speed of the Super Constellation. It could cruise at twelve thousand metres, double the ceiling for the Super Constellation, and fly above most clouds and bad weather.

Fysh and Nell flew to San Francisco on the *City of Canberra*, then set off for Sydney's Kingsford Smith airport. Nell joked that Qantas was using her as a guinea pig again, but this flight was a lot more comfortable than in the old biplane days out of Longreach. On 2 July 1959, the 707 landed in Sydney to a red-carpet reception. The trip from San Francisco to Sydney had taken just sixteen hours ten minutes actual flying time, obliterating the previous record of twenty-seven hours thirty minutes set by a Pan American DC-7C. Qantas passengers could now reach America in half the time for no increase in the fare.

As the new president of the IATA, Fysh hosted the association's seventeenth annual general meeting in Sydney in October 1961. His 21-year involvement with Tasman Empire Airways had ended in April when the New Zealand government bought out Australia's shareholding; TEAL would be renamed Air New Zealand in 1965. Through IATA, Fysh was dealing with airlines that between them owned 3339 aircraft: 382 jets with another 900 on order, 480 turboprops and 2477 piston-engines. High on the AGM agenda was the continual reduction of airfares so that air transport was brought within the reach of ordinary working people. The promise of travel in supersonic jets was also discussed, along with ways to match their economy of operation with current jets – and ways to diffuse the sonic boom so all the windows under the flight path of these futuristic machines wouldn't be blown apart.

For the overseas visitors, Fysh organised a day of entertainment with an Australian bush theme at Lane Cove National Park. Sheep were rounded up with border collies and shorn, and there were whip-cracking shows and wood-chopping contests, as well as boomerang-throwing displays by Indigenous people, who also exhibited their bark paintings. Two actors played swagmen, sitting down beside the Lane Cove River to make a fire and boil a billy, just as Fysh and McGinness had done on the route from Longreach to Darwin. And, in something of a coup, Fysh had managed to borrow the suit of armour worn by Ned Kelly at the siege of Glenrowan eighty-one years before, and an actor wearing it was escorted through the crowd by two uniformed policemen. At a dinner the following evening, Fysh introduced two of Australia's great pioneer aviators, Harry Miller and the newly knighted P. G. Taylor, to the audience. The event was a big success, and within a few weeks Fysh wrote in his diary that he was now '67 not out, and still going strong having ridden through troublesome times ... many storms, risks and difficulties. Except for some difficult periods [I have] achieved more peace of mind and better health.'[53]

At an IATA meeting in Dublin in 1962, Fysh voiced his concerns over the possibility of hijacks and recommended that a tranquilliser 'space gun' be developed to improve security on international aircraft. But it seemed Queen Elizabeth and the Duke of Edinburgh had more than adequate security when they flew in a Qantas 707 from Christchurch to Canberra during the 1963 Royal Tour. In 1965, Qantas flew the Duke and Duchess of Gloucester from London to Sydney thirty-one years after the duke had declared open the company's original overseas service in Archerfield.

Fysh and McGinness had paid less than £2000 for the first two Qantas aircraft, and now the airline's fleet was valued at more than $138 million.

FYSH TURNED SEVENTY on 7 January 1965, and his retirement from the chairman's role neared. He was the only one of the original Qantas directors still alive. He recalled, 'For some political subtlety, unfathomed by me, I was pressed to resign, but declined to do so. It was agreed that my resignation should be by mutual consent and honour was resolved by this compromise, which did not mean anything anyhow.'[54]

With John and Wendy both married and starting their own families, Fysh and Nell had moved into a smart flat on Rosemont Avenue in Woollahra. They still travelled overseas frequently for IATA meetings and other Qantas business. At a cocktail party in Istanbul, Fysh was asked if this was his first visit to Turkey, and he replied that he'd been there once before – about fifty years earlier, running like hell up a beach with bullets whizzing all around him.[55] At another time he visited Ramla's cemetery and paid his respects at the graves of the two German officers he had shot out of the sky in 1918.

Fysh farewelled the IATA in Vienna in October 1965, fittingly in his first ever balloon ride. Together with other airline chiefs from around the world, Fysh floated for an hour above the palace in Eisenstadt and the splendour of its surrounding forests. He was filled with an immense feeling of detachment and tranquillity – and satisfaction at a life well lived. Returning to the roots of aviation as the balloon sailed serenely above the greenery, Fysh could reflect on a career unparalleled in the airline industry.

Back in Sydney, his relationship with Turner hadn't improved, despite the chief executive's £100,000 windfall. In November, Fysh held a party to launch *Qantas Rising*, the first volume in his three-part autobiography, and Turner wasn't invited. The chief executive was desperate to become the company chairman, and he never hid his ambition. His personality had started to grate on other Qantas executives. 'It is no secret,' the *Sydney Morning Herald* reported, 'that relations between Fysh and Turner have never been particularly warm and that some members of the present board would by no means relish the presence of the forceful and wealthy chief executive in the chairman's seat …'[56]

Two weeks later, Turner missed out on the role he had coveted. The federal government announced that from 30 June 1966, Roland Wilson would succeed Fysh as chairman of Qantas.[57] Fysh was chasing trout in a remote part of the Snowy Mountains when he heard of the appointment. His own role with Qantas had proved to be irreplaceable: Wilson's appointment was only as a part-time chairman.

Fysh wrote a long, confidential letter to Wilson, offering to continue as a director of Qantas although he wasn't expecting to be appointed. He hoped, however, to continue as a director of the Wentworth Hotel, and he asked for an office with phone and typing facilities so he could start on the second volume of his memoirs. 'In any event,' he wrote, 'after 45 years with Qantas and QEA it will be impossible to break off suddenly.'[58]

Chapter 24

I always remember Hud as being extremely kind and extremely humble. He had been a soldier on Gallipoli and risked his life many times in the air, but he was never one to beat his chest about it. He was always modest and understated. I always had a sense that he loved me and was keenly interested in my future and the future of Australia.

FYSH'S GRANDSON DAVID MILES[1]

FYSH STILL HADN'T BOUGHT a family home despite Nell's pleas, and they continued to rent their flat in Rosemont Avenue, Woollahra. Even though he had once spent millions on bringing the most advanced aviation technology to Australia, he was often at loggerheads with his landlord over minor matters. Despite increasing the rent, the landlord was slow getting the hot water system fixed, and more than once the chairman of Qantas had to endure a cold shower, while Nell sometimes had to heat water in pots on the stove for washing-up. The tile staircase was dirty, slippery and in desperate need of some carpet, and the landlord had knocked back Fysh's request to glass in a verandah. 'Can you help us tenants out of this predicament?' Fysh pleaded.[2]

But Fysh had bought a twenty-acre (eight-hectare) piece of the Australian bush in Dural, on the north-western rural outskirts of Sydney. For a man who had always loved Australia's great outdoors, it became his second home. He and Nell were besotted at first sight of the property's golden wattle and its stunning view across the Nepean Valley to the Blue Mountains; they called it

'Golden Ridge'.[3] Fysh fenced the land, erected a small cottage and, together with his offspring and friends, cleared some of the bush, leaving enough of it in its natural state to marvel at the eagles, whipbirds, butcherbirds, kookaburras and wrens. Fysh fattened cattle and grew roses, oranges, lemons, grapefruit, figs, peaches, plums and apples. He and Nell organised family picnics at Golden Ridge, and at night Fysh would sometimes lie in the grass and gaze at the bright stars in the clear skies, wondering how long it would take before mankind flew there, too.

Fysh collected stamps and played golf and lawn bowls at Royal Sydney, but Golden Ridge and fly-fishing in the Snowy Mountains remained his great passions. 'I did a lot of fishing with Hud,' his grandson David Miles recalled.

> He took me down to the Snowy Mountains in 1965 when I was nine years old. We stayed on Lake Jindabyne. He was desperate for me to catch a fish and he taught me the art and ethos of fly-fishing; that there was much more to it than just throwing a spinner lure out. We were out there for three or four days, and I still hadn't caught anything no matter how hard I tried. He knew how keen I was to catch something, and he said, 'How 'bout we just use the spinner once but don't tell anybody?' We did, and I caught a beautiful four-pound trout which we were both very proud of. Hud just loved that mountain country and being out in the water with nature. He was never a man motivated by money or success. His great driver in life was to build an airline that could really help Australia and its people.[4]

WHEN INFORMED OF HIS impending knighthood a decade earlier, Fysh had reflected on his life and the motivations for his soaring achievements. 'Quite a few people have asked me why I have been so successful,' he told the audience at a Qantas function. 'Material success and success in the eyes of the public perhaps, yes to a degree, but while riding that tiger called "Qantas" I have not had time to dismount awhile and think as I should on the things that matter most. Religion and ethics, for instance. What is truth? I am still not sure yet ... Yet these things mean more to me in the possible attaining of them than any material reward. These were the principles taught me in youth.' Fysh said that he had a 'very poor schooling' and wasn't educated for a professional career, but that he had benefited from character training through sports and the military, and it had always stood him in good stead. 'And I mean really stood,' he said, 'for all those things on which is built a house which will last and weather the storm ... character, reputation, loyalty, confidence, honesty and all those sorts of things.'[5]

For a long time Fysh felt lost without the daily excitement of being the boss of an international airline. He sympathised with all those in the workforce who had reluctantly come to the end of their working life, but he consoled himself that it gave him more time to read and to study, and peer into the mysteries of life. On Sunday afternoons, Fysh and Wendy would often go for long walks through Trumper Park near the Fyshs' home. He told her that while he struggled with his spirituality, he had found a method of living in the story of Jesus – even though he didn't take kindly to the pompous words and ceremonies that he had encountered in many church services. One didn't need a formal religious structure, he told Wendy, to be 'steady in life and to live ethically'.[6]

Later Fysh would write, 'I cannot reconcile myself to the atheist position. The mystery of the universe, our lack of knowledge as to where it all started, and will end, is too great. I know of nothing to take the place of the Divine moral and ethical teachings of Christ. Each night of my life, as I pray to the unknown God, I ponder these matters and on many more: the Divinity or the futility of human life ...'[7]

ON 25 FEBRUARY 1966, Fysh prepared for prostate surgery. 'Today I stood at midday in the brilliant sun if not well, I was still a free man,' he wrote. 'Now, one month ahead in awful inhuman degeneration of functions – 30 days unable to leak at will. What will be the end? At best a terrible ... change of life.'[8] His mood only brightened a little when Australia's governor-general, Sir

William Slim, sent him thanks for a copy of *Qantas Rising*. 'What a splendid life you have had,' Slim wrote. 'How you have served Australia – and a lot of us, too.'[9]

Fysh's publisher at Angus and Robertson, John Ferguson, remembered him with great fondness as a man with a dry sense of humour who was 'way ahead of his time' and a gentleman to deal with. At a function the publishers held for Fysh there was spontaneous applause when Fysh arrived. The assembled guests included Ferguson's father George Adie Ferguson, the grandson of Angus and Robertson's co-founder, George Robertson.

While Fysh recovered from his surgery, the Qantas vice-chairman Robert Law-Smith, a World War II squadron leader, wrote to say that the board was happy to provide Fysh with an office, telephone and typing facilities. Law-Smith added that Fysh's long-time secretary Ida Isaacs could stay until 31 December 1966, helping him with his memoirs and with access to company records. The Qantas board, though, would run the Wentworth Hotel and could do without his services.[10]

Fysh was disappointed at not being asked to continue as a Qantas director, because that would have opened the way for him to be on the hotel board. 'The unfortunate fact is that owing to the decline in the value of money I will be left with quite an inadequate income to live reasonably on even after really severely cutting all expenses,' he told Law-Smith. 'After selling all but five acres [two hectares] of my Dural property and receiving the Commonwealth Pension of £14,450 in July, I will be able to muster £56,000, which I estimate on a reasonable average will bring me in £3,360 per annum. Now rent, household allowances and wife's dress allowance (cut to half), taxation, running a car, two Clubs only and expenses, plus annual holidays is estimated to absorb exactly that amount which, of course, is only a portion of what one's expenses will be … Obviously, I should be working in with Qantas or a Qantas association if this is possible.'[11] Working for the government hadn't paid as well as it could have. Fysh said he was devastated that after all his years of pioneering and running Qantas 'in financial circumstances so completely out of line with the rewards from similar service in private industry', he was being left with an uncertain financial future, especially 'when one sees the sumptuous wealth displayed on every hand by successful people not in Government positions'.[12]

Fysh then appealed to the new Minister for Civil Aviation, Reginald Swartz,[13] a former prisoner of war on the notorious Burma–Thailand Railway. Fysh asked Swartz if he could be retained in an advisory capacity on the Qantas board for an annual fee equivalent to that of a director. 'I find I will definitely need to augment my income in order for my wife and I to live in reasonable comfort, after heavily curtailing expenses,' Fysh explained. 'I retire after 45 years creative service having forgone the fortune, expensive homes and cars etc. which the great majority of my friends and associates have as a result of their private enterprise. I can't help being conscious of being the main lasting factor in building what is today a great Commonwealth enterprise.'[14] A week later, on 20 May, with Fysh still chairman of the board, Boeing made its first proposal for Qantas to acquire three of its huge new 747 Jumbo jets at a cost of $US18,381,000 each.

While preoccupied with worry about what would happen when he soon left the helm of the airline he had built, Fysh wrote to the Duke of Norfolk, at Arundel Castle in England, about the running of the duke's race meeting at Epson, which Fysh had visited on his last trip to England. He could always see the wry side of life. He complained lightheartedly that he had placed a £1 bet on a winning horse called Caterina at odds of seven to four. The bookmaker, 'a big, swarthy, dark, greying haired man', refused to pay. 'There was also a gullible little fat man who did the smooth talk,' Fysh complained, 'and, finally, the penciller who was, I think, one-legged, very abusive and a bad type. I realise that these conditions could either be regarded as warranting a good laugh or, alternatively, as disastrous to the honour of England, when viewed by a visitor from Australia where racing is properly regulated, and bookmakers must themselves record the bet on the back of the card. English racing, of which you are a Steward, owes me £1 …'[15]

Fysh was finally allowed to remain as a Qantas consultant after his official departure as chairman on 30 June 1966. On the twentieth, Minister Swartz hosted a lunch for him in Canberra, and the Qantas board organised a dinner for Fysh and Nell at the Australian Club. At another staff send-off, Qantas workers chipped in $472.47 for Fysh to buy a grandfather clock in England as a memento of his forty-six years of service.[16] He was given other farewell functions in Australia and overseas, and in Launceston, where he was born seventy-one years earlier, he was awarded the Freedom of the City.

Australia's new prime minister, Harold Holt, sent him a letter on 27 June 1966: 'I feel I should write to you in formal terms how much I have in mind that this is your last week in the position you have filled so admirably. It is in fact the culmination of the great public service which you have rendered to this country for what truly constitutes a great many years in the life of any man. We thank you ...'[17] Not long after that, the federal government granted Qantas permission to place deposits on four of the Boeing 747s, costing a total of $A123 million.[18] Qantas also ordered four of the Anglo-French supersonic aircraft, the Concorde, though those plans were soon curtailed.

Fysh continued writing his memoirs and the history of Qantas – as well as a book on fly-fishing[19] – from his office in Sydney's ANZ Bank building, but in February 1968, the Qantas board voted that he would have to relinquish his secretary and office at the end of the year. After continued appeals by Fysh, the board agreed to continue paying him as a consultant at a yearly fee of $2000 with an occasional pay rise.

For all of Fysh's business success over the years and his years of service to Australia, he and Nell didn't live extravagantly. They rented a comfortable flat, Nell always dressed well, and they treated themselves to a good meal occasionally at their local restaurant, Pruniers. In his twilight years, Fysh feared running out of his retirement savings. Now without a Qantas driver, he bought a Chrysler Valiant but had to chase $27.25 in damages after another motorist ran into his vehicle at Edgecliff.[20] Then he had to hunt down $33.75 in a reimbursement from the Chrysler dealers in Darlinghurst after the car needed mechanical repairs while he was on a fishing trip.[21]

Qantas chairman Roland Wilson distanced himself from Fysh so that any of Fysh's communications had to go through Robert Law-Smith. Fysh told Law-Smith that he was grateful for Qantas's generosity to him and for the help given to Arthur Baird and George Harman before that. Fysh wrote,

> They both needed it, but in their cases the arrangement, I think, ended in their deaths. For some peculiar reason fate ruled that I should live on … My only complaint – having nothing to do with the Board – is that despite my record … I have never been able to get into a financial position approaching most of my associates and contemporaries, having had a fortune snatched from under my nose while new millionaires arise almost daily … I have never been one to make money a god, but I would have liked to help more than I can some needy members of my family.[22]

On 7 February 1969, Fysh wrote again to Law-Smith.

> I have had a month of it at home here in the flat and, of course, have missed the old facilities in the ANZ Bank building very sorely. At my age, when I needed a bit of sympathetic consideration, I am afraid it has put an end to any further serious work. No one can understand why I had to leave the city office, in view of my long and creative service in Qantas … The Australian government, when it took over Qantas, proceeded to cash in on the pioneering work of a few private individuals … who have never been suitably recompensed … My shares in the old Qantas and then in QEA – which I was a very active leader in – should have been worth $2 million today. Pioneers, of course, traditionally do not make money … You will remember the great struggle I had to stay on as long as I did in the face of an arbitrary notice to quit.[23]

In July 1969, Fysh and Nell flew to Cloncurry for a function organised by the Australian Inland Mission. They sat with Frith Fysh, Graham's son and Frith's wife Elizabeth. The young couple would eventually take over the Acacia Downs station once owned by Ainslie Templeton, Qantas's first shareholder.

Eight men had landed on the moon by the Australian winter of 1971, and not long after dawn on Monday, 16 August, Sydney was treated to the arrival of an aircraft that appeared out of this world. Qantas's first Jumbo jet, the huge 747B called the *City of Canberra*, cruised over Rose Bay[24] where the Empire flying boats had been based when Fysh was the most powerful man in Australian aviation. The Jumbo, capable of carrying 356 passengers, had been handed over to Qantas staff at the Boeing plant in Seattle, then flown nonstop from Honolulu. At 7.10 a.m. it had crossed the eastern coast of Australia and was joined by two escort aircraft as it circled Sydney before touching down to a rapturous welcome at Mascot half an hour later.

About a month after that, Qantas started its 747 passenger service from Sydney to Singapore. Amid the tremendous interest in these extraordinary aircraft, Fysh appeared on the ABC documentary series *Those Were the Years*, talking about the formation of the airline and the days when he would try to sell joy rides to wary station owners in the outback.[25] He posed for photographs beside the Jumbo and a replica of his old Avro 504K, the first Qantas plane he had bought, in order to illustrate the gargantuan growth of both the airline and the aircraft since 1920. The University of Tasmania presented Fysh with an honorary degree as a Doctor of Engineering.

Fifty years after Fysh carried Alexander Kennedy in the first scheduled Qantas passenger flight, and after having completed the last two volumes of his memoirs, he and Nell visited western Queensland at the end of 1972 in what the press called a 'sentimental journey'.[26] 'Hud' and 'Darl' posed together for photographs in front of the very same hangar where Fysh had spent a decade building Qantas from its humble origins.

FYSH WAS SEVENTY-SEVEN, and for some months he had made little jokes to his family about his 'dodgy ticker', masking the discomfort he often suffered. The pain in his chest wasn't so bad when he creaked his bones to their full height, but when he sat in a chair it was murder. He saw a doctor several times about the pain but was told just to take it easy. Finally his doctor said he would probably need a hernia operation and sent him for an X-ray.

The Craven A people had marketed their cigarettes as the 'brand you can trust', but they had lied. The X-ray showed a cancer as big as an orange sitting on Fysh's lung and pressing against his heart.

After the diagnosis Fysh remained at home on Rosemont Avenue and underwent treatment in his stride. He became frailer and sipped on Bonnington's Irish Moss cough syrup as his voice grew more raspy. He joked that he only drank the cough mixture because it had a little alcohol in it, but it helped calm his smoker's cough.

He was desperate to finish writing a biography of his extraordinary grandfather Henry Reed,[27] which he did despite his illness, exploring Reed's remarkable business life and relationship with God. Fysh also wrote a letter to Wendy about his own life, thanking her for all she had done for the family and for always being there in his time of need. 'He told me he didn't expect any of his two children or seven grandchildren to burden themselves with trying to match Hud's achievements,' Wendy recalled. 'He just wanted them to be happy in whatever they chose to do in life. Dad told me he'd just been lucky in getting into aviation in its infancy. He was being very modest, of course. It wasn't luck that helped him run Qantas for almost 50 years.'

Fysh told Wendy that he had learned a long time ago that abiding by principles of justice and honesty, rather than the pursuit of fame or money, should be the guiding light in a person's life. It was a legacy he hoped to pass on to his grandchildren.

'To be a good sort of understanding character, and in kindness and helpfulness to others is so much more important,' he wrote, 'and I would much rather see anyone of mine more apt in this than in making a million, being a movie star, or running some vast industry.' Fysh signed the letter 'Huddy'.[28]

On Saturday, 6 April 1974, Fysh collapsed at home and was unable to get back on his feet. Nell called an ambulance to take him to the Scottish Hospital in the neighbouring suburb of Paddington. Stretcher-bearers carried Fysh down the stairs from his flat, and in a weak, hoarse voice he told them, 'You know fellas, I did a lot of this, too – carrying blokes on stretchers down the slopes of Gallipoli.'[29]

After Fysh was stabilised and made comfortable in hospital, he looked weakly around the bare walls of his room and joked with the matron that he had plenty of fight left in him, telling her and his family that he had no intention of giving out in a room without a single picture. His family went home for the night so Fysh could get some sleep. But in the quiet, still darkness, the man who had once defied Turks and Germans in battle, who had fought one dogfight after another over Palestine and then battled for almost half a century guiding Australia's airline through stormy times, rolled out of bed and fractured his shoulder on the hard hospital floor. Hudson Fysh died that night.

A private funeral service was held three days later. On 11 April a large gathering of mourners at St Andrew's Cathedral honoured him with a public memorial. Don Anderson,[30] who had succeeded Roland Wilson as Qantas chairman, said that Fysh was 'a great Australian and an eminent figure in international aviation' who would never be forgotten.[31]

Despite the place he held in Australian business life, Fysh did not die a rich man: his estate was valued for probate at $95,817. While Bob Menzies had once suggested a statue for Fysh, the *Canberra Times* told its readers that his memorial was actually Qantas itself, 'airborne somewhere around the world every minute of the year'.[32]

Fysh was cremated, and his family placed his ashes in the cemetery of St Jude's Anglican Church in Dural, not far from Golden Ridge and the birds and trees that Fysh loved so much. Nell was laid to rest next to him in 1983.

John Fysh organised a simple headstone for his father, with gold lettering on black marble.

Sir Wilmot Hudson Fysh KBE DFC
7-1-1895 – 6-4-1974
SOLDIER. AVIATOR. COUNTRYMAN. WRITER.
CO-FOUNDER OF QANTAS. A MAN OF COURAGE,
VISION, AND INTEGRITY.
A LOYAL SERVANT OF HIS COUNTRY.

Epilogue

ON 2 MAY 2022, QANTAS CEO Alan Joyce announced that the airline would begin direct flights from Sydney to London and New York from 2025. He said the new routes, using Airbus A350-1000s, would be the longest passenger flights in the world. A Qantas test flight direct from New York to Sydney in 2019 took nineteen hours and sixteen minutes.

The life and career of Hudson Fysh, who flew the first scheduled passenger flight for Qantas in 1922 and ran the company for almost half of its life, is celebrated at the Qantas Founders Museum in Longreach, where he lived for ten years. The museum opened in 1996 in the heritage-listed Qantas Hangar, which Qantas built a century ago, and which is located on Hudson Fysh Drive. Historic Qantas aircraft are on display there, including one of the company's first 707s. One of the tallest structures in Longreach is the decommissioned Qantas 747 parked under a new 'Airpark roof'. The museum also displays a Super Constellation, a Consolidated Catalina, a replica Avro 504K, a replica de Havilland DH.50, a replica de Havilland DH.61, and a former Qantas Douglas DC-3. A sound and light show projects a narrated history of Qantas onto these aircraft at night.

One of the museum's directors is David Miles, Fysh's eldest grandson, who shared a love of the Australian bush with him. David is a former CEO of Chase Trust Bank in Tokyo, and a former Chief Operating Officer of the JPMorgan Investment Bank based in Sydney. He now runs a small Angus cattle breeding operation three hours west of Sydney and is also a director and treasurer of Australian Doctors International, a not-for-profit that provides volunteer medical aid to the people of Papua New Guinea, eighty years after Fysh's Qantas planes began sending aid there. Fysh and Paul McGinness were both inducted into the Australian Aviation Hall of Fame in 2013.

Qantas claimed to have lost more than $A22 billion because of border closures during the COVID-19 pandemic before launching a huge marketing push to win back customers. The company, which was gradually privatised between 1993 and 1997, now employs thirty thousand staff and flies fifty million passengers every year. Crucial to its planned fightback at the end of 2021 was the return of its massive 'super-jumbos', the Airbus A380s, after they spent almost two years in storage during the global pandemic. The first of these enormous machines to return to Australia touched down in Sydney on 9 November 2021. Appropriately it was the aircraft registered as VH-OQB – the *Hudson Fysh*.

Acknowledgements

Hudson Fysh was a remarkable Australian who lived a huge and fascinating life and left Australia with the enduring legacy of affordable, reliable, comfortable travel to the most distant parts of the earth. He risked his life many times in the service of his country during the Great War and after it dedicated most of his life to improving the lot of his countrymen.

By peering into his astonishing story, I felt I got to know Fysh well and I grew to like him very much. My sincere thanks to his children John Fysh and Wendy Miles who kindly shared their deeply personal memories of a father they adored. Thanks also to David Miles and Alastair Fysh who generously gave of their time and recollections of the kindly old grandfather they called 'Hud'.

Elizabeth Fysh and her marvellous book on Fergus McMaster, *When Chairmen Were Patriots*, gave me a fascinating introduction into the hardships the Qantas pioneers faced and the grit they needed to overcome them.

I owe a great debt of gratitude to Tom Harwood of the Qantas Founders Museum in Longreach, and the team there who do such an outstanding job preserving and promoting the history of Qantas and Australian aviation. Karen Nelson, the former CEO of the Founders Museum, gave me some valuable insights. I am also grateful to Fysh's former publisher John Ferguson, from Angus and Robertson, and his daughter Kathryn for their kind assistance.

Thanks also to the staff at the Mitchell Library in Sydney and the John Oxley Library in Brisbane, custodians of so many of Australia's aviation treasures.

Fysh's three-volume autobiography and John Gunn's three-volume history of Qantas were an essential part of my research.

This book would not have been possible without the unstinting support of the marvellous team at HarperCollins: Jude McGee, Brigitta Doyle, Lachlan McLaine, Matt Howard, Lara Wallace, Nicolette Houben and my editors Kevin McDonald and Kate Goldsworthy, whose advice and assistance were priceless.

And, of course, I am forever grateful to my wife Colleen, whose superb work at the Hinkler Hall of Aviation in Bundaberg launched me into the wild blue yonder that is Australia's beguiling history of flight.

Endnotes

Preface

1 John Fysh, interview with the author.

2 Sir Wilmot Hudson Fysh, born 7 January 1895, at The Gables, 52 High Street, East Launceston, Tasmania; died 6 April 1974, Paddington, Sydney.

3 Paul Joseph McGinness, born 14 February 1896 Framlingham, Victoria; died 25 January 1952, Perth, Western Australia.

4 'Good day, God, My Name's Smith', ABC Commercial, 1971.

5 Later Sir Fergus McMaster, born 3 May 1879, Morinish near Rockhampton, Queensland; died 8 August 1950 at his home in Rees Avenue, Clayfield.

6 'Good day, God, My Name's Smith', ABC Commercial, 1971.

7 'Jumbo jets outdate airports', *The Canberra Times*, 3 November 1967, p. 1.

8 'Qantas "Looks Closely" At Boeing 747', *Ibid*, 4 November 1966, p. 4.

Chapter 1

1 Margaret S. E. Reed, *Henry Reed: An Eventful Life Devoted to God and Man*, Morgan and Scott, 1906, p. 4.

2 Henry Reed, born 28 October 1806 in Doncaster, England; baptised 23 December; died 10 October 1880 in Mount Pleasant, Tasmania.

3 Hudson Fysh, *Qantas Rising*, Angus & Robertson, 1965, p. 5.

4 Margaret S. E. Reed, *Henry Reed: An Eventful Life*, p. 4.

5 *Colonial Times and Tasmanian Advertiser* (Hobart), 27 April 1827, p. 2.

6 Reed, *Henry Reed: An Eventful Life*, p. 9.

7 John Ward Gleadow, born 1801 Hull, Yorkshire, England; died 25 August 1881 Launceston, Tasmania.

8 John Batman, born 21 January 1801, died 6 May 1839.

9 Also known as Elizabeth Callaghan (1802-1852).

10 On 29 March 1828.

11 Brian Hoad, 'The Hack who Became Master', *The Bulletin*, 10 December 1977, p. 73.

12 Henry Reynolds, *Fate of a Free People: A Radical Re-examination of the Tasmanian Wars*, Penguin, 1995, p. 78.

13 Hudson Fysh, *Qantas Rising*, p. 5.

14 Harry Bean, 'A Remarkable Man - Henry Reed', *Western Tiers* (Deloraine, Tasmania), 27 August 1996, p. 25.

15 *Ibid*.

16 Reed, *Henry Reed: An Eventful Life*, p. 40.

17 Ian Welch, 'Henry Reed, Australian Pan-Protestant Evangelical and Businessman', (Working Paper), Australian National University Canberra, 2014.

18 On 22 October 1831. Maria Susanna Grubb (1814-1860).

19 'Victoria's Religious History', *The Mercury* (Hobart), 29 April 1936, p. 10.

20 Reed in Launceston, to Thomas Umphelby, 60 Collins Street, Melbourne, 16 October 1877, from *The Mercury* (Hobart), 29 June 1925, p. 4.

21 Reed, *Henry Reed: An Eventful Life*, pp. 26-27.

22 Harry Bean, 'A Remarkable Man – Henry Reed', *Western Tiers* (Deloraine, Tasmania), 27 August 1996, p. 25.

23 Sir George Cayley, (27 December 1773 – 15 December 1857).

24 More than two centuries later, the disc is held at London's Science Museum as the first modern conception of an aeroplane.

25 Richard P. Hallion, *Taking Flight:*

Inventing the Aerial Age, from Antiquity Through the First World War, Oxford University Press, 2003, p. 111.

26 'Grecian Temple, Dunorlan Park', historicengland.org.uk

27 Census for Henry Reed and family, Tunbridge Wells, Kent, 1871, *Census Returns of England and Wales, 1871*. Kew, Surrey, England: The National Archives, London, England.

28 On 17 March 1863.

29 Margaret Frith, born 1827 Enniskillen, Fermanagh, Northern Ireland; died 8 September 1924, Evandale, Tasmania.

30 Mary Reed, born 28 April 1868 Dunorlan Park, Tunbridge Wells, England; died 24 May 1943, Kew, Melbourne.

31 On 4 July 1533.

32 *Writings Of Tindal, Frith, and Barnes*, Vol. 2, Religious Tract Society, 1830, pp. 9-10.

33 Hudson Fysh, *Qantas Rising*, p. 5.

34 Margaret Reed, *Henry Reed: An Eventful Life*, preface.

35 Information from Tunbridge Wells Museum.

36 *Ibid*.

37 Hudson Fysh, *Qantas Rising*, p. 8.

38 Lawrence Hargrave (29 January 1850 – 6 July 1915).

39 Frederick Wilmot Fysh, born 12 November 1866, Launceston; died 29 August 1920, Launceston.

40 On 28 March 1894. 'Family Notices', *Launceston Examiner*, 7 May 1894, p. 1.

41 Later Sir Philip Oakley Fysh, born 1 March 1835 in Highbury, London; died 20 December 1919 in Sandy Bay, Hobart.

42 Frederick Lewis Fysh, born 13 February 1837, Islington, Middlesex; died 25 March 1900, Launceston, Tasmania.

43 James Hudson Taylor (21 May 1832 – 3 June 1905).

Chapter 2

1 'Guillaux Loops at 5000ft', *Geelong Advertiser*, 6 July 1914, p. 3.

2 Henry Frith Fysh, born 2 May 1896, Launceston; died 24 March 1922, Launceston.

3 Mary Geraldine Fysh, born 2 May 1896, Launceston; died November 1987 Lady Gowrie Nursing Home, Gordon, Sydney.

4 Margaret Dorothea Boileau (Peggie) Fysh (later Mrs Alexander), born 20 July 1899, Launceston; died 4 January 1977 East Kew, Melbourne.

5 Graham Stuart Fysh, born 9 January 1903 Launceston; died 27 March 1976, Brisbane.

6 Hudson Fysh, *Qantas Rising*, p. 10.

7 Author's interview with Fysh's daughter Wendy Miles, May 2022.

8 *Ibid*.

9 Hudson Fysh, *Qantas Rising*, p. 12.

10 *Ibid*., p. 4.

11 Interview with Wendy Miles.

12 Sir John Stokell Dodds (1848–1914)

13 Hudson Fysh, *Qantas Rising*, p. 14.

14 Charles Harvard Gibbs-Smith, *The Invention of the Aeroplane, 1799–1909*, Taplinger, 1966.

15 Leonard Harford Lindon, born April 1858, Bolton Lancashire; died 12 May 1953 in South Yarra, Melbourne.

16 'The Church of England Grammar School, Geelong', *Church of England Messenger for Victoria and Ecclesiastical Gazette for the Diocese of Melbourne*, 3 January 1896, p. 5.

17 'Roll Of Honor', *Daily Telegraph* (Launceston), 7 December 1918, p. 10.

18 James Valentine (Jim) Fairbairn, born 28 July 1897 Wadhurst, Sussex, England; died 13 August 1940 in an aircraft crash, Canberra.

19 Charles Allan Seymour Hawker, born 16 May 1894 Bungaree homestead, near Clare, South Australia; died 25 October 1938 in an aircraft crash, Mount Dandenong, Victoria.

20 Hudson Fysh, *Qantas Rising*, p. 18.

21 Gordon Armytage 'Beau' Fairbairn, born 29 June 1892; died 6 November 1973, Geelong.

22 Hudson Fysh, *Qantas Rising*, p. 17.

23 Francis Ernest Brown, born 12 March 1869 Bristol, England; died 1 June 1939 Gloucestershire, England.

24 'Brown, of Geelong', *Age*, 7 October 1939, p. 9.

25 Hudson Fysh, *Qantas Rising*, p. 20.
26 Mary Soames, *Clementine Churchill: The Biography of a Marriage*, Mariner Books, 2003.
27 Hudson Fysh, World War 1 service record, National Archives of Australia, B2455, Item No: 4001684.
28 'Guillaux at Geelong', *Ballarat Star*, 6 July 1914, p. 4.
29 Herbert John Louis Hinkler (8 December 1892 – 7 January 1933).
30 'Wizard Stone Injured: His Monoplane Wrecked', *Brisbane Courier*, 2 June 1914, p. 7.
31 'Guillaux The Great', *Advertiser* (Adelaide), 20 July 1914, p. 15.

Chapter 3

1 General John Monash to his wife Victoria, 16 May 1915, published *Argus*, 10 July 1915, p. 19.
2 'Austria Declares War', *Argus* (Melbourne), 30 July 1914, p. 9.
3 *Weekly Times*, 18 July 1914, p. 53.
4 *Argus*, 4 August 1914, p. 9.
5 The announcement was made at 11 p.m. on 4 August, London time.
6 The crew of the *Pfalz* was sent to Holsworthy internment camp in NSW, and Captain Kuhlken and his officers to Berrima where they remained until the end of the war. Inspection of the *Pfalz* revealed preparations had already been undertaken to convert the ship into a merchant raider, with deck plates drilled to take 4-inch guns stored in the hold. *Pfalz* was converted to an Australian troop transport and renamed HMAT *Boorara*.
7 'Enlistment statistics, First World War', awm.gov.au.
8 *Ibid*.
9 Bill Gammage, *The Broken Years: Australian Soldiers in the Great War*, Melbourne University Publishing, 2010.
10 Hudson Fysh, World War 1 service record, National Archives of Australia, B2455, Item No: 4001684.
11 Henry Frith Fysh, World War 1 service record, National Archives of Australia, B2455, Item No: 4001683.
12 Charles Bean, *The Official History of Australia in The War of 1914-1918*, Volume I, Angus & Robertson, 1941, p. 95.
13 'First convoy of Australian troops in World War I', anzacportal.dva.gov.au.
14 A.B. Paterson, 'The Transports', *Sydney Morning Herald*, 8 December 1914, p. 8.
15 Hudson Fysh, *Qantas Rising*, p. 23.
16 Kenzy Fahmy, 'The Trees of Maadi', csa-living.org.
17 'Aussies In Maadi (Meadi) Camp, Egypt 1914-19', diggerhistory.info.
18 A.J. Sweeting, 'Walker, Sir Harold Bridgwood (1862–1934)', *Australian Dictionary of Biography*, MUP, 1990.
19 Hudson Fysh, *Qantas Rising*, p. 27.
20 Group Captain Keith Isaacs, RAAF (ret.), 'Wings over Gallipoli', Defence Force Journal, Gallipoli 75th Anniversary 1915–1990, No. 81, March–April 1990, p. 7.
21 Later Major General Ewen George Sinclair-Maclagan (1868–1948).
22 John Kirkpatrick (enlisted as John Simpson), born 6 July 1892, South Shields, England; died 19 May 1915, Gallipoli Peninsula, Turkey.
23 Hudson Fysh, *Qantas Rising*, p. 31.
24 *Ibid*., p. 30.

Chapter 4

1 Fysh, *Qantas Rising*, p. 42.
2 Murdoch to Andrew Fisher, 23 September 1915, Murdoch Papers, MS 2823, National Library of Australia.
3 *Ibid*.
4 Later Sir Cyril Brudenell Bingham White born 23 September 1876, St Arnaud, Victoria; killed in an air crash 13 August 1940 Canberra.
5 Monash to his wife Vic, 20 December 1915, *War letters of General Monash: Volume 1, 24 December 1914 – 4 March 1917*, awm.gov.au.
6 *Ibid*.
7 Sir Ross Macpherson Smith, born 4 December 1892 in Semaphore, Adelaide, killed in an air crash at Brooklands Airfield, Weybridge Surrey, 13 April 1922.

8 From a letter written by William Adams at Bourke, NSW to his son after Ross Smith visited there in his Vickers Vimy on 13 February 1920; from Peter McMillan, Terry Gwynn-Jones; *The Greatest Flight*, Turner Publishing, 1995.
9 Friedrich Siegmund Georg Freiherr Kress von Kressenstein (24 April 1870 – 16 October 1948).
10 John Robinson Royston, born 29 April 1860 Durban, South Africa; died 25 April 1942 Durban.
11 It stayed there until Royston's death 26 years later.
12 Fysh, *Qantas Rising*, p. 39
13 *Ibid.*, p. 41.
14 Bryan Cooper & John H. Batchelor, *Fighter: A History of Fighter Aircraft*, Macdonald & Co., 1973.
15 Also known as No. 67 (Australian) Squadron, Royal Flying Corps (RFC), to distinguish it from Britain's No. 1 Squadron.
16 Major General W.G. Salmond, Commanding Royal Air Force. Middle East - from Leslie William Sutherland's Account of Operations of the 1st Squadron, A.F.C., 40th Wing, R.F.C., 1917-1919, p. 70, State Library of NSW, World War 1 diaries, Item: 04.
17 Captain Reginald Francis Baillieu, MC, born Queenscllff, Victoria, 17 March, 1896; died 18 January 1965 Trawalla, Victoria.
18 F.M. Cutlack, *Official History of Australia in the War of 1914–1918, Volume VIII – The Australian Flying Corps in the Western and Eastern Theatres of War*, pp. 58-9.
19 Later Air Vice Marshal Francis Hubert McNamara, born 4 April 1894, Rushworth, Victoria; died Amersham, Buckinghamshire, 2 November 1961.
20 David Wilson, *The Brotherhood of Airmen*, Allen & Unwin, 2014, pp. 11–13.
21 Alan Stephens, *The Royal Australian Air Force*, Oxford University Press, 2009, pp. 14–15.
22 'No. 30122', *The London Gazette (Supplement), 8 June 1917, pp. 5701–5703.*
23 Wilson, *The Brotherhood of Airmen*, pp. 11–13.
24 Stephens, *The Royal Australian Air Force*, pp. 14–15
25 Cutlack, *Official History*, pp. 58-9.
26 Eustace Slade Headlam (20 May 1892 – 25 May 1958) A left-handed batsman and slow left arm orthodox bowler, he played one first-class match for Tasmania in 1911/12. He won the Tasmanian Open golf championship in 1913 and 1919 and the Tasmanian amateur championship five times between 1912 and 1927.
27 Cutlack, *Official History*, pp. 69-70.
28 On 20 April 1917.
29 Later Air Marshal Sir Richard Williams, born 3 August 1890 Moonta Mines, South Australia; died 7 February 1980 in St George's Hospital, Kew.
30 Adrian Lindley Trevor Cole, born 19 June 1895 at Glen Iris, Melbourne; died 14 February 1966 at the Heidelberg Repatriation Hospital, Melbourne.
31 F.M. Cutlack, *Official History of Australia in the War of 1914–1918*, Volume VIII, p. 63.
32 Later Captain Stanhope Irving Winter-Irving MC, born 13 November 1890 Malvern, Victoria; died 22 October 1967 Nagambie, Victoria.
33 'Personal and Anecdotal', *Sunday Mail*, 10 May 1931, p. 2.
34 Henry Frith Fysh, World War 1 service record, National Archives of Australia, B2455, Item No: 4001683.
35 F.M. Cutlack, *Official History*, p. 66.

Chapter 5

1 Fysh, *Qantas Rising*, p. 128.
2 Ronald Albert Austin, born 21 June 1893, Eilyer Station, Lake Bolac, Victoria; died 20 July 1965, Mortlake Victoria.
3 Hudson Fysh, *Qantas Rising*, p. 61.
4 F.M. Cutlack, *Official History*, Volume VIII, p. 74.
5 Gerhardt Felmy, born December 12, 1891, Berlin; died December 8, 1955, Dreieichenhain. He was a major general by the end of World War II. His older brother, Hellmuth Felmy was a high-ranking Nazi who was convicted of war crimes in 1948 but served just three years of a 15-year sentence.

6 Hudson Fysh, World War 1 service record, National Archives of Australia, B2455, Item No: 4001684.
7 winsleyatwar.wordpress.com.
8 Thomas Taylor, born 17 August 1894, Lucknow, Victoria; died 2 July 1953, Edithvale, Melbourne.
9 Later Air Commodore Francis William Fellowes Lukis, born 27 July 1896 Balingup, Western Australia; died 18 February 1966 Melbourne.
10 Australian War Memorial, caption on photograph, Accession Number P01034.031
11 Allan Murray Jones, born 25 February 1895 Caulfield, Melbourne, died Double Bay, Sydney, 8 December 1963.
12 Keith Isaacs, 'Jones, Allan Murray (1895–1963)', *Australian Dictionary of Biography*, Vol. 9, Melbourne University Press, 1983.
13 F.M. Cutlack, *Official History*, Volume VIII, p. 72.
14 Archibald Henry Searle, born 27 December 1887, Bendigo, Victoria; died 13 July 1917 near Beersheba.
15 Hudson Fysh, *Qantas Rising*, p. 48.
16 John Herbert Ellerslie Butler, born 19 January 1894, Bellerive, Hobart; died 30 April 1924, Royal Prince Alfred Hospital, Newtown, Sydney.
17 Oliver Mathew Lee, born 22 June 1886 in Deloraine, Tasmania; died 27 September 1966, Deloraine.
18 On 19 March 1918.
19 Alfred William Leslie Ellis, born 14 October 1894, Steiglitz, Victoria; died 22 January 1948, Sydney.
20 Ernest Andrew 'Pard' Mustard, born 21 September 1893, Oakleigh, Melbourne; died 10 October 1971, Coolangatta, Queensland.
21 T.E. Lawrence, *Seven Pillars of Wisdom*, Jonathan Cape, 1935.
22 Hudson Fysh, *Qantas Rising*, p. 51.
23 Sydney Wentworth Addison, born 31 January 1887 Hobart, Tasmania.
24 Leslie William Sutherland's Account of Operations of the 1st Squadron, A.F.C., 40th Wing, R.F.C., 1917-1919, p. 26, State Library of NSW, World War 1 diaries, Item: 04. E
25 *Ibid.*, p. 35.
26 Cutlack, pp. 109-110.
27 Also known as Ramleh.
28 Hudson Fysh, *Qantas Rising*, p. 53.
29 Leslie William Sutherland's Account, p. 35.
30 *Ibid*, p. 39.
31 Norman Ellison, *Flying Matilda: Early Days in Australian Aviation*, Angus and Robertson, 1957, p. 40.
32 *Ibid*, p. 39.
33 On 22 August 1918.
34 John Mercer Walker, born January 1888, Ballarat Victoria; died 22 August 1918 Ramla, Palestine.
35 Harold Alexander Letch, born 3 May 1894, Donnybrook Victoria; died 22 August 1918 Ramla, Palestine.
36 Ernest Cecil Stooke, born 28 May 1894, Hawthorn, Melbourne; died 19 August 1918 Ramla, Palestine.
37 Louis Paul Krieg, born 24 October 1893 Dimboola, Victoria; died 19 August 1918 Ramla, Palestine.
38 Wilfred Arthur Baird (originally his surname was listed as Beard), born in Benalla, Victoria, 6 November 1889; died 7 May 1954, at his home in Yarranabbe Rd, Darling Point, Sydney.
39 Hudson Fysh, *Qantas Rising*, p. 52
40 Pauline Cottrill, *The Man Australia Forgot*, Qantas Founders Museum, 2012, p. 26.
41 *Ibid.*, p. 27.
42 Cottrill, *The Man Australia Forgot*, p. 39.
43 Tom Harwood, 'Who Was Paul McGinness?', qfom.com.au.
44 Cottrill, *The Man Australia Forgot*, p. 64.
45 *Ibid.*, p. 102; McGinness to his mother, 16 October 1916.
46 Paul Joseph McGinness, World War 1 service record, National Archives of Australia, B2455, Item: 1944353.
47 *Ibid.*
48 Ellison, *Flying Matilda*, p. 47
49 Cottrill, *The Man Australia Forgot*, p. 134.
50 Cutlack, p. 135.
51 *Ibid.*, p.141.
52 *Ibid.*, p. 142.
53 *Ibid.*, pp. 143-4.
54 Sir Samuel McCaughey, born 1 July 1835 at Tullyneuh, near Ballymena, County Antrim, Ireland; died 25 July 1919, Yanco, NSW.

55 Ellison, *Flying Matilda*, p. 40.
56 Howard Bowden Fletcher, born 22 November 1890, Newcastle, NSW; died 10 April 1967, Southport, Queensland.
57 Air Vice Marshal Amyas Eden 'Biffy' Borton (20 September 1886 – 15 August 1969).
58 Hudson Fysh, World War 1 service record, National Archives of Australia, B2455, Item No: 4001684'.
59 From the diary of Clive Conrick, 30 August 1918 and private correspondence, Pat Conrick (edit), *The Flying Carpet Men*, P. Conrick, 1993.
60 Information supplied by Fysh's son John Fysh.
61 Cutlack, p. 147.
62 Fysh, *Qantas Rising*, p. 65.
63 Cutlack, p. 152.
64 *Ibid.*, p. 171.
65 Hudson Fysh, *Qantas Rising*, p. 65.
66 Cottrill, *The Man Australia Forgot*, p. 154.
67 Fysh, *Qantas Rising*, p. 128.
68 Leslie William Sutherland, born 17 December 1892, Murrumbeena, Victoria; died24 October 1967, Moruya, NSW.
69 Fysh, *Qantas Rising*, p. 101.
70 *Ibid.*, p. 63.
71 James Mallett Bennett, 14 January 1894, St Kilda, Melbourne, died 13 April 1922 Surrey, England.
72 Walter Henry Shiers, born 17 May 1889 at Norwood, Adelaide; died 2 June 1968 Adelaide.
73 'Aerial Travelling', *Barrier Miner* (Broken Hill, NSW), 5 March 1919, p. 4
74 'The Country Drive that Created an Airline', qfom.com.au
75 'Aerial Services Limited', *Brisbane Courier*, 1 January 1919, p. 7.
76 'Motoring', *Sunday Times*, 2 February 1919, p. 16.
77 Jean Claude Marduel, born Lyon, France, 9 November 1877; died 10 July 1939, Paris.
78 Now the site of RAAF Base Richmond.
79 'Aviation Development', *Age* (Melbourne), 20 March 1919, p. 6.

Chapter 6

1 'To London by Air', *Daily Mail*, Brisbane, 9 August 1919, p. 7.
2 'Soldiers Arrive', *Mercury* (Hobart), 19 April 1919, p. 4.
3 *Ibid.*
4 'First World War 1914–18', awm.gov.au
5 Nelson, E. (2004), 'Homefront hostilities: the first world war and domestic violence in Victoria', PhD thesis, History Department, The University of Melbourne.
6 'To London by Air', *Daily Mail*, Brisbane, 9 August 1919, p. 7.
7 'Australian Flight', *West Australian* (Perth), 10 June 1919, p. 5.
8 'Death Of Our Greatest Pastoralist', *Sunday Times* (Sydney), 27 July 1919, p. 4.
9 Peter Hohnen, 'McCaughey, Sir Samuel (1835–1919)', *Australian Dictionary of Biography*, Volume 5, Melbourne University Press, 1974.
10 Lieutenant General James Gordon Legge, born 15 August 1863, Hackney, London; died 18 September 1947 Oakleigh, Melbourne.
11 'Australian Air Force', *Daily Mercury* (Mackay), 13 November 1919, p. 3.
12 Recollections of Fysh's daughter Wendy Miles, 'Three adventurers follow pioneer outback road journey to a T', couriermail.com.au, 19 August 2009.
13 David Lindsay, born on 20 June 1856 at Goolwa, South Australia; died 17 December 1922, Darwin.
14 'The Aerial Survey', *Townsville Daily Bulletin*, 2 June 1919, p. 4.
15 Cottrill, *The Man Australia Forgot*, p. 169.
16 'To London by Air', *Daily Mail*, Brisbane, 9 August 1919, p. 7.
17 *Ibid.*
18 Fysh, *Qantas Rising*, p. 69.
19 Ainslie Neville Templeton, born 4 October 1873, Columbra Station, Clermont; died 8 September 1959, Brisbane.
20 George Thomas Gorham; born July 1876, Sevenoaks, Kent, England; died 10 August 1948, Longreach, Queensland.
21 Cottrill, *The Man Australia Forgot*, p. 170.

22 'Late G. Gorham', *Longreach Leader*, 27 August 1948, p. 13.
23 George Thomas Gorham, World War 1 service record, National Archives of Australia, B2455, Item No: 4774766.
24 Hudson Fysh, *Qantas Rising*, p. 71.

Chapter 7

1 'Preparing the Way for World Fliers', *The Herald* (Melbourne), 21 August 1920, p. 14.
2 'Taroom', *Brisbane Courier*, 9 June 1868, p. 3.
3 Emily Caroline Creaghe diary 1883, John Oxley Library, State Library of Queensland.
4 Francis Birtles, *Lonely Lands: Through the Heart of Australia*, N.S.W. Bookstall Company, 1909, p. 74.
5 'Preparing the Way for World Fliers', *The Herald* (Melbourne), 21 August 1920, p. 14.
6 Fysh, *Qantas Rising*, p. 75.
7 Cottrill, *The Man Australia Forgot*, p. 182.
8 'Flying To Australia', *Herald* (Melbourne), 22 September 1919, p. 1.
9 Cottrill, *The Man Australia Forgot*, p. 184.
10 'Preparing the Way for World Fliers', *The Herald* (Melbourne), 21 August 1920, p. 14.
11 Cottrill, *The Man Australia Forgot*, p. 186.
12 Recollections of McGinness's daughter Pauline Cottrill, from 'Living the Dream' documentary produced by A Shorething TV, 2012.
13 Cottrill, *The Man Australia Forgot*, p. 185.

Chapter 8

1 'The Crossing of Australia', *Examiner* (Launceston) 17 October 1919, p. 5.
2 Hubert Ernest de Mey Warren, born 2 March 1885 Prahran, Victoria; died 19 October 1934 in an aircraft crash over Bass Strait.
3 'Rev. H. E. Warren Memorial', *Sydney Morning Herald*, 15 December 1934, p. 16.
4 Hudson Fysh diary, 1 October 1919, State Library of NSW, MLMSS 2413.
5 Fysh, *Qantas Rising*, p. 85.
6 *Ibid.*, p. 86
7 David Lindsay, born 20 June 1856 at Goolwa, South Australia; died 17 December 1922, Darwin.
8 Suzanne Edgar, 'Lindsay, David (1856–1922)', *Australian Dictionary of Biography*, Volume 10, Melbourne University Press, 1986.
9 Cottrill, *The Man Australia Forgot*, p. 190.
10 George Campbell Matthews, born 6 July 1883 Stranraer, Scotland; died 27 January 1958, Melbourne.
11 Dominic Thomas Kay, born 20 October 1886, Springmount, Victoria; died 19 May 1963, Toowong, Brisbane
12 'When The Aeroplanes Arrive', *Northern Territory Times and Gazette* (Darwin), 22 November 1919, p. 16.
13 Fysh to Legge, 30 October 1919, From Fergus McMaster papers, quoted in John Gunn, *The Defeat of Distance*, University of Queensland Press, 1985, p. 15.
14 Kellaway to Fysh, *Ibid.*
15 Fysh, *Qantas Rising*, p. 86.
16 Sir Keith Macpherson Smith, born 20 December 1890 Adelaide; died 19 December 1955, Sydney.
17 Roger Douglas, born 5 June 1894 at Charters Towers, Queensland; died 1 3 November 1919, Surbiton, England.
18 James Stuart Leslie Ross, born 20 October 1895, Moruya, New South Wales; died 13 November 1919, Surbiton, England.
19 Alexander Kennedy, born 11 November 1837, Dunkeld, Scotland; died at his home, 'Loretto', Chasely Street, Auchenflower, Brisbane 12 April 1936. The Auchenflower home is now part of the Wesley Hospital complex.
20 Marion Kennedy (nee Murray), born 20 January 1847, Mauchline, Ayrshire, Scotland; died 1 September 1936, 'Loretto', Chasely Street, Auchenflower, Brisbane.
21 'A Farsighted Scot – Alexander Kennedy', qfom.com.au.
22 Cottrill, *The Man Australia Forgot*, p. 210.
23 Later Air Vice Marshal Henry Neilson Wrigley, born 21 April 1892, Collingwood, Melbourne, died 14 September 1987, Fairfield, Melbourne.

24 Later Air Commodore Arthur William Murphy, born 17 November 1891, Kew, Melbourne; died 21 April 1963, Essendon, Melbourne.

25 Valdemar Richard Rendle, born 12 November 1897, Brisbane; died 8 November 1962, Brisbane.

26 Sir George Hubert Wilkins, born 31 October 1888, Mount Bryan East, South Australia; died 30 November 1958, Framingham, Massachusetts.

27 Cedric Ernest 'Spike' Howell, born 17 June 1896, Adelaide, South Australia; died 10 December 1919, St George's Bay, Corfu, Greek Islands.

28 George Henry Fraser, born 1880, Macorna, Victoria; died 10 December 1919, St George's Bay, Corfu, Greek Islands.

29 'The Australian Flight, A Terrible Tragedy', *Advertiser* (Adelaide), 16 December 1919, p. 7.

30 'The 1919 Great Air Race', epicflightcentenary.com.au.

31 Sir Ross Macpherson Smith, *14,000 Miles Through the Air*, Macmillan, 1922, p. 70.

32 'The Great Flight', *Northern Territory Times and Gazette* (Darwin), 13 December 1919, p. 5.

33 H.G. Castle, 'The First to Australia', from *Wonders of World Aviation*, Vol. 1, ed. Clarence Winchester, Amalgamated Press, London, 1938.

34 'Good day, God, My Name's Smith', ABC Commercial, 1971.

35 'Australia at Last', *Register* (Adelaide), 11 December 1919, p. 7.

36 Fysh, *Qantas Rising*, p. 87.

Chapter 9

1 'Sir Ross Smith: Prize of £10,000 Presented', *Age* (Melbourne), 28 February 1920, p. 13.

2 'Major-General Legge's Praise', *West Australian*, 13 December 1919, p. 9.

3 'Ross Smith Knighted', *Telegraph* (Brisbane), 24 December 1919, p. 6.

4 'Gossip', *Townsville Daily Bulletin*, 10 January 1920, p. 6.

5 'Personal', *The Northern Miner* (Charters Towers), 17 December 1919, p. 4.

6 Hugh McMaster, *Fickle Fortune*, Self-published, 1949, p. 55.

7 'Cobb's Creek, Northern Territory', epicflightcentenary.com.au.

8 'Ross Smith's Luck', *Sun* (Sydney), 17 December 1919, p. 5.

9 'Flying Again', *Newcastle Sun*, 20 December 1919. p. 5.

10 Fysh, *Qantas Rising*, p. 86.

11 Fysh, 'Buffalo Hunting on Foot', *The Bulletin*, 30 September 1920, p. 40.

12 'Ross Smith At Charleville', *Herald* (Melbourne), 23 December 1919, p. 1.

13 'Aircraft repairs at the Railway Workshops', blog.qm.qld.gov, 10 January 2020.

14 Raymond John Paul Parer, born 18 February 1894, South Melbourne; died 5 July 1967, Greenslopes, Brisbane.

15 John Cowe McIntosh, born February 1892, Aberdeen Scotland; killed in aircraft crash 28 March 1921, near Pithara, Western Australia.

16 'Sir Ross Smith: Prize of £10,000 Presented', *Age* (Melbourne), 28 February 1920, p. 13.

17 'Lieut. Parer's Flight', *Transcontinental* (Port Augusta, SA), 10 March 1922, p. 4.

18 'Here & There Mouse as Mascot', *The Evening News* (Boulder, WA), 5 January 1922, p. 1.

19 'Parer and McIntosh', *The Argus* (Melbourne), 1 September 1920, p. 9.

20 Fysh, *Qantas Rising*, p. 89.

21 *Ibid.*, p. 93.

22 *Graziers' Review*, 16 April 1924, p. 53.

23 Later Sir Fergus McMaster, born 3 May 1879, Morinish near Rockhampton, Queensland; died 8 August 1950 at home in Rees Avenue, Clayfield.

24 'Cloncurry Notes', *The Northern Miner* (Charters Towers), 29 October 1919, p. 3.

25 John Gunn, *Defeat of Distance*, University of Queensland Press, 1985, p. 1.

26 *Ibid.*, p. 2.

27 Elizabeth Fysh, *When Chairmen were Patriots*, Boolarong Press, 2020, p. 61.

28 Gunn, *Defeat of Distance*, p. 17.

29 *Ibid.*, p. 62.

30 Cottrill, *The Man Australia Forgot*, p. 222.

31 Thomas John Lynott, born 1848, Bendigo, Victoria; died 17 November 1925 Borroloola, Northern Territory.
32 Fysh, *Qantas Rising*, p. 90.
33 Jeannie Gunn, *We of the Never Never*, Hutchinson, 1908.
34 Fysh, *Qantas Rising*, p. 90.
35 'Darwin to Cloncurry by Motor', *The Brisbane Courier*, 26 July 1920, p. 4.
36 *Ibid.*

Chapter 10

1 'Aviation: Queensland Scheme – Aerial Service to Darwin', *The Sydney Morning Herald*, 26 October 1920, p. 8.
2 'The Prince of Wales', *Newcastle Morning Herald and Miners' Advocate*, 24 July 1920, p. 7.
3 'Aeroplane Falls', *The Sydney Morning Herald*, 17 June 1920, p. 5.
4 Officially the Royal Queensland Show, and originally called the Brisbane Exhibition.
5 'Views on the Route of the Aerial Survey Party from Queensland to Darwin', *Weekly Times* (Melbourne), 28 August 1920, p. 32.
6 The hotel was demolished after Brisbane's 1974 floods.
7 Cottrill, *The Man Australia Forgot*, p. 227.
8 Elizabeth Fysh, *When Chairmen were Patriots*, Boolarong Press, p. 67.
9 Gunn, *Defeat of Distance*, p. 22.
10 *Ibid.*, p. 23.
11 John Thomson, born Liverpool, England, 5 November 1864; died Brisbane 20 September 1935.
12 Noel Adsett, 'John Thomson', heritage.saintandrews.org.au
13 Gunn, *Defeat of Distance*, p. 23.
14 Alan Walter Campbell, born 27 June 1880, Apple Tree Gully, near Inverell, NSW; died 8 December 1972, Clayfield, Brisbane.
15 Thomas James O'Rourke, born 1 October 1870, Morinish, Queensland; died 6 May 1938 Winton.
16 McMaster's recollections, 'Sir Fergus McMaster's Version of the Beginning', Qantas Founders Museum, qfom.com.au, 27 August 2020.
17 Fysh, *Qantas Rising*, p. 98.
18 'A Farsighted Scot – Alexander Kennedy', qfom.com.au
19 Qantas still has the historical artefact.
20 Cottrill, *The Man Australia Forgot*, p. 229.
21 Tom Harwood, '100 Years of What?', 13 November 2020, qfom.com.au.
22 Nigel Borland Love, born 16 January 1892, South Kurrajong, NSW; died 2 October 1979 Killara, Sydney.
23 Fysh, *Qantas Rising*, p. 99.
24 'Personal', *Daily Telegraph* (Launceston), 31 August 1920, p. 4.
25 'Preparing the Way for World Fliers: The Men that Mapped the Route', *The Herald* (Melbourne) 21 August 1920, p. 14.
26 'Aviation', *The Sydney Morning Herald*, 26 October 1920, p. 8.
27 *Ibid.*
28 'Aviation: Queensland Scheme – Aerial Service to Darwin', *The Sydney Morning Herald*, 26 October 1920, p. 8.
29 *Ibid.*
30 McMaster to Campbell, 30 September 1920, from Gunn, *Defeat of Distance*, pp. 26–7.
31 A.D. Allen, 19 Castlereagh St, Sydney to A.W. Campbell, 26 October 1920, from Gunn, *Defeat of Distance*, p. 25.
32 Gunn, *Defeat of Distance*, p. 26.
33 McMaster to Campbell, 30 September 1920, from Gunn, *Defeat of Distance*, p. 26.
34 'Commercial Aviation', *Morning Bulletin* (Rockhampton), 5 February 1921, p. 7.
35 A.J. Jackson, *Avro Aircraft since 1908*, Putnam, 1990, p. 184.
36 'Aviation: Queensland Scheme – Aerial Service to Darwin', *The Sydney Morning Herald*, 26 October 1920, p. 8.
37 *Ibid.*
38 Cottrill, *The Man Australia Forgot*, p. 231.
39 Gunn, *Defeat of Distance*, p. 24.
40 'Commercial Aviation', *The Western Champion and General Advertiser for the Central-Western Districts* (Barcaldine), 23 October 1920, p. 6.
41 Prospectus, 14 October 1920, Gunn, *Defeat of Distance*, p. 28.
42 'Aviation', *The Daily Mail* (Brisbane), 28 October 1920, p. 6.

43 Tom Harwood, '100 Years of What?', 13 November 2020, qfom.com.au.
44 Fysh, *Qantas Rising*, p. 97.
45 winton.qld.gov.

Chapter 11

1 'Aerial Services.', *The Daily Mail* (Brisbane), 29 December 1920, p. 5.
2 It was passed on 11 November 1920, and became law on 11 February 1921.
3 'Advertising ', *The Catholic Press* (Sydney), 2 December 1920, p. 47.
4 Gunn, *Defeat of Distance*, p. 15.
5 'Commercial Flying', *The Sydney Morning Herald*, 13 December 1920, p. 10.
6 'Aerial Services.', *The Daily Mail* (Brisbane), 29 December 1920, p. 5.
7 'The Hotel Metropole', *The Sydney Morning Herald*, 14 January 1890, p. 5.
8 John Flynn, born 25 November 1880 at Moliagul, Victoria; died 5 May 1951, Royal Prince Alfred Hospital, Sydney.
9 Cottrill, *The Man Australia Forgot*, p. 233.
10 Graeme Bucknall, 'Flynn, John (1880–1951)', *Australian Dictionary of Biography*, Volume 8, Melbourne University Press, 1981.
11 Second Lieutenant John Clifford Peel, born 17 April 1894 at Inverleigh, Victoria; killed 19 September 1918, near St Quentin Canal, France.
12 On 2 November and 20 November 1917.
13 'Lieutenant Clifford Peel – Providing a Blueprint for the RFDS', flyingdoctor.org.au, 12 June 2017.
14 'Big Airplane', *The Sydney Morning Herald*, 18 January 1921, p. 8.
15 Fysh, *Qantas Rising*, p. 100.
16 Charles Martin Castle Knight, born 1 July 1890, Rockhampton, Queensland; died 15 February 1938, Muttaburra Hospital, Queensland, from a gunshot wound to the head.
17 J. H. Butler, National Archives Australia, Series: B3455, Item: 3176221
18 'Air Record', *The Daily Mail* (Brisbane), 10 July 1920, p. 7.
19 Herbert Darton 'Herb' Avery, born 19 May 1889, Berridale, NSW; died 16 July 1953, Brisbane.
20 'Aeroplane Smash at Longreach', *The Western Champion and General Advertiser for the Central-Western Districts* (Barcaldine), 18 September 1920, p. 9.
21 'Two Aeroplanes Arrive', *Singleton Argus*, 1 February 1921, p. 2.
22 'Aerial Derby to Sopwith-Gnu Machine — Lieutenant Love Wins Handicap', *Sunday Times* (Sydney), 28 November 1920, p. 3.
23 'Aviation in Queensland', *The Brisbane Courier*, 24 January 1921, p. 6.
24 Letter to McMaster from State Government Insurance Office, 19 January 1921, quoted in Gunn, *Defeat of Distance*, p. 30.
25 'In the Back Country.', *Herald* (Melbourne), 7 March 1921, p. 4.
26 Fysh, *Qantas Rising*, p. 102.
27 Cottrill, *The Man Australia Forgot*, p. 240.
28 'Commercial Aviation in Queensland', *The Sydney Morning Herald*, 3 February 1921, p .8.
29 Fysh, *Qantas Rising*, p. 102.
30 'Two Planes Unexpectedly Visit St. George', *Balonne Beacon* (St. George, Queensland), 5 February 1921, p. 2.
31 *Ibid.*
32 *Ibid.*
33 'In the Back Country.', *Herald* (Melbourne), 7 March 1921, p. 4.
34 *Ibid.*
35 *Ibid.*
36 'Queensland *News*', *Morning Bulletin* (Rockhampton), 5 February 1921, p. 9.
37 Acacia cambageii
38 Fysh, *Qantas Rising*, p. 103.
39 'In the Back Country.', *Herald* (Melbourne), 7 March 1921, p. 4.
40 Fysh, *Qantas Rising*, p. 103.
41 *Ibid.*
42 Dr Frederick Archibald Hope Michod, born 22 August 1872, Richmond, Surrey; died 18 March 1938, St Martin's Hospital, Ann St, Brisbane.
43 'Aviation in Central Queensland', *The Western Champion and General Advertiser for the Central-Western Districts* (Barcaldine), 12 February 1921, p. 5.
44 'Aviation in Queensland', *The Brisbane Courier*, 8 February 1921, p. 7.

45 *Ibid.*
46 The notes are on display at the Qantas Heritage Centre at Sydney Airport.
47 Elizabeth Fysh, *When Chairmen Were Patriots*, Boolarong Press, 2020, pp. 76–7.
48 Harriet Whistler Riley, born 20 July 1898 Winton, Queensland; died 16 May 1956.
49 'They argued about Waltzing Matilda', *The Sun* (Sydney), 26 August 1945, p. 4.
50 Frederick Whistler Riley, born 7 December 1856, Gheringhap, Victoria; died 9 November 1914, Winton, Queensland.
51 Frederick Adams Whistler Riley, born 18 August 1902, Winton, Queensland, died 4 August 1974, Townsville, Queensland.
52 'Western Aerial Service', *The Evening Telegraph* (Charters Towers), 19 February 1921, p. 2.
53 Fysh, *Qantas Rising*, p. 104.
54 Gunn, *Defeat of Distance*, p. 34, quoting McMaster's Narrative.
55 Fysh, *Qantas Rising*, p. 104.
56 'Commercial Aviation', *The Western Champion and General Advertiser for the Central-Western Districts* (Barcaldine), 23 October 1920, p. 6.
57 Sociedad Colombo Alemana de Transportes Aéreos, founded on 5 December 1919.
58 'A Farsighted Scot – Alexander Kennedy', qfom.com.au, 30 October 2020.
59 Qantas Provisional Directors Report, 1921, Gunn, *The Defeat of Distance*, p. 36.
60 Cottrill, *The Man Australia Forgot*, p. 245.
61 'Birthplace of QANTAS', experiencewinton.com.au.

Chapter 12

1 'Turkey-Shooting By Aeroplane', *The Herald* (Melbourne), 11 April 1921, p. 2.
2 Fysh, *Qantas Rising*, p. 105.
3 Gunn, *The Defeat of Distance*, p. 37.
4 *Ibid.*, p. 37, quoting McMaster Papers.
5 Interview with Fysh's grandson David Miles, Sydney, May 2022.
6 Arthur Stanislaus Rodgers, born 20 March 1876 at Geelong, Victoria; died Melbourne 4 October 1936.
7 Sir Donald Charles Cameron, born 1 9 November 1879, Brisbane, died 19 November 1960, Brisbane.
8 Gunn, *The Defeat of Distance*, p. 38.
9 John Ronald Shafto 'Ron' Adair OBE (22 May 1893 – 27 June 1960).
10 Frank Leonard Roberts, OBE, born 18 August 1896, Bendigo, Victoria; died 24 June 1993, Baxter, Victoria.
11 'Sheep Deal by Aeroplane', *The Capricornian* (Rockhampton), 25 September 1920, p. 20.
12 Fysh, *Qantas Rising*, p. 107
13 *Ibid.*, p. 110.
14 'Aviation in the Central West', *The Evening Telegraph* (Charters Towers, Qld), 19 March 1921, p. 4, quoting *The Winton Herald*.
15 'Turkey-Shooting by Aeroplane', *The Herald* (Melbourne), 11 April 1921, p. 2.
16 Fysh, *Qantas Rising*, p. 110.
17 Fysh to McMaster, from Gunn, *The Defeat of Distance*, p. 38, quoting McMaster Papers.
18 The Aeroplane Tragedy', *Western Mail* (Perth, WA), 28 April 1921, p. 30.
19 'Australia's Trade', *Daily Commercial News and Shipping List* (Sydney), 28 July 1920, p. 7.
20 'Torrential Rain', *The Daily Mail* (Brisbane), 5 March 1921, p. 8.
21 Hudson Fysh, 'Above the Plains', *The Sydney Morning Herald*, 26 May 1922. p. 8.
22 Gunn, *The Defeat of Distance*, p. 39.
23 'Linking The Interior', *The Daily Telegraph* (Sydney) 3 March 1921, p. 6.
24 *Ibid.*
25 Cottrill, *The Man Australia Forgot*, p. 250
26 *Ibid.*, p.251.
27 Gunn, *The Defeat of Distance*, p. 39.
28 'Record by a Baby Avro', *The Sydney Morning Herald*, 6 April 1921, p. 12.
29 'Record Non-Stop Flight', *Brisbane Courier*, 12 April 1921, p. 7.
30 Fysh to McMaster, from Cottrill, p. 259.
31 'Per Aeroplane', *The Sydney Stock and Station Journal*, 1 April 1921, p. 3.
32 Cottrill, *The Man Australia Forgot*, p. 60.
33 Rose Lilian Joliffe, born 19 October 1920, Longreach, Queensland; died 16 June 2004, Barcaldine, Queensland.

34 Alec Walter Joliffe (1888-1943).
35 'Aeroplane as Ambulance', *The Brisbane Courier*, 14 April 1921, p. 6.
36 *Ibid.*, *The Capricornian* (Rockhampton), 23 April 1921, p. 28.
37 'Aeroplane Outback', *The Sydney Morning Herald*, 15 April 1921, p. 9. The little baby Qantas rescued all those years ago married Jack Chilcott at Longreach in 1942 and lived to the grand old age of 83.
38 'Aeroplane as Ambulance', *Queensland Times* (Ipswich), 14 April 1921, p. 5.
39 'Good day, God, My Name's Smith', ABC Commercial, 1971.

Chapter 13

1 Fysh, *Qantas Rising*, p. 108.
2 *Ibid.*, p. 109.
3 The machine was eventually rebuilt and is now on display at the Hinkler Hall of Aviation in his home town of Bundaberg.
4 Charles James Anthony Brabazon, born 19 August 1869, Tummaville, Queensland; died 26 September 1944, Winton, Queensland.
5 Henry G. Lamond to Fysh, 30 June 1948, Hudson Fysh Papers, State Library of NSW, MLMSS 2413.
6 Gunn, *The Defeat of Distance*, p. 42, quoting Chairman's Report at the first Qantas Annual General Meeting.
7 Edgar Charles Johnston, born 30 April 1896 East Perth, Western Australia; died 24 May 1988, Malvern, Melbourne.
8 Fysh to McMaster, 8 June 1921, from Gunn, *The Defeat of Distance*, p. 45.
9 'Aviation Outback', *The Herald* (Melbourne), 29 June 1921, p. 1.
10 Horace Clowes Brinsmead, born 2 February 1883 at Hampstead, London, died 11 March 1934 Melbourne.
11 'Aviation Outback', *The Herald* (Melbourne), 29 June 1921, p. 1.
12 *Ibid.*
13 Fysh, *Qantas Rising*, p. 124.
14 'Aerial Doctors', *Border Watch* (Mount Gambier, SA), 29 July 1921, p. 3.
15 *Ibid.*
16 'Herbert River Notes', *Cairns Post*, 25 July 1921, p. 7.
17 'Aeroplane Lands Unexpectedly', *Townsville Daily Bulletin*, 1 August 1921, p. 4.
18 Herbert River Notes', *Cairns Post*, 13 August 1921, p. 8.
19 Fysh, *Qantas Rising*, p. 112.
20 'Nor'-West Aerial Service', *Sunday Times* (Perth, WA), 6 November 1921, p. 9.
21 'General *News*', *The Western Champion* (Barcaldine, Qld.), 27 August 1921, p. 7.
22 Gunn, *The Defeat of Distance*, p. 47.
23 *Ibid.*
24 Thomas A. Stirton (1860–1926).
25 *Pastoral Review*, 16 December 1926, p. 1095.
26 Fysh, *Qantas Rising*, p. 116.
27 *Ibid*, p. 113.
28 James Aitchison Johnston Hunter (1882-1968).
29 Gunn, *The Defeat of Distance*, p. 49.
30 Fysh, *Qantas Rising*, p. 113.
31 'Good day, God, My Name's Smith', ABC Commercial, 1971.
32 Gunn, *The Defeat of Distance*, pp. 49-50,
33 *Ibid.*
34 Fysh, *Qantas Rising*, p. 114.
35 *Ibid.*
36 Hansard, 22 November 1921.
37 Gunn, *The Defeat of Distance*, p. 51.
38 *Ibid.*
39 Hansard, 22 November 1921.
40 Samille Mitchell, 'Triumph to tragedy: Remembering those who died in Australia's first scheduled air service', abc.net.au, 12 September 2018.
41 'Landing Grounds Wanted', *Queensland Times* (Ipswich), 7 December 1921, p. 5.
42 'The Aerial Mail: Major Brearley's Mishap,' *The West Australian* (Perth), 28 December 1921, p. 6.
43 'Is Flying Safe? Australia's Aerial Routes. Linking The Western Railheads', *The Brisbane Courier*, 29 May 1922, p. 19.
44 *Ibid.*
45 *Ibid.*
46 'Aviation', *The Western Champion* (Barcaldine), 24 December 1921, p. 7.
47 Fysh, *Qantas Rising*, p. 116.
48 Carlyon Mark Mozart Foy (1889-1956).
49 'Aviation', *The Western Champion*, 24 December 1921, p. 7.

50 *Ibid.*
51 *Ibid.*
52 Fysh, *Qantas Rising*, p. 116.
53 Herbert Joseph 'Jimmy' Larkin, born 8 October 1894, South Brisbane; died 20 June 1972, St Martin, Guernsey, Channel Islands.
54 Reginald Stanley Larkin, born 10 June 1898, Norwood, Adelaide; died 12 February 1983, Ferndown, England.
55 Ann G. Smith, 'Larkin, Herbert Joseph (1894–1972), *Australian Dictionary of Biography*, Volume 9, Melbourne University Press, 1983.
56 'Feud That Lay Behind the Shepherd Inquiry', *Smith's Weekly* (Sydney), 16 February 1929, p. 1.
57 Gunn, *The Defeat of Distance*, p. 53.
58 Fysh, *Qantas Rising*, p. 119.
59 Gunn, *The Defeat of Distance*, pp. 54-5.
60 *Ibid.*
61 Fysh, *Qantas Rising*, p. 122.
62 *Ibid.*

Chapter 14

1 'Outback Aerial Service', *The Daily Mail* (Brisbane), 5 November 1922, p. 1.
2 Major Harry Turner Shaw OBE (1889-1973).
3 'Parer's Aeroplane Crashes', *The Herald* (Melbourne), 7 February 1922, p. 1.
4 Fysh to McMaster, 3 February 1922, Gunn, *The Defeat of Distance*, p. 57.
5 The board decided on 7 January 1922 to give Clarkson his notice of termination and he left on 22 April.
6 Fysh, *Qantas Rising*, p. 127.
7 *Ibid.*
8 Fysh to McMaster, 7 March 1922, Gunn, *The Defeat of Distance*, p. 58.
9 Marcus Griffin, born 20 March 1889, Richmond, NSW, died 24 March 1949, Hunters Hill, Sydney.
10 Gunn, *The Defeat of Distance*, p. 59.
11 C.Vickey (4 Bridge St, Sydney) to McMaster, 24 February 1922, Gunn, p. 59.
12 Fysh, *Qantas Rising*, p. 123.
13 'About People', *Examiner* (Launceston), 27 March 1922, p. 6.
14 'Gallant Airmen Killed', *Argus* (Melbourne), 15 April 1922, p. 11
15 'Sir Keith Smith: Story of the Tragedy', *West Australian* (Perth), 2 June 1922.
16 Fysh, *Qantas Rising*, p. 120.
17 Peter Abelson, 'House and Land Prices in Sydney: 1925 to 1970', *Urban Studies*, Vol. 22, No. 6 (December 1985), Sage Publications, pp. 521-534.
18 'Aerial Derby', *The St George Call* (Kogarah, NSW), 5 May 1922, p. 5.
19 *Ibid.*, *National Advocate* (Bathurst, NSW), 8 May 1922, p. 1.
20 *The Sun* (Sydney), 12 Apr 1922, p. 10.
21 Fysh, *Qantas Rising*, p. 120.
22 Elizabeth Eleanor (Nell) Dove, born 24 March 1896, Rockhampton, Queensland; died 26 July 1983, Sydney.
23 Interview with Wendy Miles.
24 Interview with Fysh's relative Elizabeth Fysh.
25 'Death of the Rev. W.W. Dove', *The Maitland Mercury and Hunter River General Advertiser*, 26 Mar 1867, p. 3.
26 'Rockhampton Police Court', *Morning Bulletin* (Rockhampton, Qld), 9 February 1894, p. 6.
27 *Ibid.*, 28 March 1892, p. 5.
28 'Aeroplane Services', *The Western Champion* (Barcaldine), 27 May 1922, p. 17.
29 'The Suicide Club', Smithsonian National Postal Museum, postalmuseum.si.edu.
30 Charles Augustus Lindbergh (February 4, 1902 – August 26, 1974).
31 'Aeroplane Services', *The Western Champion* (Barcaldine), 27 May 1922, p. 17.
32 Horatio Clive Miller, born 30 April 1893, Ballarat, Victoria; died Dalkeith, Perth 27 September 1980.
33 'Aviation', *The Queenslander* (Brisbane), 7 October 1922, p. 32.
34 Fysh, *Qantas Rising*, p. 126.
35 *Ibid.*, p. 125.
36 Godfrey Wigglesworth, born 3 October 1885, Pannal, Yorkshire; died 25 June 1935, Townsville, Queensland. Fysh and others wrote of him as 'Geoffrey' Wigglesworth.
37 Fysh, *Qantas Rising*, p. 125.
38 'Aviation', *The Queenslander* (Brisbane), 7 October 1922, p. 32.

39 Interview with Wendy Miles.
40 Cottrill, *The Man Australia Forgot*, p. 280.
41 McMaster to McGinness, 28 October 1922, Gunn, *The Defeat of Distance*, p. 66.
42 'Aerial Mail', *The Sun* (Sydney), 3 November 1922, p. 3.
43 'Aerial Mails', *Morning Bulletin* (Rockhampton, Qld), 3 November 1922, p. 8.
44 *Ibid.*
45 *Ibid.*
46 Fysh, *Qantas Rising*, p. 133.
47 'By Aerial Mail', *Morning Bulletin* (Rockhampton), 15 November 1922, p. 7.
48 'Outback Aerial Service', *The Daily Mail* (Brisbane), 5 November 1922, p. 1.

Chapter 15

1 Philippa Coates, 'My grandmother was the first paying passenger to fly Qantas', afr.com, 28 February 2020.
2 Ivy Lillian McLain, born 24 October 1899, Normanton, Queensland; died 7 September 1991, Brisbane.
3 Cecil Jack Hazlitt, born 6 October 1897, South Yarra, Melbourne; died 15 June 1993.
4 'Australian Biography: Jack Hazlitt', interview recorded 27 February 1992, National Film and Sound Archive of Australia.
5 Frederic (also Frederick) George Huxley, born 31 August 1892, Currie, King Island, Tasmania; died 6 April 1960, Malvern, South Australia.
6 'At a Drome 40 Miles from Sydney ', *The Sun* (Sydney), 4 August 1940, p. 3.
7 'Flyers With a Story', *King Island News* (Currie, King Island), 15 October 1980, p. 11.
8 Cottrill, *The Man Australia Forgot*, p. 280.
9 *Ibid.*
10 Fysh, *Qantas Rising*, p. 225.
11 Gunn, *The Defeat of Distance*, p. 71.
12 'Record Air Flight', *Daily Mercury* (Mackay, Qld), 8 February 1923, p. 7.
13 'The Aerial Mail,' *Toowoomba Chronicle and Darling Downs Gazette*, 1 December 1922, p. 5.
14 Melda Elaine 'Peg' Glasson, born 12 September 1922, Toowoomba, Queensland, died July 2020, Gold Coast, Queensland.
15 Melda Olive 'Hilda' Lane, born 2 October 1896, Winton, Queensland; died 22 April 1974, Holland Park, Brisbane.
16 'The Aerial Mail,' *Toowoomba Chronicle and Darling Downs Gazette*, 1 December 1922, p. 5.
17 'Happy Anniversary – A Little Late', Qantas Founders Museum, qfom.com.au, 28 July 2016.
18 'The Aerial Mail,' *Toowoomba Chronicle and Darling Downs Gazette*, 1 December 1922, p. 5.
19 Thomas Quarles Back, born 23 September 1892, Norwich, England; died 20 February 1986, War Veterans Hospital, Caboolture, Queensland. Fysh wrote of him as Tom Bach.
20 Fysh, *Qantas Rising*, p. 146.
21 'Western Flying', *Morning Bulletin* (Rockhampton), 2 June 1923, p. 9.
22 'Personal Items', *Daily Standard* (Brisbane), 21 February 1923, p. 6.
23 Gunn, *The Defeat of Distance*, p. 69, quoting McMaster papers.
24 Fysh, *Qantas Rising*, p. 135.
25 Gunn, *The Defeat of Distance*, p. 74.
26 'Large Passenger Aeroplane!', *The Daily News* (Perth, WA), 25 January 1923, p. 9.
27 *Ibid.*
28 H.C. 'Horrie' Miller, *Early Birds: Magnificent Men of Australian Aviation Between the Wars*, Rigby Australia, 1968, p. 104.
29 *Ibid.*
30 Great Britain, Royal Aero Club Aviators' Certificates, 1910-1950, No. 1892.
31 'The Vickers-Vulcan 'Plane', *The Longreach Leader*, 9 March 1923, p. 7.
32 'The Vulcan Plane', *The Longreach Leader*, 29 March 1923, p. 10.
33 'Large Passenger Aeroplane!', *The Daily News* (Perth, WA), 25 January 1923, p. 9.
34 Gunn, *The Defeat of Distance*, p. 70.
35 *Ibid.*
36 'Western Flying', *Morning Bulletin* (Rockhampton), 2 June 1923, p. 9.

37 'Aeroplane As Ambulance', *Horsham Times* (Vic), 2 March 1923, p. 5.
38 In 1946, William Ballinger's son, Kenneth Hulton Ballinger, married Peg Glasson, who as a baby had flown with her mother on the Charleville–Winton flight in 1922.
39 Fysh, *Qantas Rising*, p. 139.
40 Miller, *Early Birds*, p. 104.
41 *Ibid.*
42 Gunn, *The Defeat of Distance*, p. 71.
43 'Plane Crashes', *The Sun*, 16 March 1923, p. 7.
44 'Handed Over', *The Daily Mail* (Brisbane), 26 March 1923, p. 2.
45 Miller, *Early Birds*, p. 103.
46 Fysh, *Qantas Rising*, p. 141.
47 Tom Harwood, 'Frank McNally – An Accidental Bonus', qfom.com.au
48 Miller, *Early Birds*, p. 104.
49 Fysh, *Qantas Rising*, p. 142.
50 *Ibid*, p. 190.
51 Frederick William Haig (originally Schultz), born 29 July 1895, South Yarra, Melbourne; died 2 March 1984, Rosebud Hospital, Victoria.
52 'Western Flying', *Morning Bulletin* (Rockhampton), 2 June 1923, p. 9.
53 Tom Harwood, 'Frank McNally – An Accidental Bonus', qfom.com.au
54 *Ibid.*
55 Registered G-AUDE.
56 Fysh, *Qantas Rising*, p. 147.
57 Gunn, *The Defeat of Distance*, p. 77.
58 *Ibid.*
59 Arthur Whitehair Vigers, born 20 January 1890, Isleworth, Middlesex, UK; died September 1968, Bunbury, Western Australia.
60 Fysh, *Qantas Rising*, p. 151.
61 Gunn, *The Defeat of Distance*, p. 79.
62 'Western Flying', *Morning Bulletin* (Rockhampton), 2 June 1923, p. 9.
63 Fysh, *Qantas Rising*, p. 149.

Chapter 15

1 Interview with the author.
2 'Wedding Bells', *The Wingham Chronicle and Manning River Observer* (NSW), 11 December 1923, p. 2.
3 Butler died five months later on 30 April 1924 aged 30 from lymphatic leukaemia. He was buried at Waverley Cemetery after a military funeral. His comrades in the AFC erected a memorial over the foot of his grave, consisting of iron propellor blades formed into a cross.
4 Wedding Bells', *The Wingham Chronicle and Manning River Observer* (NSW), 11 December 1923, p. 2.
5 'Aerial Service ', *The Daily Mail* (Brisbane), 4 January 1924, p. 6.
6 *Ibid.*
7 Interview with Wendy Miles.
8 *Ibid.*
9 'Advertising ', *The Longreach Leader* (Qld), 4 January 1924, p. 12.
10 Fysh, *Qantas Rising*, p. 151
11 *Ibid.*, p. 203.
12 'The Pride of the Central West: Q.A.N.T.A.S', *The Longreach Leader*, 14 March 1924, p. 13.
13 Fysh, *Qantas Rising*, p. 161.
14 Interview with Elizabeth Fysh, Graham's daughter-in-law.
15 'A Record Flight', *The Brisbane Courier*, 7 July 1924, p. 5.
16 Fysh, *Qantas Rising*, p. 163.
17 *Ibid*, p. 157.
18 Cottrill, *The Man Australia Forgot*, p. 305.
19 Percival Harold 'Skipper' Moody, born 21 May 1893, Albany, Western Australia; died 19 July 1978, Sydney Hospital, from myocardial infarction.
20 'Hamilton Island', *The Proserpine Guardian* (Qld), 28 July 1950, p. 4.
21 Lester Joseph Brain AO, born 27 February 1903, Forbes, New South Wales; died 30 June 1980, Sydney.
22 Edward P. Wixted, *The North-West Aerial Frontier 1919-1934*, Boolarong Publications, 1985, p. 15.
23 'Round Australia Flight', *The Longreach Leader*, 8 August 1924, p. 14.
24 'Aviation', *The Longreach Leader*, 26 September 1924, p. 5.
25 Fysh to McMaster, 11 October 1924, from Gunn, *The Defeat of Distance*, p. 86.
26 Fysh, *Qantas Rising*, p. 166.
27 He revealed his real name as James Porter when applying for the old age pension.

28 Fysh, *Qantas Rising*, p. 184.
29 Charles Augustin Yorke (Cay) Johnston, born 6 August 1892, Brisbane; died 2 November 1983, Townsville, Queensland.
30 Fysh, *Qantas Rising*, p. 174.
31 *Sydney Morning Herald*, 2 June 1925, p. 8.
32 Interview with Wendy Miles.
33 Family Notices, *The West Australian* (Perth), 30 December 1925, p. 1.
34 Sir Neville Reginald Howse (1863-1930).
35 A. J. Hill, 'Howse, Sir Neville Reginald (1863–1930)', *Australian Dictionary of Biography*, Volume 9, Melbourne University Press, 1983
36 Fysh, *Qantas Rising*, p. 176.
37 On 20 October 1926, from Fysh, *Qantas Rising*, p. 177.
38 John Lawrence Baird, 1st Baron Stonehaven, (27 April 1874 – 20 August 1941).
39 'Lord and Lady Stonehaven at Longreach.', *The Western Champion* (Barcaldine), 31 July 1926, p. 9.
40 Fysh, *Qantas Rising*, p. 179.
41 *Mercury* (Hobart), 8 July 1926, p. 7.
42 'Another Aviation Feat', *The Longreach Leader*, 13 August 1926, p. 10.
43 *Ibid.*
44 'Cobham's Arrival in Melbourne', *Brisbane Courier*, 16 August 1926, p. 7.
45 'Historic Vice-Regal Flight', *The Register* (Adelaide), 21 September 1926, p. 9.
46 Registered as G-AUFA
47 'The Iris', *The Brisbane Courier*, 19 August 1926, p. 7.
48 Alan Douglas Davidson, born 15 December 1899; died 24 March 1927, Tambo, Queensland.
49 Arthur Herbert Affleck, born 3 July 1903, Brighton, Melbourne; died 1 September 1966 while cruising off Vancouver, Canada, in the liner *Orsova*.
50 Charles Cauvet Matheson, born 14 August 1898, Trafalgar, Victoria; died 5 June 1955, Hawthorn, Victoria.
51 Charles William Anderson Scott, born 13 February 1903, Westminster, London; died 15 April 1946, Bad Arolsen, Germany.
52 Fysh, *Qantas Rising*, p. 198.
53 *Ibid.*, p. 197.
54 'Archerfield', *Toowoomba Chronicle and Darling Downs Gazette*, 22 January 1931, p. 9.
55 Peter Dunn, 'History of Archerfield Aerodrome', ozatwar.com.
56 G-AUER
57 Fysh, *Qantas Rising*, p. 196.
58 Qantas Manager's Report, 6 April 1927, from Gunn, *The Defeat of Distance*, p. 94.
59 Russell Brooke Tapp, born 23 August 1898, Barton Regis, Gloucestershire; died 2 March 1984, Longreach.
60 'Snake In Cockpit', *Johnstone River Advocate and Innisfail News* (Qld), 8 March 1929, p. 7.
61 G-AUED.
62 Fysh, *Qantas Rising*, p. 195.
63 'Aeroplane Crash at Tambo', *Northern Standard* (Darwin), 5 March 1927, p. 2.
64 *Ibid.*
65 Fysh, *Qantas Rising*, p. 195.
66 Qantas Manager's Report, 6 April 1927, from Gunn, *The Defeat of Distance*, p. 94.
67 'Aviation: Federal Grant', *The Sydney Morning Herald*, 22 August 1927, p. 11.
68 Fysh, *Qantas Rising*, p. 173.

Chapter 17

1 Lise Mellor, 'Kenyon St Vincent Welch', The University of Sydney School of Medicine Online Museum, sydney.edu.au
2 '*Lone Australian Airman Completes Epic Flight', The Daily Telegraph* (Sydney), 23 February 1928, p. 2.
3 'Was Hinkler Forced Down by Engine Trouble?' *Daily Telegraph* (Sydney), 25 February 1928, p. 2.
4 Charles Thomas Philippe Ulm, born 18 October 1898, Middle Park, Melbourne; disappeared on a flight to Hawaii after leaving Oakland, California on 3 December 1934. The two Americans on the flight were radio operator James Warner and navigator Harry Lyon.
5 'Mantle of Safety', *The West Australian* (Perth), 24 March 1925, p. 6.
6 'Medical Service in Remote Parts', *Daily Standard* (Brisbane), 1 November 1927, p. 2.
7 *Ibid.*

8 Alfred Hermann Traeger, born 2 August 1895 at Glenlee, Dimboola, Victoria; died 31 July 1980, Rosslyn Park, Adelaide.
9 That system was replaced by voice communication in 1937.
10 Kenyon St Vincent Welch, born 22 February 1885, Greenwich, Sydney; died Sydney 1961.
11 'Kenyon St Vincent Welch', The University of Sydney School of Medicine Online Museum, sydney.edu.au.
12 'The first "flying doctor" trip', National Archives of Australia.
13 Kenyon St Vincent Welch', The University of Sydney School of Medicine Online Museum, sydney.edu.au.
14 Gunn, *Defeat of Distance*, p. 106.
15 Henry George Nutson (1907-1928).
16 'Late Mr. G. Nutson.', *The Western Champion* (Barcaldine, Qld), 27 October 1928, p. 5.
17 Gunn, *Defeat of Distance*, p. 106.
18 'Qantas Biplane Smashed', *Northern Standard* (Darwin), 7 September 1928, p. 5.
19 Late Mr. G. Nutson.', *The Western Champion* (Barcaldine, Qld), 27 October 1928, p. 5.
20 Fysh, *Qantas Rising*, p. 225.
21 Gunn, *Defeat of Distance*, p. 107.
22 'Air Mail Services', *The Daily Telegraph* (Sydney), 25 September 1928, p. 5.
23 'Air Mail Services', *The Daily Telegraph* (Sydney), 25 September 1928, p. 5.
24 Fysh to McMaster, 23 November 1928, from Gunn, *Defeat of Distance*, p. 107.
25 Fysh, *Qantas Rising*, p. 242.
26 Albert Plesman (7 September 1889 – 31 December 1953)
27 Interview with Wendy Miles.
28 Fysh, *Qantas Rising*, p. 223.

Chapter 18

1 Hudson Fysh, 'Qantas – A Modern Pioneer', Truth (Brisbane), 6 July 1930, p. 3.
2 Fysh, *Qantas Rising*, p. 224.
3 'Valuable Link', *The Brisbane Courier*, 19 April 1929, p. 18.
4 Fysh's report to Qantas, 31 January 1929, from Gunn, *The Defeat of Distance*, p.109.
5 Sir Thomas William Glasgow (1876-1955).
6 Qantas Directors' Meeting, 7 March 1929, from Gunn, *The Defeat of Distance*, *Ibid.*, pp.109-10.
7 Fysh, *Qantas Rising*, p. 226.
8 Philip Henry Compston (1905–1969).
9 Frederick William Stevens (1898–1966)
10 Keith Vincent Anderson, born 6 July 1898, Perth; died about 12 April 1929 in Tanami, Central Desert Region, Northern Territory.
11 Henry Smith 'Bobby' Hitchcock, born 11 October 1891, Broken Hill, New South Wales; died about 12 April 1929 in Tanami, Central Desert Region, Northern Territory.
12 Leslie Hubert Holden, born on 6 March 1895, East Adelaide; killed in an air crash 18 September 1932, Byron Bay, NSW.
13 'Captain Holden's Story', *Daily Standard* (Brisbane), 15 April 1929, p. 1.
14 'Kookaburra', *The Young Chronicle* (NSW), 23 April 1929, p. 2.
15 'Captain Brain's Story', *Toowoomba Chronicle and Darling Downs Gazette*, 25 April 1929, p. 7.
16 'Anderson's Last Words Scratched on Plane', *The Register News*-Pictorial (Adelaide), 2 May 1929, p. 3.
17 'Southern Cross', *The Canberra Times*, 17 May 1929, p. 4.
18 'Moir and Owen Safe', *Lithgow Mercury*. 27 May 1929, p. 1.
19 Fysh, *Qantas Rising*, p. 229.
20 Albert Ernest Green (21 December 1869 – 2 October 1940).
21 'Commercial Air Force', The *Advertiser* (Adelaide), 14 December 1929, p. 31.
22 George Urquhart 'Scotty' Allan, born 2 February 1900 in Bellshill, Scotland; died 12 August 1996 in Sydney.
23 George (Scotty) Allan interviewed by Dr Amy McGrath, 1 January 1980, nla.obj-215152982
24 'Start from Sydney', *The Telegraph* (Brisbane), 2 January 1930, p. 9.
25 George (Scotty) Allan interviewed by Dr Amy McGrath.
26 'Australia Served by Air', *Queensland Times* (Ipswich), 1 January 1930, p. 2.

27 *Ibid*.
28 *Ibid*.
29 John Henry Arthur Treacy, born 13 July 1895 in Wagga Wagga, NSW, died 30 July 1984, Maclean, NSW.
30 'In Five Hours', *The Central Queensland Herald* (Rockhampton), 27 March 1930, p. 22.
31 Keith Allison Virtue, born 23 June 1909 in Lismore; died 7 February 1980 on the course at Brisbane Golf Club, Yeerongpilly.
32 George Albert Robinson, born 1880 Macleay River, NSW; died 1953, Byron Bay, NSW.
33 Fysh, *Qantas Rising*, p. 232.
34 Australia Served by Air', *Queensland Times* (Ipswich), 1 January 1930, p. 2.
35 'Golf Notes', *The Longreach Leader*, 28 March 1930, p. 24.
36 'Qantas Re-Organisation', *The Longreach Leader*, 17 April 1930, p. 11.
37 'Secretary to Qantas', *The Herald* (Melbourne), 10 May 1930, p. 8.
38 'Qantas activities', *The Queenslander*, 2 May 1929, p. 26.
39 Fysh, *Qantas Rising*, p. 236.
40 'Longreach Notes', *Townsville Daily Bulletin*, 21 May 1930, p. 11.
41 Fysh, *Qantas Rising*, p. 236.
42 'Qantas Headquarters for Brisbane', *Daily Standard* (Brisbane), 24 June 1930, p. 5.
43 'Qantas Entertains', *The Longreach Leader*, 27 June 1930, p. 17.
44 Fysh to R.D. Miller, 23 June 1931, Hudson Fysh Papers, State Library NSW, MLMSS 2413.
45 R.D. Miller to Fysh, 27 August 1931, *Ibid*.
46 'In the Social Circle', *The Week* (Brisbane), 27 June 1930, p. 9.
47 Interview with Wendy Miles.
48 Fysh, *Qantas Rising*, p. 266.
49 Gunn, *The Defeat of Distance*, p. 121
50 'Qantas Commemorates', *The Longreach Leader*, 31 October 1930, p. 5.
51 'Qantas', *Daily Standard* (Brisbane), 9 January 1931, p. 11.
52 *Ibid*.
53 Fysh, *Qantas Rising*, p. 234-5.
54 'Imperial Air Connection— Brisbane Route', *The Brisbane Courier*, 21 November 1930, p. 12.
55 Guy Alexander Taunton, born 13 November 1880, Rockhampton, Queensland; died 15 July 1946, Newcastle.
56 'Business Man Dies in Office', *Newcastle Morning Herald and Miners' Advocate*, 16 July 1946, p. 2.
57 'Qantas Commemorates', *The Longreach Leader*, 31 October 1930, p. 5.
58 *Ibid*.
59 'Qantas Commemorates', *Ibid*, 3 1 October 1930, p. 5.
60 John Fysh interviewed by Dr Amy McGrath, 28 August 1979, NLA, ID 4763547.
61 'Lower Wages', *Mudgee Guardian and North-Western Representative*, 5 May 1930, p. 3.
62 Fysh, *Qantas Rising*, p. 241.
63 'Not Going to Loaf', *Sunday Mail* (Brisbane), 17 May 1931, p. 3.
64 Cottrill, *The Man Australia Forgot*, p. 307.
65 Fysh, *Qantas Rising*, p. 244.
66 'Bowral's Idol', *Recorder* (Port Pirie, SA), 5 November 1930, p. 4.
67 Donald Bradman, *Farewell to Cricket*, Hodder & Stoughton, 1950, p. 35.
68 'Looters Raid Plane Wreck', *The Canberra Times*, 30 October 1958, p. 1.
69 'Imperial Airways Launch Big Venture', *Daily Standard* (Brisbane), 1 April 1931, p. 11.

Chapter 19

1 'Flying Fysh's Dreams', *The Daily Telegraph* (Sydney), 5 September 1932, p. 6.
2 'Aviation', *The Queenslander* (Brisbane), 9 April 1931, p. 18.
3 Fysh, *Qantas Rising*, p. 245.
4 Gunn, *The Defeat of Distance*, p. 132.
5 'England to Australia – Ten Days', *The World's News* (Sydney), 29 October 1930, p. 18.
6 'Air Mail Lands: History Made at Archerfield', *The Brisbane Courier*, 29 April 1931, p. 11.
7 'To England ', *Sunday Mail* (Brisbane), 17 May 1931, p. 3.

8 'Air Mail *Hippomenes* Reaches Darwin', *The Telegraph* (Brisbane), 19 May 1931, p. 8.
9 'Activities in Aviation', *Sunday Mail* (Brisbane), 17 May 1931, p. 3.
10 'Great Depression', National Museum Australia, nma.gov.au.
11 *Aircraft* (magazine), 1 July 1931.
12 'Suspending Air Services', *News* (Adelaide), 19 June 1931, p. 5.
13 Fysh, *Qantas Rising*, pp. 240-1
14 *Ibid.*, p. 265.
15 *Ibid*, p. 269.
16 Hudson Fysh, *Taming the North*, Angus & Robertson, 1933.
17 *James* Allan *Mollison* (19 April 1905 – 30 October 1959).
18 Cecil Arthur Butler (1902-1980), born 8 June 1902 at Sparkhill, England; died 13 April 1980, Wahroonga, Sydney.
19 Joseph Benedict Chifley, born 22 September 1885, Bathurst, NSW; died 13 June 1951, Canberra.
20 Fysh, *Qantas Rising*, p. 245.
21 *Ibid.*, p. 243.
22 'Dutch Mail Crashes', *Northern Standard* (Darwin), 8 December 1931, p. 5.
23 'Southern Star Crashes in English Orchard', *Herald* (Melbourne), 22 December 1931.
24 Gunn, *The Defeat of Distance*, p. 139.
25 *Ibid.*
26 *Ibid.*
27 Templeton to Fysh, 3 December 1931, from Gunn, *The Defeat of Distance*, pp. 140-1.
28 Joseph Aloysius Lyons (15 September 1879 – 7 April 1939).
29 *Sir George* Foster *Pearce* (14 January 1870 – 24 June 1952).
30 Johnston to Fysh, 13 January 1932, from Gunn, *The Defeat of Distance*, pp. 143-4.
31 Ulm to Lyons, 13 January 1932, *Ibid.*, pp. 144-5.
32 *Ibid.*
33 'Air Service', *The Sydney Morning Herald*, 19 January 1932, p. 9.
34 'Here To Discuss Air Link,' *The Telegraph* (Brisbane), 3 February 1932, p. 10.
35 Fysh, *Qantas Rising*, pp. 245-6.
36 Fysh to Brearley, 27 February 1932, from Fysh, *Qantas Rising*, p. 246.
37 Memories of Jean Fysh (nee McMaster), from Elizabeth Fysh, *When Chairmen Were Patriots*, p. 142.
38 Fysh to McMaster, 17 December 1932, from Elizabeth Fysh, *Ibid.*
39 Fysh, *Qantas Rising*, p. 251.
40 Albert Ernest Rudder, born 14 December 1868, Kempsey, NSW; died 2 June 1956, Sydney.
41 Fysh, *Qantas Rising*, p. 253.
42 'The Astraea', *The Sydney Morning Herald*, 19 June 1933, p. 9.
43 Herbert G. Brackley (4 October 1894 - 15 November 1948).
44 'Log of the Astraea', *The Telegraph* (Brisbane), 26 June 1933, p. 9.
45 Robert Archer Prendergast (1901-1934).
46 'The Astraea', *The Sydney Morning Herald*, 19 June 1933, p. 9.
47 'Log of the Astraea', *The Telegraph* (Brisbane), 26 June 1933, p. 9.
48 'Giant Of Air: Big Welcome at Archerfield', *Daily Standard* (Brisbane), 24 June 1933, p. 5.
49 'Big Escort', *The Sun* (Sydney), 26 June 1933, p. 7.
50 'Astraea's Arrival Here Today', *The Herald* (Melbourne), 29 June 1933, p. 5.
51 'Brisbane-London', *The Brisbane Courier*, 3 July 1933, p. 12.
52 'Notes on Aviation', *The Telegraph* (Brisbane), 3 July 1933, p. 9.
53 'From the Log of the Astraea III', *The Brisbane Courier*, 18 August 1933, p. 9.
54 Hudson Fysh, *The log of the Astraea: the story of a journey from Australia to England by air with Imperial Airways*, Qantas Ltd, 1933.
55 'From the Log of the Astraea', *The Brisbane Courier*, 16 August 1933, p. 13.
56 'From the Log of the Astraea II', *Ibid.*, 17 August 1933, p. 15.
57 *Ibid.*
58 'From the Log of the Astraea III', *Ibid.*, 18 August 1933, p. 9.
59 'From the Log of the Astraea', *Ibid.*, 16 August 1933, p. 13.
60 'From the Log of the Astraea III', *Ibid.*, 18 August 1933, p. 9.
61 'The Log of the Astraea', *The Courier-Mail* (Brisbane), 31 August 1933, p. 11.

62 'Desert Air Fort', *Ibid.*, 6 September 1933, p. 15.
63 'From Basra to Bagdad', *Ibid.*, 7 September 1933, p. 16.
64 *Ibid.*
65 'Across the Mediterranean,' *Ibid.*, 8 September 1933, p. 17.
66 *Ibid.*
67 'At the End of the Journey ', *Ibid.*, 9 September 1933, p. 14.
68 Fysh to McMaster, 30 August 1933, Hudson Fysh Papers, State Library of NSW.
69 Parliamentary Debates, 1933.
70 Gunn, *The Defeat of Distance*, p. 182.
71 'The Astraea', *The Sydney Morning Herald*, 27 June 1933, p. 8.
72 'Idea Ridiculed in Europe', *The Courier-Mail* (Brisbane), 14 September 1933, p. 13.
73 Fysh to George Woods Humphery, 17 October 1933, from Fysh, *Qantas Rising*, p. 258.
74 'Kingsford Smith Reaches Melbourne', *Townsville Daily Bulletin*, 16 October 1933, p. 4.
75 'Ulm At Derby', *The Evening News* (Rockhampton, Qld), 20 October 1933, p. 1.
76 Fysh to George Woods Humphery, 17 October 1933, from Fysh, *Qantas Rising*, p. 258.
77 *Ibid.*, 19 October 1933, *Ibid*, p. 259.
78 *Ibid.*
79 'Qantas: British Tribute', *The Telegraph* (Brisbane), 4 January 1934, p. 1.
80 'Aust. Air Mail Propaganda', *The Labor Daily* (Sydney), 25 November 1933, p. 6.
81 Sidney Albert Dismore (1900–1945).
82 William Alexander Watt (1871–1946).
83 'Qantas Empire Airways Formed', *The Courier-Mail* (Brisbane), 27 January 1934, p. 14.
84 Fysh, *Qantas Rising*, p. 255.
85 Hudson Fysh, *Qantas at War*, Angus & Robertson, 1968, p. 22
86 Fysh, *Qantas Rising*, p. 264.
87 Sir Ivan Nello Holyman (1896-1957).
88 Sir Macpherson Robertson (6 September 1859 – 20 August 1945)
89 'Airmail Tenders', *The Argus* (Melbourne), 20 April 1934, p. 7.
90 'Kingsford Smith's Disappointment', *The Sydney Morning Herald*, 21 April 1934, p. 16.
91 'Wedding', *Townsville Daily Bulletin.* 24 July 1934. p. 6.

Chapter 20

1 Fysh, *Qantas Rising*, p. 268.
2 Interview with Wendy Miles.
3 *Ibid.*
4 'Inspecting Air Route', *The Longreach Leader*, 16 June 1934, p. 15.
5 'Hudson Fysh At Darwin', *The Telegraph* (Brisbane), 18 June 1934, p. 15.
6 'Australian Air Service', *The Courier-Mail* (Brisbane), 4 July 1934, p. 15.
7 Norman Morris Chapman, born 28 November 1899, Albury, NSW; killed in crash 3 October 1934, near Winton, Qld.
8 'Air Crash: Chapman's Death Accidental', *Weekly Times* (Melbourne), 1 January 1927, p. 7.
9 'Qantas Aeroplane Tragedy', *Worker* (Brisbane), 10 October 1934, p. 6.
10 *Ibid.*
11 'Charred Remains Found in Burnt 'Plane', *The Evening News* (Rockhampton), 5 October 1934, p. 6.
12 *Ibid.*
13 Fysh to Mrs Chapman, 24 November 1934, from Gunn, *Defeat of Distance*, p. 205.
14 'Scott's Great Win in Air Race', *The Argus* (Melbourne), 24 October 1934, p. 7.
15 'Albury and the Uiver', uivermemorial.org.au.
16 Sir Cedric Oban Turner, born 13 February 1907, Dubbo, NSW; died 16 November 1982 Sydney.
17 *The Sun* (Sydney), 15 November 1934, p. 23.
18 'The Air Disaster', *Townsville Daily Bulletin*, 19 November 1934, p. 4.
19 *Ibid.*
20 *Ibid.*
21 Hudson Fysh, *Qantas at War*, Angus & Robertson, 1968, p. 6.
22 'Search Abandoned', *The Central Queensland Herald* (Rockhampton), 20 December 1934, p. 40.

23 Fergus McMaster missed the ceremony because of dust storms at Longreach.
24 'Important Link in Empire Air Chain', *Queensland Times* (Ipswich), 11 December 1934, p. 7.
25 Fysh, *Qantas at War*, p. 11.
26 *Ibid.*
27 Fysh to Woods Humphery, 31 December 1934, Hudson Fysh Papers, State Library NSW, MLMSS 2413.
28 *Ibid.*
29 Johnston to Fysh, 29 January 1935, from Gunn, *Defeat of Distance*, p. 238.
30 Fysh, *Qantas at War*, pp. 17–18.
31 Later Lieutenant-Colonel Harold Pedro Joseph 'Bunnie' Phillips (6 November 1909 – 27 October 1980).
32 Edwina Cynthia Annette Mountbatten, Countess Mountbatten of Burma, (née Ashley; 28 November 1901 – 21 February 1960).
33 Louis Francis Albert Victor Nicholas Mountbatten, 1st Earl Mountbatten of Burma (25 June 1900 – 27 August 1979).
34 Lady Pamela Hicks, *Daughter of Empire: My Life as a Mountbatten*, Simon & Schuster, 2014, p. 25.
35 Herbert Bindley Hussey, born Port Elliot, South Australia, 11 February 1896; died Christmas Day 1958, Brisbane.
36 'First Air Passenger for London', *T he Telegraph* (Brisbane), 17 April 1935, p. 10.
37 'First Passenger', *The Central Queensland Herald* (Rockhampton), 25 April 1935, p. 47.
38 'Investors in real estate', *The Courier-Mail* (Brisbane), 27 August 1935, p. 9.
39 Fysh, *Qantas at War*, p. 32.
40 *Ibid.*, p. 29.
41 Sir Robert Archdale 'Archie' Parkhill (27 August 1878 – 2 October 1947).
42 'Aerial Funeral', *Queensland Times*, 4 September 1936, p. 7.
43 'Mining Notices', *Kalgoorlie Miner*, 22 February 1935, p. 1.
44 'Family Ousted', *The Daily News* (Perth), 9 July 1942, p. 4.
45 'Farmer's Wife's Game Struggle', *Mirror* (Perth), 11 July 1942, p. 16.
46 Cottrill, *The Man Australia Forgot*, p. 311.
47 *Ibid.*, p. 312.
48 Fysh, *Qantas at War*, p. 59.
49 William Bloxham Purton, born 15 November 1912, Hobart, Tasmania; died 28 February 1942 in flying boat crash near Java.
50 Fysh, *Qantas at War*, p.96.
51 Wing Commander Patrick Windsor Lynch-Blosse, born 11 April 1900 Glamorgan, Wales; killed 9 May 1942 serving in the RAF near Berlin.
52 Fysh, *Qantas at War*, p. 72.
53 John Gunn, *Challenging Horizons: Qantas 1939-1954*, University of Queensland Press, 1987, p. 318.
54 Fysh, *Qantas at War*, p. 75.
55 Elizabeth Fysh, *When Chairmen Were Patriots*, p. 160.
56 Fysh to Johnston, 12 November 1938, Hudson Fysh Papers, State Library NSW.
57 Fysh, *Qantas at War*, p. 77.
58 *Ibid*, p.87.
59 'West-Bound Flying Boats: Hundredth Service To-day', *The Sydney Morning Herald*, 9 March 1939, p. 10.

Chapter 21

1 'Mr. Menzies Broadcasts to Nation', *The Sydney Morning Herald*, 4 September 1939, p. 11.
2 Sebastien Roblin, 'Stuka: The Nazi Dive Bomber That Terrified Millions', 19fortyfive.com, 25 December 2021.
3 'Mr. Menzies Broadcasts to Nation', *The Sydney Morning Herald*, 4 September 1939, p. 11.
4 Cottrill, *The Man Australia Forgot*, p. 281.
5 McGinness to McMaster, 23 January 1940, Hudson Fysh Papers, State Library NSW, MLMSS 2413.
6 McMaster to Fysh, 26 January 1940, State Library NSW, MLMSS 2413.
7 Fysh to McMaster, 30 January 1940, *Ibid*.
8 Fysh to Menzies, 23 January 1941, *Ibid*
9 Brain to Edgar Johnston, 5 February 1941, from Gunn, *Challenging Horizons*, University of Queensland Press, 1987, p. 30.

10 Albert Aubrey (Aub) Koch, born 2 October 1904, Ulverstone, Tasmania; died 21 June 1975, Mount Eliza, Victoria.
11 Walter Leslie Runciman, 2nd Viscount Runciman, (26 August 1900 – 1 September 1989).
12 'Shot Down by Japanese', The *Age* (Melbourne), 4 February 1942, p. 1.
13 'Japs Blitzed Bomber in Jungle After Smashing Flying Boat', *The Telegraph* (Brisbane), 13 February 1942, p. 2.
14 'Captain Among 5 Saved ', *The Sun* (Sydney), 4 February 1942, p. 4.
15 Charles Henry Cecil Swaffield, born 31 August 1902 Rockhampton, Queensland; died 20 February 1942, Brisbane.
16 'Nine Lose Lives in Plane Crash Near Brisbane', *Morning Bulletin* (Rockhampton), 2 February 1942, p. 5.
17 ozatwar.com
18 'Flying Boat Circe Loss Announced', *Townsville Daily Bulletin*, 1 April 1942, p. 2.
19 Arthur Brownlow Corbett (18 February 1877 – 20 March 1970).
20 Corbett to Fysh, 1 May 1942, from Gunn, *Challenging Horizons*, p. 67.
21 Air Vice Marshal William Dowling Bostock (5 February 1892 – 28 April 1968).
22 Fysh to Bostock, 14 May 1942, State Library NSW, MLMSS 2413.
23 Father John Corbett Glover (1909-1949).
24 'Wants Indian Ocean Airway ', *The Sun* (Sydney), 10 July 1942, p. 3
25 Fysh to McMaster, 24 July 1942, State Library NSW, MLMSS 2413.
26 *Ibid.*, 20 August 1942, *Ibid.*
27 Sir Reginald Myles Ansett, born 13 February 1909, Inglewood, Victoria; died 23 December 1981, Mount Eliza, Victoria.
28 Fysh to Holyman, 20 June 1942, State Library NSW, MLMSS 2413.
29 Holyman to Fysh, 27 June 1942, *Ibid.*
30 Arthur Samuel Drakeford, born 26 April 1878, Fitzroy, Melbourne; died 9 June 1957, Moonee Ponds, Victoria.
31 Drakeford to Fysh, 3 July 1942, State Library NSW, MLMSS 2413.
32 Brigadier General Alexander Gore Arkwright Hore-Ruthven, 1st Earl of Gowrie (6 July 1872 – 2 May 1955)
33 Corbett to Fysh, 31 August 1942, State Library NSW, MLMSS 2413.
34 Fysh memo to executives, 23 December 1942, *Ibid.*
35 Fysh, *Qantas at War*, p. 177.
36 *Ibid.*, p. 183.
37 'Qantas Double Sunrise', qfom.com.au
38 Fysh, *Qantas at War*, p. 205.
39 Hudson Fysh, *Wings to the World: The Story of Qantas 1945-1966*, Angus & Robertson, 1970, p. 5.
40 Arthur Samuel Drakeford (26 April 1878 – 9 June 1957).
41 Lieutenant-Colonel John Theodore Cuthbert Moore-Brabazon (8 February 1884 – 17 May 1964).
42 Fysh to Lord Knollys, 26 April 1945, Hudson Fysh Papers, State Library of NSW, MLMSS 2413.
43 Later Sir Daniel McVey, born 24 November 1892 at Carronshore, Stirlingshire, Scotland; died 24 December 1972, East Melbourne.
44 Fysh notes on UK trip, 23 June to 9 August 1945, Hudson Fysh Papers, State Library of NSW, MLMSS 2413.
45 Paul Joseph McGinness, Service Number 251538, National Archives of Australia, 1939-1948, Series A9300, Item ID: 5243949.
46 *Ibid.*
47 Cottrill, *The Man Australia Forgot*, p. 318.
48 Fysh, *Qantas at War*, p. 193.
49 'Airlines Bill Passed', *The Sydney Morning Herald*, 3 August 1945, p. 5.

Chapter 22

1 Interview with the author.
2 P.J. Birtles, 'The de Havilland Comet Series: 1–4', *Aircraft in Profile*, Volume 5. Doubleday, 1970, p. 124
3 *Ibid.*, p.16
4 'Air Monopoly Plan Held Invalid', *The Sydney Morning Herald*, 15 December 1945, p. 5.
5 Fysh, *Wings to the World*, p. 7 quoting S. Ralph Cohen, *IATA, the First Three Decades*, Head Office of the International Air Transport Association, 1949.

6 Fysh, *Wings to the World*, p. 8.
7 *Ibid.*, p. 62.
8 Fysh, *Wings to the World*, p. 18.
9 'Qantas Overseas Service', *Cairns Post*, 4 April 1946, p. 9.
10 'Airlines General Manager', *The Argus* (Melbourne), 4 May 1946, p. 1.
11 'First T. A. A. Flight', *Age* (Melbourne), 9 September 1946, p. 3.
12 'We Do Not Want to Wait for British Planes', *The Sydney Morning Herald*, 31 August 1946, p. 2.
13 Fysh, *Wings to the World*, p. 30.
14 *Ibid.*
15 *Ibid.*
16 Cable No. 331, Chifley to Dominions Office, 10 September 1946, British Airways archives; RAF Museum, Hendon, London.
17 Read in House of Representatives, 4 December 1946.
18 Pastor S.O. Gross to Fysh, 21 March 1947, Hudson Fysh Papers, State Library of NSW, MLMSS 2413.
19 'Government's Plan to Buy Out Qantas', *The Central Queensland Herald* (Rockhampton), 29 May 1947, p. 28.
20 'Exit Qantas', *The Bulletin*, 15 October 1947, p. 13.
21 *Ibid.*
22 John Stackhouse, *The Longest Hop*, Focus Publishing, 1997, p. 25.
23 Interview with Wendy Miles, Sydney, May 2022.
24 'Catalina Destroyed', *Age* (Melbourne), 24 June 1949, p. 3.
25 'Crude Time-Bomb Was Used to Blow Up Catalina Flying Boat', The *Newcastle Sun*, 29 August 1949, p. 1.
26 'Acquitted on Plane Explosion Charge', *The Daily Telegraph* (Sydney), 16 July 1950, p. 51.
27 'Hudson Fysh, aviation expert, looks into future air development', *The Sun* (Sydney), 20 October 1949, p. 24.
28 *Ibid.*
29 Interview with Wendy Miles, Sydney, May 2022.
30 Lieutenant General Charles George Norman Miles (10 November 1884 – 18 February 1958).
31 Fysh, *Wings to the World*, p. 123.
32 Elizabeth Fysh, *When Chairmen Were Patriots*, p. 189.
33 'McMaster Left £32,398', Brisbane *Telegraph*, 20 May 1952, p. 20.
34 Fysh, *Wings to the World*, p. 124.
35 'Seven Killed in Airliner Crash Off New Guinea', Daily *Advertiser* (Wagga Wagga, NSW), 17 July 1951, p. 1.
36 'Air crash in New Guinea', *Cairns Post*, 15 December 1951. p. 5.
37 'Obscure funeral for hero', The Mail (Adelaide), 2 February 1952, p. 5.
38 Fysh, *Wings to the World*, p. 128
39 'We'll put something on the moon', *T he Daily Telegraph* (Sydney), 6 June 1953, p. 22.

Chapter 23

1 Fysh diary, 24 April 1957, Hudson Fysh Papers, State Library of NSW, MLMSS 2413.
2 'Million Cheer Progress of Queen Through Sydney', Brisbane *Telegraph*, 3 February 1954, p. 2.
3 '140 honored in Govt. House ballroom', *The Daily Telegraph* (Sydney), 7 February 1954, p. 6.
4 'Coronation Honours List', Daily *Examiner* (Grafton), 1 June 1953, p. 3.
5 'We'll put something on the moon', *The Daily Telegraph* (Sydney), 6 June 1953, p. 22.
6 Fysh diary, 20 February 1950, Hudson Fysh Papers, State Library of NSW, MLMSS 2413.
7 Fysh manuscript of *Wings to the World*, p. 334, Hudson Fysh Papers, State Library of NSW, MLMSS 2413. The book was later published without many of Fysh's more provocative comments.
8 *Ibid*, p. 338.
9 Interview with Fysh's daughter, Wendy Miles, Sydney, May 2022.
10 *Ibid.*
11 Fysh manuscript of *Wings to the World*, p. 442, Hudson Fysh Papers, State Library of NSW, MLMSS 2413.
12 George Percival Norman Watt (1890–1983)
13 John Gunn, *High Corridors, Qantas 1954-1970*, University of Queensland Press, 1988, p. 3.

14 Fysh manuscript of *Wings to the World*, p. 339, Hudson Fysh Papers, State Library of NSW, MLMSS 2413.
15 *Ibid.*, p. 338.
16 Sir Roland Wilson, born 7 April 1904 at Ulverstone, Tasmania; died 25 October 1996, Canberra.
17 Selwyn Cornish, 'Wilson, Sir Roland (1904–1996)', *Australian Dictionary of Biography*, published online 2021.
18 Gunn, High Corridors, p. 2
19 John Gunn, High Corridors, p. 59.
20 *Ibid.*
21 Hudson Fysh, *Wings to the World: The Story of Qantas 1945-1966*, Angus & Robertson, 1970, p. 138.
22 Athol Gordon Townley, born 3 October 1905, Hobart, died 24 December 1963, East Melbourne.
23 *Ibid.*
24 'Advertising', *The Australian Jewish Times* (Sydney), 16 September 1955, p. 16.
25 Fysh to Townley, 9 June 1955, file SP1844 (1943-63), *Ibid.*
26 Qantas press release, 24 June 1955, Hudson Fysh Papers, State Library of NSW, MLMSS 2413.
27 Fysh to Turner, 28 June 1955, *Ibid.*
28 'History – at jet speed', *The Argus* (Melbourne), 11 September 1956, p. 13.
29 Fysh to Turner, 3 January 1957, Hudson Fysh Papers, State Library of NSW, MLMSS 2413.
30 Fysh to Robert Law-Smith, 17 January 1957, *Ibid.*
31 Fysh to Turner, 3 January 1957, *Ibid.*
32 Gunn, *High Corridors, Qantas 1954-1970*, p. 140.
33 Fysh diary, 24 April 1957, Hudson Fysh Papers, State Library of NSW, MLMSS 2413.
34 'Qantas To Restore Services in Week', *The Canberra Times*, 22 April 1957, p. 1.
35 Fysh handwritten notes, 3 July 1957, Hudson Fysh Papers, State Library of NSW, MLMSS 2413.
36 'Building Opened as Plane Circles ', *The Sydney Morning Herald*, 29 October 1957, p. 5.
37 Hudson Fysh, *Ethics and Other Things*, Qantas, 1955.
38 Fysh manuscript of *Wings to the World*, pp. 444–45, Hudson Fysh Papers, State Library of NSW, MLMSS 2413.
39 John Fysh in Foreword to, Fysh, *Qantas Rising*, pp. v–vi.
40 Gunn, *High Corridors, Qantas 1954-1970*, p. 142.
41 Harold Charles Gatty, born 5 January 1903, Campbell Town, Tasmania; died 30 August 1957, Fiji.
42 Sir Shane Dunne Paltridge (11 January 1910 – 21 January 1966).
43 'Inauguration Qantas Round World Service', *The Canberra Times*, 15 January 1958, p. 8.
44 'The Snowy Mountains', Australian Fly-Fishing Museum, affm.net.au.
45 On 18 January 1975.
46 Fysh manuscript of *Wings to the World*, pp. 444–45, Hudson Fysh Papers, State Library of NSW, MLMSS 2413.
47 *Ibid*, pp. 444–45.
48 *Ibid.*, pp. 456–57.
49 Fysh to Law-Smith and Sir Roland Wilson, 23 June 1961, Hudson Fysh Papers, State Library of NSW, MLMSS 2413.
50 '£100,000 to «Sick» Man', *The Canberra Times*, 18 August 1962, p. 9.
51 Fysh diary, 8 February 1963, Hudson Fysh Papers, State Library of NSW, MLMSS 2413.
52 Fysh, *Wings to the World*, p. 157.
53 Fysh diary entry, 9 January 1962, Hudson Fysh Papers, State Library of NSW, MLMSS 2413.
54 Fysh, *Wings to the World*, p. 186.
55 Fysh, *Qantas Rising*, p. 34.
56 'Who'll take over Qantas?', *The Sydney Morning Herald*, 19 November 1965, p. 6.
57 'New chief of Qantas named', *The Canberra Times*, 4 December 1965, p. 3.
58 Fysh to Wilson, 31 December 1965, from Gunn, *High Corridors, Qantas 1954-1970*, p. 341.

Chapter 24

1 Interview with David Miles, Sydney, May 2022.
2 Fysh correspondence with J Fielding & Co, Buckingham St, Surry Hills, Hudson Fysh Papers, State Library of NSW, MLMSS 2413.

3 Fysh, *Wings to the World*, p. 191.
4 Interview with David Miles, Sydney, May 2022.
5 Fysh address, 28 July 1953, Hudson Fysh Papers, State Library of NSW, MLMSS 2413.
6 Interview with Wendy Miles, Sydney, May 2022.
7 Fysh, *Qantas Rising*, p. 190.
8 Fysh diary 23 February 1966, Hudson Fysh Papers, State Library of NSW, MLMSS 2413.
9 Slim to Fysh, 26 February 1966, *Ibid.*
10 Law-Smith to Fysh, 28 March 1966, from Gunn, High Corridors, p. 356.
11 Fysh to Law-Smith, 19 April 1966, *Ibid*, p. 357.
12 *Ibid.*
13 Sir Reginald William Colin Swartz (14 April 1911 – 2 February 2006).
14 Fysh to Swartz, 13 May 1966, from Gunn, High Corridors, p. 357.
15 Fysh to the Duke of Norfolk, 26 May 1966, Hudson Fysh Papers, State Library of NSW, MLMSS 2413.
16 Qantas memo, 15 February 1967, *Ibid.*
17 Holt to Fysh, 27 June 1966, *Ibid.*
18 'Jumbo jets outdate airports', *The Canberra Times*, 3 November 1967, p. 1.
19 Hudson Fysh, *Round the Bend in the Stream*, Angus & Robertson, 1968.
20 Fysh to J. Roche, Cowper St, Randwick, 20 December 1966, Hudson Fysh Papers, State Library of NSW, MLMSS 2413.
21 Fysh to Harden & Johnston Ltd, 22 February 1967, *Ibid.*
22 Fysh to Law-Smith, 18 July 1968, *Ibid.*
23 *Ibid*, 7 February 1969, *Ibid.*
24 `33 Years of Aviation History', *The Sydney Morning Herald*, 17 August 1971, p. 14.
25 'Colour TV for NZ by 1973', *The Canberra Times*, 18 October 1971, p. 15.
26 'Flight to where it began', *The Canberra Times*, 3 November 1972, p. 7.
27 Hudson Fysh, *Henry Reed: Van Diemen's Land Pioneer*, Cat & Fiddle Press, 1973.
28 Fysh to Wendy Miles, 10 January 1973, letter provided by Wendy Miles.
29 Interview with Wendy Miles, Sydney, May 2022.
30 Sir Donald George Anderson (1 March 1917 – 30 November 1975).
31 'Service for Air Pioneer', *The Sydney Morning Herald*, 9 April 1974, p. 8.
32 'Hudson Fysh: Father of Australia's International Airline', *The Canberra Times*, 9 April 1974, p. 2.

Bibliography

BOOKS

Charles Bean, *The Official History of Australia in The War of 1914-1918, Volume I,* Angus & Robertson, 1941.

Francis Birtles, Lonely Lands: Through the Heart of Australia, N.S.W. Bookstall Company, 1909.

P.J. Birtles, 'The de Havilland Comet Series: 1–4', *Aircraft in Profile, Volume 5.* Doubleday, 1970

Bryan Cooper & John H. Batchelor, *Fighter: A History of Fighter Aircraft,* Macdonald & Co., 1973.

Pat Conrick (edit), The Flying Carpet Men: From the diary of Clive Conrick, P. Conrick, 1993.

Pauline Cottrill, *The Man Australia Forgot,* Qantas Founders Museum, 2012

Jim Eames, *Courage in the Skies*, Allen & Unwin, 2019.

Norman Ellison, *Flying Matilda: Early Days in Australian Aviation*, Angus and Robertson, 1957.

Elizabeth Fysh, *When Chairmen Were Patriots*, Boolarong Press, 2020.

Hudson Fysh, *Taming the North*, Angus & Robertson, 1933.

Hudson Fysh, Qantas Rising, Angus & Robertson, 1965.

Hudson Fysh, *Qantas at War,* Angus & Robertson, 1968.

Hudson Fysh, *Wings to the World: The Story of Qantas 1945-1966*, Angus & Robertson, 1970.

Hudson Fysh, *Henry Reed: Van Diemen's Land Pioneer,* Cat & Fiddle Press, 1973.

Bill Gammage, *The Broken Years: Australian Soldiers in the Great War,* Melbourne University Publishing, 2010.

Charles Harvard Gibbs-Smith, *The Invention of the Aeroplane, 1799–1909,* Taplinger, 1966.

John Gunn, *The Defeat of Distance*, University of Queensland Press, 1985.

John Gunn, *Challenging Horizons: Qantas 1939-1954*, University of Queensland Press, 1987.

John Gunn, *High Corridors, Qantas 1954-1970,* University of Queensland Press, 1988.

Richard P. Hallion, *Taking Flight: Inventing the Aerial Age, from Antiquity Through the First World War,* Oxford University Press, 2003.

Lady Pamela Hicks, *Daughter of Empire: My Life as a Mountbatten*, Simon & Schuster, 2014.

A.W. Jose, *Official History of Australia in the War of 1914–1918,* published 1920 to 1942.

T.E. Lawrence, *Seven Pillars of Wisdom*, Jonathan Cape, 1935.

Hugh McMaster, *Fickle Fortune,* Self-published, 1949.

Peter McMillan, Terry Gwynn-Jones; *The Greatest Flight*, Turner Publishing, 1995.

H.C. 'Horrie' Miller, *Early Birds: Magnificent Men of Australian Aviation Between the Wars,* Rigby Australia, 1968.

Margaret S. E. Reed, *Henry Reed: An Eventful Life Devoted to God and Man,* Morgan and Scott, 1906.

Henry Reynolds, *Fate of a Free People: A Radical Re-examination of the Tasmanian Wars,* Penguin, 1995.

Sir Ross Macpherson Smith, *14,000 Miles Through the Air,* Macmillan, 1922.

Mary Soames, *Clementine Churchill: The Biography of a Marriage*, Mariner Books, 2003.

John Stackhouse, *The Longest Hop*, Focus Publishing, 1997

Alan Stephens, *The Royal Australian Air Force*, Oxford University Press, 2009.

David Wilson, *The Brotherhood of Airmen*, Allen & Unwin, 2014.

Edward P. Wixted, *The North-West Aerial Frontier 1919-1934*, Boolarong Publications, 1985

Writings Of Tindal, Frith, and Barnes, Vol. 2, Religious Tract Society, 1830.

INTERNET RESOURCES

3squadron.org.au

19fortyfive.com

'Aircraft repairs at the Railway Workshops', blog.qm.qld.gov

anzacportal.dva.gov.au

Australian Dictionary of Biography

Australian Fly-Fishing Museum, affm.net.au.

Australian War Memorial, awm.gov.au.

csa-living.org

diggerhistory.info

dustyheaps.blogspot.com

epicflightcentenary.com.au

espace.library.uq.edu.au
experiencewinton.com.au.
historicengland.org.uk
'John Thomson', heritage.saintandrews.org.au
John Oxley Library
Libraries Tasmania
National Archives, London
National Library of Australia
National Museum Australia, nma.gov.a
ozatwar.com
qantas.com.au
Qantas Founders Museum, qfom.com
qantasnewsroom.com.au
Smithsonian National Postal Museum, postalmuseum.si.edu.
South Australian Aviation Museum, saam.org.au
State Library of NSW
Sydney Morning Herald
Trove
Tunbridge Wells Museum
uivermemorial.org.au
University of Sydney School of Medicine Online Museum, sydney.edu.au
winsleyatwar.wordpress.com
winton.qld.gov